W9-BXH-241

BIOTIC DIVERSITY IN SOUTHERN AFRICA:

Concepts and conservation

Edited by: B J Huntley

Oxford University Press
Cape Town
1989

ST. PHILIP'S COLLEGE LIBRARY

Oxford University Press

Walton Street, Oxford OX2 6DP, United Kingdom

OXFORD   NEW YORK   TORONTO
DELHI   BOMBAY   CALCUTTA   MADRAS   KARACHI
PETALING   JAYA   SINGAPORE   HONG KONG   TOKYO
NAIROBI   DAR ES SALAAM   CAPE TOWN
MELBOURNE   AUCKLAND

AND ASSOCIATED COMPANIES IN
BERLIN   IBADAN

ISBN 0 19 570549 1

© Oxford University Press 1989

OXFORD is a trademark of Oxford University Press

All rights reserved. No part of this publication may be reproduced, stored in a retrieval system, or
transmitted in any form or by any means, electronic, mechanical, photocopying, recording or
otherwise, without the prior permission of the copyright owner.

Published by Oxford University Press Southern Africa,
Harrington House, Barrack Street, Cape Town, 8001, South Africa

Set in 10/12 pt Times by BellSet
Printed and bound by Clyson Press, Maitland, Cape.

Contents

Part 3: The survey, evaluation and monitoring of biotic diversity 135

Part 4: Conservation status of terrestrial ecosystems and their biota 185

Authors

Begg G W (Dr), Environmental Advisory Services, P O Box 37122, Overport, 4067

Berruti A (Dr), Durban Natural History Museum, P O Box 4085, Durban, 4000

Breen C M (Prof), Director: Institute of Natural Resources, University of Natal, P O Box 375, Pietermaritzburg, 3200

Bond W J (Dr), Botany Department, University of Cape Town, Rondebosch, 7700

Botha P Roelf (Prof), Chairman: Council for the Environment, Private Bag X447, Pretoria, 0001

Buxton C D (Dr), Department of Ichthyology and Fisheries Science, Rhodes University, P O Box 94, Grahamstown, 6140

Cooper J (Mr), Percy FitzPatrick Institute of African Ornithology, University of Cape Town, Rondebosch, 7700

Cowling R M (Dr), Botany Department, University of Cape Town, Rondebosch, 7700

Cunningham A B (Dr), Institute of Natural Resources, University of Natal, P O Box 375, Pietermaritzburg, 3200

Davies B R (Prof), Freshwater Research Unit, Department of Zoology, University of Cape Town, Rondebosch, 7700

Ferrar A A (Mr), Co-ordinator: Nature Conservation Research, Foundation for Research Development, CSIR, P O Box 395, Pretoria, 0001

Geldenhuys C J (Mr), Saasveld Forestry Research Centre, Private Bag X6515, George, 6530

Gibbs Russell G E (Dr), Botanical Research Institute, Private Bag X101, Pretoria, 0001

Grant W S (Dr), Department of Genetics, University of the Witwatersrand, P O Wits, 2050

Hall A V (Prof), Bolus Herbarium, University of Cape Town, Rondebosch, 7700

Hardy M B (Mr), Department of Agriculture and Water Supply, Natal Region, Private Bag X9059, Pietermaritzburg, 3200

Harrison J A (Mr), Southern African Bird Atlas Project (SAFRING), University of Cape Town, Rondebosch, 7700

Heydorn A E F (Dr), Division of Earth, Marine and Atmospheric Sciences and Technology, CSIR, P O Box 320, Stellenbosch, 7600

Hilton-Taylor C (Mr), Department of Botany, University of Stellenbosch, 7600

Hobbs J C A (Mr), Environmental Impact Management, ESCOM, P O Box 1091, Johannesburg, 2000

Hockey P A R (Dr), Percy FitzPatrick Institute of African Ornithology, University of Cape Town, Rondebosch, 7700

Hoffman M T (Mr), Botany Department, University of Cape Town, Rondebosch, 7700

Huntley B J (Mr), Foundation for Research Development, CSIR, P O Box 395, Pretoria, 0001

King J M (Dr), Freshwater Research Unit, Department of Zoology, University of Cape Town, Rondebosch, 7700

Le Roux A (Miss), Jonkershoek Nature Conservation Station, Private Bag X5014, Stellenbosch, 7600

MacDevette D R (Mr), Natal Parks Board, P O Box 662, Pietermaritzburg, 3200

Macdonald I A W (Mr), Percy FitzPatrick Institute of African Ornithology, University of Cape Town, Rondebosch, 7700

Mentis M T (Prof), Department of Botany, University of the Witwatersrand, P O Wits, 2050

O'Keeffe J H (Dr), Institute for Freshwater Studies, Rhodes University, Grahamstown, 6140

Shelton P A (Dr), Sea Fisheries Research Institute, Private Bag 2, Rogge Bay, 8012

Siegfried W R (Prof), Director: Percy FitzPatrick Institute of African Ornithology, University of Cape Town, Rondebosch, 7700

Skelton P H (Dr), J L B Smith Institute of Ichthyology, Private Bag 1015, Grahamstown, 6140

Smuts G L (Dr), Anglo American Corporation of South Africa Limited, P O Box 61587, Marshalltown, 2107

Tainton N M (Prof), Faculty of Agriculture, University of Natal, P O Box 375, Pietermaritzburg, 3200

Woods D R (Prof), Deputy Vice-Chancellor, University of Cape Town, Rondebosch, 7700

Zacharias P J K (Mr), Faculty of Agriculture, University of Natal, P O Box 375, Pietermaritzburg, 3200

Preface

The past decade has witnessed a growing concern throughout the world regarding the imminent mass extinction of animals and plants, especially in the humid tropics.

This concern has stimulated an unprecedented growth in the field of biotic diversity research and a desire to quantify all aspects of environmental degradation. Yet suddenly scientists find that they do not have an even vaguely accurate idea of how many species exist. Current estimates range from five to thirty million, of which only 1,4 million have been classified and named. Statistics on tropical forest destruction also vary widely, but the best estimates indicate that 11 million hectares are deforested annually. This is equivalent to an area the size of the Kruger National Park being transformed from rain forest to cattle pastures every two months. Projections of species loss range from nine to fifty per cent of all tropical species by the year 2000 — as many as 100 species becoming extinct every day!

Despite heated debate regarding the credibility of such statistics, the need for precise and accurate data is open to question. What is beyond doubt is that we are currently losing species at least 1 000 times faster than normal evolutionary rates. An extinction spasm is occurring, the like of which has not been experienced since the mass extinctions believed to have been triggered by enormous asteroid strikes some 65 million years ago.

It is therefore not surprising that biotic diversity is currently the vogue concept in nature conservation, having emerged from obscurity to international prominence in the space of a few years. Like many newly fashionable concepts, it runs the risk of falling into disrepute through misuse or over-exposure. It is therefore appropriate that biotic diversity be clearly defined at the start of this book, even if individual authors reflect subtle differences in their interpretation and use of the term.

At the XVII General Assembly of the International Union for the Conservation of Nature and Natural Resources (Costa Rica, February 1988), the Species Survival Commission adopted the following definition of biotic diversity:

The variety and variability of all living organisms. This includes the genetic variability within species and their populations, the variety of species and their life forms, the diversity of the complexes of associated species and of their interactions, and of the ecological processes which they influence or perform.

The considerable interest in, and importance attached to, the maintenance of biotic diversity has resulted in increased concern for effective environmental management, reaching far beyond the traditional realm of nature conservation. Major international organizations, such as the World Bank, the World Health Organization, the World Commission for Environment and Development, and the Food and Agriculture Organization of UNESCO, have initiated massive and ambitious programmes of research, development and technology transfer to enhance our ability to maintain the biotic resources on which humankind's future survival and prosperity depends.

For many years, southern African scientists and environmental managers have realized that biological richness should not be measured only in terms of elephants, rhinos, succulents or cycads. Co-ordinated efforts to assess the adequacy of conservation measures for our diverse fauna and flora were initiated during the International Biological Programme in the late 1960s and early 1970s. These led to the publication of the first series of South African Red Data Books and the initiation of numerous national co-operative scientific programmes. While individual workers have made important original contributions to the theory and practice of biotic diversity conservation in southern Africa, much of the momentum developed over the past decade has resulted from the co-operative efforts of closely knit teams of researchers and managers from different disciplines and organizations.

Some of the studies reported on in this book have developed over an extended period and draw on vast data bases, whereas others explore new fields and are of necessity speculative. Despite the somewhat uneven cover in depth and breadth that results, this assemblage of papers provides a synthesis of insight, information and understanding of biotic diversity at a regional level that is currently unmatched anywhere on the African continent.

In order to provide as comprehensive a review as possible, authors were urged to include recent and often unpublished information, or results available only in departmental reports. The inclusion of this 'grey literature' is thus deliberate. Access to such material is available from the individual authors.

In this work 'southern Africa' is defined as the region comprising the republics of South Africa, Bophuthatswana, Ciskei, Transkei and Venda, and the kingdoms of Lesotho and Swaziland. 'South Africa' refers to the Republic of South Africa and its included self-governing homelands.

Another term requiring definition is that of conservation status. This signifies the extent to which populations, species or communities have been modified by the influence of industrial man, and the degree to which they may be expected to maintain their diversity and ecological processes in the medium term (10 to 100 years). Although conservation status as a concept is essentially judgemental, various authors have attempted to lend objectivity to it by means of a variety of quantitative measures.

This book comprises five parts. Part 1 examines the dynamic nature of biotic diversity. William Bond reviews the different methods of measuring biotic diversity and concludes that the Linnaean species concept, and through it the mere number of species in a taxon or nature reserve, is a poor guide to establishing conservation priorities. Various measures of biotic diversity are used to describe the patterns and evolution of floristic diversity by Richard Cowling and co-authors, who find no answer to the question 'Why and how did so many species evolve?' in the southern African flora. The changes wrought by man to the region's landscape, ecosystems and species composition are analysed by Ian Macdonald, who considers southern Africa to be one of the most man-modified regions on Earth.

The nature of human dependence on biotic diversity is considered in Part 2. Mike Mentis provides a new and iconoclastic interpretation of the concept 'conservation'. He argues that conservation is not knowingly executable or achievable and that as a consequence the World Conservation Strategy is based on untestable concepts. The current and direct uses of and dependence on wild plant resources by rural people in southern Africa is reviewed by Tony Cunningham. Both plant species and traditional knowledge are being lost through environmental and social change. The influence of plant diversity patterns on livestock production, and of grazing management on diversity, is discussed by Neil Tainton and colleagues, while David Woods examines the exciting frontiers being explored, if not yet conquered, in the field of genetic engineering and biotechnology.

Part 3 provides a comprehensive review of southern African experience in the survey, evaluation and monitoring of biotic diversity. Chapters detail the preparation and use of Red Data Books (Tony Ferrar), approaches to rare plant surveys (Tony Hall), bird atlases (James Harrison), and the application of molecular methods in the assessment of biotic diversity (Stewart Grant). These papers illustrate the tremendous advances made in the quantification of the conservation status of species in southern Africa during the past decade.

Reviews of the conservation status of terrestrial ecosystems and their biota are included in Part 4. Roy Siegfried presents the remarkable results of an assessment of the percentage of various taxonomic groups found in southern Africa's 582 nature reserves, which collectively cover 5,8 per cent of the region. These reserves include populations of 92%, 92%, 97% and 93% of the region's amphibian, reptilian, avian and mammalian faunas respectively. The floristic diversity and conservation status of the fynbos, succulent karoo and Nama-karoo biomes are described in detail by Craig Hilton-Taylor and Annelise le Roux. These biomes include some of the most diverse, yet least adequately conserved, floristic communities in southern Africa.

The indigenous forests of southern Africa, although extremely limited in area, are in general very well conserved (Coert Geldenhuys and Dave MacDevette). A similar situation prevails on the continental and oceanic islands, although invasive animals and plants pose a significant threat to indigenous species on the islands, as described by John Cooper and Aldo Berruti. Perhaps the most severely threatened of all terrestrial ecosystems in southern Africa are the wetlands, yet Charles Breen and George Begg show that they have not been studied in any detail and that little is known of their biotic diversity.

Part 5 reviews the conservation status of river, coastal and marine ecosystems and their biota. Rivers, until recently badly neglected by conservation researchers and managers alike, have been the subject of detailed study in the past five years, as described by the main workers in this field, Jay O'Keeffe and co-authors. Estuaries, too, have enjoyed considerable, if belated, attention although little information is available on their biotic diversity (Alan Heydorn). The level of protection afforded coastal ecosystems, in particular intertidal communities, is described as inadequate by Phil Hockey and Colin Buxton, although this situation could be markedly improved by minor legislative changes. The interdependence of different pelagic populations and their vulnerability to excessive exploitation is illustrated in Peter Shelton's paper, which concludes the reviews of the conservation status of southern African ecosystems.

The book ends with two chapters on policies which could provide the basis for the protection of biotic diversity. Roelf Botha and Brian Huntley describe current efforts by the Council for the Environment to develop a new conceptual base to environmental conservation policy in South Africa. Emphasis is placed on equality of access to use and enjoy the region's natural resources and development opportunities, while maintaining environmental quality and biotic diversity. In the final chapter, Butch Smuts and Jonathan Hobbs argue that the motivation for large corporations and the private sector to become involved in environmental management is based on sound ethical, social, political, economic and ecological considerations.

This volume is by no means an exhaustive 'state of the art' review of biotic diversity and its conservation in southern Africa. Many important advances, amongst others, in the role of botanical and zoological gardens, the management of protected areas, and the contribution of non-governmental organizations and private landowners to conserving biotic diversity, have been achieved in recent years, yet are excluded for reasons of space.

Acknowledgements

This book is based on invited review papers delivered at a conference titled 'Conserving Biotic Diversity in Southern Africa' held at the University of Cape Town, 13 to 15 June 1988. The conference was sponsored by the National Programme for Ecosystem Research of the Foundation for Research Development, CSIR. The National Programme is a co-operative venture of scientists, resource managers, administrators and the general public concerned with research in the natural environment. Funds for the research conducted within the National Programme are provided by the individual participating organizations, while substantial additional support is made available to university and museum-based research from central funds contributed by the Department of Environment Affairs. The support of all participants and sponsors is gratefully acknowledged.

The editing of this book was greatly facilitated by the valuable inputs of an active team of referees. In particular, I would like to thank Martin Cody, Michael Gilpin, Brian Norton, Mark Plotkin, Carleton Ray and Ken Tinley for their detailed reviews.

It is also a great pleasure to thank Fifi Bierman, Marie Breitenbach and Tisha Greyling for their assistance in the efficient organization of the conference, Gillian Ress, and Sandie Vahl of Oxford University Press for their invaluable support in meeting the publication deadline while ensuring a product of high quality.

In conclusion I would like to thank copyright holders for permission to reproduce the following copyright material:

Figure   **1.2**   Bond W J, Midgeley J and Vlok J, When is an island not an island? Insular effects and their causes in fynbos islands. *Oecologia* **77**.

Figure   **1.3:**   By D Goodman in M Soule (ed), *Viable Populations for Conservation,* Cambridge University Press, Cambridge.

Figure   **7.1:**   From *Molecular Biology of Bacterial Viruses* by G S Steut. Copyright © 1963 by W H Freeman and Company. Used by permission.

Figure   **7.8**   Watson J D, Hopkins N H, Roberts J W, Steitz J A and Weiner A M. *Molecular Biology of the Gene, Volume 1.* © 1987. The Benjamin/Cummings Publishing Company.

Figure   **18.2:**   From *In the mangroves of Southern Africa* by P Berjak et al (1977).

Brian J Huntley
Pretoria, April 1989

INTRODUCTION

Challenges to maintaining biotic diversity in a changing world

B J Huntley

THE NUMBER AND VALUE OF BIOTA

The future is not normally what it used to be. Sunter (1987)

Nowhere is this apparently glib statement more relevant than in the field of biotic diversity. Neither Darwin nor Wallace could have envisaged that the biological wealth they were revealing to the world at the start of the Industrial Revolution would be heading for destruction within the next century. Yet the Atlantic coastal forests of Brazil, which inspired the young Darwin in 1832 with 'wonder, astonishment and sublime devotion' and in which Warming developed the first principles of tropical ecology, have been reduced to one per cent of their former extent.

The tropical rain forests of South America, Africa, Asia and Australia have entered an episode of extinction more dramatic than that at the end of the Cretaceous, 65 million years ago. Even conservative estimates indicate that the rate of species loss is now 1 000 to 10 000 times the rate before human intervention (Wilson 1988a). As Myers (1988) has so aptly stated, the current demise of species is occurring within a mere 'twinkling of an evolutionary eye'.

Biologist seem ill equipped to measure, let alone resolve, the problem. About 1,4 million living species have been given Latin binomials (Parker 1982) but 200 years after Linnaeus laid the foundations to the science of classifying and naming plants and animals, we are no closer to answering the basic question of 'How many species are there?'. As recently as 1982, estimates ranged from three to five million species. Yet a single paper in an obscure journal (Erwin 1982) suddenly raised the estimate to 30 million. Could biologists really be so hopelessly wrong in their understanding of the diversity of living organisms?

In an elegant review of the empirical and theoretical studies that could be used to sharpen the global species estimate, May (1988) concluded that 'we cannot explain from first principles, why the global total is of the general order of 10^7 rather than 10^4 or 10^{10}'. It is unlikely that we shall significantly improve our estimate of species numbers within the next decades. The task is enormous, and the world has fewer than 1 500 systematists competent to classify tropical biota, which probably account for at least 90%, or over 20 million species, of the undescribed taxa. Most conservationists argue that the exact number of living species is not important. What does matter is how to identify and implement mechanisms that will slow the rate of species loss. In particular it is necessary to identify those species that are useful to humanity, for both present and future generations.

Conservationists have been hard-pressed to find convincing arguments in support of protecting the multitude of wild species. Molecular biologists argue that biotechnology will adequately maintain the genetic diversity of the 30 species of crop plants that provide 95% of global food needs. Why do we need to preserve hundreds of thousands of species, most of which are not even known to science? Can anyone truly believe that civilization is worse off for the loss of the dodo, the passenger pigeon or the quagga?

Norton (1988) considers the value of biotic diversity within three fields — commodity value, amenity value, and moral value. None of these is easy to measure in monetary terms — although Prestcott-Allen and Prestcott-Allen (1986) provide the interesting estimate that wild species contributed $27 billion to the United States economy in 1980 — 4,5% of the GDP. Farnsworth et al (1985) have estimated that over 3,5 billion people (80% of the Third World population) rely on a wide diversity of plants for medicinal purposes. Yet the list of species brought in from the wild that have truly improved the quality of life of humanity at large is embarrassingly short. For most species 'The burden of proof lies on the preservationist to find or create values for a species or else regard it as useless' (Norton 1986a).

The most plausible argument for the inherent usefulness of all species is the parable of the 'Rivet Poppers' (Ehrlich and Ehrlich 1981). It is an analogy that should be related far more frequently. It reads as follows:

As you walk from the terminal toward your airline, you notice a man on a ladder busily prying rivets out of its wing. Somewhat concerned, you saunter over to the rivet popper and ask him just what the hell he's doing.

'I work for the airline — Growthmania Intercontinental,' the man informs you 'and the airline has discovered that it can sell these rivets for two dollars a piece.'

'But how do you know you won't fatally weaken the wing doing that?' you inquire.

'Don't worry,' he assures you. 'I'm certain the manufacturer made this plane much stronger than it needs to be, so no harm's done. Besides, I've taken lots of rivets from this wing and it hasn't fallen off yet. Growthmania Airlines needs the money; if we didn't pop the rivets Growthmania wouldn't be able to continue expanding. And I need the commission they pay me — fifty cents per rivet!'

'You must be out of your mind!'

'I told you not to worry; I know what I'm doing. As a matter of fact, I'm going to fly on this flight also. So you can see there's absolutely nothing to be concerned about.'

Any sane person would, of course, go back into the terminal, report the gibbering idiot and Growthmania Airlines to the FAA, and make reservations on another carrier. You never have to fly on an airliner. But unfortunately all of us are passengers on a very large spacecraft — one on which we have no option but to fly. And, frighteningly, it is swarming with rivet poppers behaving in ways analogous to that just described.

CONSERVATION BIOLOGY: THE NEW TOOL KIT

Thirty years ago it was sufficient for a wildlife manager to be able to ride a horse, handle a gun and fix fences. The most successful protected areas in southern Africa were built on the tradition of the game ranger on horseback. Wardens were drawn from the ranks of ex-servicemen and repentant hunters. But despite these apparent shortcomings, much was achieved — Kruger National Park, Hluhluwe Game Reserve, Etosha, Hwange, Gorongosa — all are monuments to the tenacity of their early guardians. The 1960s and 1970s saw the introduction of wildlife management as a formal postgraduate degree course in many universities, particularly in the United States, but also in southern Africa. The courses in South Africa have produced over 200 graduates in the past twenty years, but it was not until the early 1980s that conservation entered the arena of hard science, and even then it did so somewhat hesitantly. Conservation biology, as this new, self-conscious discipline has been styled, draws on genetics, parasitology, community ecology, biogeography, ecology and good old-fashioned wildlife management.

The remarkable growth of interest in conservation biology is illustrated by the increase in articles published in journals included in the BIOSIS data base. Ginsberg (1987) analysed the number of citations concerning conservation and wildlife management and those that addressed both ecology and conservation. Articles in both categories had nearly doubled in the 1983 to 1986 period. More important, conservation biologists have contributed a spate of basic textbooks giving both identity and stature to their discipline: Soule and Wilcox 1980; Frankel and Soule 1981; Norton 1986b; Soule 1986, 1987; Usher 1986.

The discipline has been further strengthened by advances in related fields, most particularly the application of hierarchy theory to ecological issues (O'Neill et al 1986) and the development of landscape ecology (Godron 1986). However, many wildlife ecologists regard conservation biology with healthy disdain. Jordan (1988) notes 'the indignation I have felt stemming from the usurpation of the term conservation by the gene counters and butterfly watchers, seemingly to make population biology more palatable to applied funding agencies!'

Population biologists have certainly made somewhat bold inroads into conservation science, but one can only welcome the breath of fresh air that is blowing through the stagnant territory of plant phytosociologists, animal behaviourists and the hunting and fishing fraternity. But it is clear that a long learning curve lies ahead. Many fashionable trends will rise and fall, such as the energy-sapping SLOSS debate (Soule and Simberloff 1986) and the similar controversy surrounding the concept of Minimum Viable Populations.

What conservation biology has introduced is a new and exciting momentum to wildlife conservation, and a proliferation of workshops, syntheses and critical reviews of issues that were previously ignored or accepted at face value. In both North America (Lacy 1988) and southern Africa (Grant et al 1988) training courses in fields such as population genetics and modelling have become necessary components of the development of the newly emerging discipline. What is clear is that there can be no turning back to the traditions of dogma and sentiment that have in the past governed much of conservation decision-taking in southern Africa. A new set of concepts and technologies is available for testing and application.

GLOBAL CHANGE AND BIOTIC DIVERSITY

Superimposed on all the regional and short-term pressures on biotic diversity is the recently recognized phenomenon of global climatic change, particularly that resulting from the 'greenhouse effect'. The possible impacts of global warming on maintaining biotic diversity in the medium term — to the twenty-second century — will far outweigh any of the current crises of deforestation, desertification, expanding human populations, waste management, pollution and resource depletion. Yet even as recently as the National Forum on BioDiversity, held in Washington in 1986 (Wilson 1988b) only one of 57 papers dealt specifically with the effect of global climatic change on natural communities (Peters 1988). That the problem has now been widely recognized is illustrated by the plethora of conferences, workshops and international collaborative research programmes that have been convened since 1986. Attention was focused specifically on the topic of the Consequences of the Greenhouse Effect for Biological Diversity at a conference convened under that title in Washington in 1988 (Peters 1989).

Southern African conservation authorities would do well to note the potential impacts of global change, and to plan accordingly. Our present network of protected areas includes a very high percentage of southern Africa's biotic diversity (Siegfried, this volume), but the future may see rapid extinctions as climate change displaces individual species or whole communities out of protected areas onto radically transformed land. The extreme species richness of numerous fynbos reserves, scattered like islands in a sea of agricultural development, will be decimated if the south-western Cape changes from a winter-rainfall to a summer-rainfall region, as several general circulation models suggest may occur. Increased winter temperatures and moister conditions over the east coast would transform Hluhluwe and Umfolozi to tropical forest, eliminating the last remnants of white rhinoceros habitat from an already rapidly changing vegetation complex. The predicted warming of mean annual temperatures by three to five degrees Celsius by AD 2050 would push many communities up mountain slopes, assuming that their dispersal mechanisms permit very rapid migration. The Afro-alpine communities of the Drakensberg summits would probably be replaced by dense tussock grasslands, while the mountain fynbos communities of the Cape would have to accommodate even higher levels of species packing than at present, since a 500 m altitudinal shift would be induced by the warmer climate.

These examples are highly speculative and may never occur. But if conservationists are serious about their commitment to maintaining biotic diversity for the benefit of future generations, they cannot afford to ignore the possible impacts of global change.

SOCIO-ECONOMIC REFORM IN SOUTH AFRICA

In the long term, global warming will probably determine the future of all life on planet Earth, but in the short term, socio-political events will shape the environment of southern Africa. And without the short term, there can be no long term.

The past few years have seen the publication of numerous books on South Africa's political and economic futures (Louw and Kendal 1986; Sunter 1987; Berger and Godsell 1988; Bethlehem 1988). All describe the need for major political reform and, in particular, greatly increased sharing of development opportunities and access to natural resources. Yet none of these studies addresses the environmental consequences of socio-economic change. To redress this imbalance Huntley, Siegfried and Sunter (1989) have examined possible environmental scenarios that would emerge following different socio-economic trajectories in South Africa to the early twenty-first century.

They describe four possible scenarios. The absence or failure of reform would lead to a siege economy in a partitioned state. This would ultimately deteriorate to a protracted civil war. Both situations would result in major negative impacts on the environment. An active reform policy with rapid economic growth would pave the way for the development of the wealth needed to support environmental health and a high quality of life. But unless economic growth is accompanied by an improved relationship between the human population and its environment, and the development of a strong environmental ethic, the natural resources on which southern Africa's wealth depends could be rapidly exhausted. Critical factors could include the impact of massive urbanization (with over 750 000 people moving from the rural areas to the cities every year), the impact of acid rain resulting from the country's enormous coal-burning power stations (with emissions exceeding 120 million tonnes of carbon dioxide and one million tonnes of sulphur dioxide per year) and the pressure on farmland in the homelands (where 42% of the country's population occupies only 13% of the total land area). Thus the choice of political options during the next decade in southern Africa will be decisive in shaping the future course of environmental quality — either moving towards a Utopian state of social, economic and ecological health or descending rapidly to a regional wasteland.

CHANGES IN ENVIRONMENTAL POLICY AND PRACTICE

The problem is very big and the fuse very short. Lovejoy (1988)

Lovejoy was referring to the need to chart the biological diversity of the planet — a quick and dirty survey to identify areas rich in biotic diversity in order to protect them before they disappear.

The call reflects the current environmental dilemma. Scientists have to resort to crisis management approaches simply because sensible, affordable strategies have been allowed to slip while government decision-takers and science policy-makers back the vogue themes of molecular biology and biomedicine. Ecological research in the USA currently receives about $50 million annually, two orders of magnitude less than that received by the two more popular areas of research mentioned above. Yet while the multi-billion dollar investment in biomedical research may lengthen the life expectancy of affluent Americans by a few years or provide a cure for social diseases such as AIDS, the enormous impact of deteriorating environmental health could remain largely unresolved, affecting the quality of life of the entire planet.

While the AIDS scare attracted the attention of the media within months of the discovery of the HIV virus, some ten years separated the early scientific warnings concerning acid rain and the first substantive demonstrations of the effects of the phenomenon. Even the 'greenhouse effect' was dismissed as an overstudied 'non-crisis' problem as far back as 1969 (Mercer and Petersen 1988).

These examples illustrate the urgent need for politicians to attach greater significance to the warning signals raised by innately modest environmental scientists. They will have to adapt to making hard decisions on apparently soft information. Woodwell (1989) has brought focus to the problem: 'Scientists have been infused with an unbecoming industrial bias, to the point where they demand a special higher standard of objectivity and proof for matters of public welfare that may affect the profits from the system than for their everyday business of developing knowledge.'

What is needed, therefore, is a new positioning on environmental matters. Whether the issue is the maintenance of biotic diversity, the progression of Third World populations through the demographic transition, the agricultural impacts of global warming or the quality of one's immediate environment, a new awareness of the interaction between economic growth, environmental health and quality of life is urgently needed. More than anything else, a consciousness of intergenerational responsibility is required.

Caring for the environment of our children and grandchildren requires both socio-economic and scientific actions. In the latter context, a fundamental legacy which the present generation should guarantee the next is an inventory of knowledge. Systematists call for biological surveys and a more comprehensive check-list of the world's living resources. Ecologists call for a greater understanding of the functioning of ecosystems, atmospheric scientists call for general circulation models. All are urgently needed, and all must be made accessible to decision-makers through the medium of geographic information systems, expert systems and other advances in information technology.

PRIORITIES FOR ACTION

While there is little doubt that the primary determinant of the conservation status of species, ecosystems and landscapes in southern Africa will be the type of political dispensation that develops over the next decade, many other factors can play a significant role. More important, unless certain actions are taken now on which to build the knowledge and understanding required for the effective management of the region's extremely rich biota, the maintenance of this diversity will not be possible regardless of the type of political settlement achieved.

The key actions that are required include:

— The establishment of an independent, adequately financed team of scenario planners charged with developing and regularly up-dating environmental scenarios at national and regional scales.

— The provision of substantial support for the adaptation and validation of general circulation models for the southern African subcontinent, particularly in terms of future agro-meteorological patterns and the dispersal climatology of industrial air pollutants.

— Substantially increased support for the biological survey of southern Africa, both for the collection of material on which to describe the region's biological diversity, and for the biosystematic studies needed to explain its origin and probable future changes.

— The identification of the minimum set of protected areas and populations necessary to ensure the maintenance of southern Africa's diversity of landscapes, habitats, plants and animals in terrestrial, freshwater, coastal and marine environments.

— The establishment of a network of environmental research and monitoring stations, where the structure and functioning of natural and agricultural ecosystems can be elucidated and where the impacts of global and regional environmental change phenomena may be monitored.

This wish-list of needs may appear overwhelming, but all can be achieved, and in the short term, by southern African science. All that is needed is a commitment, on the part of those charged with the responsibility of directing taxpayers' money to taxpayers' welfare, to identify and resolve environmental problems now and in the future.

REFERENCES

BERGER P L and GODSELL B (1988). *A Future South Africa.* Human and Rousseau, Tafelberg, Cape Town. 344 pp.

BETHLEHEM R W X (1988). *Economics in a revolutionary society.* A D Donker, Johannesburg. 367 pp.

EHRLICH P R and EHRLICH A H (1981). *Extinction. The causes and consequences of the disappearance of species.* Random House, New York. 305 pp.

ERWIN T L (1982). Tropical forests: Their richness in Coleoptera and other Arthropod species. *Coleopterists Bulletin* **36(1)**, 74 – 75.

FARNSWORTH N R, AKERELE O, BINGEL A S, SOEJARTO D D and GUO Z G (1985). Medicinal plants in therapy. *Bulletin of the World Health Organization* **63**, 965 – 981.

FRANKEL O H and SOULE M E (1981). *Conservation and Evolution.* Cambridge University Press, Cambridge.

GINSBERG J R (1987).What is conservation biology? *Trends in Ecology and Evolution* **2**, 262 – 264.

GODRON M (1986). *Landscape Ecology.* John Wiley and Sons, New York.

GRANT W S, ROBINSON E R and FERRAR A A (1988). Population genetics and the management of small populations. *South African Journal of Science* **84**, 868 – 870.

HUNTLEY B J, SIEGFRIED W R and SUNTER C L (1989). *South African Environments in the 21st Century.* Human and Rousseau Tafelberg, Cape Town. (In press).

JORDAN C F (1988). Ecosystem ecology. *Conservation Biology* **2**, 137 – 138.

LACY (1988). A report on Population Genetics in Conservation. *Conservation Biology* **2**, 245 – 247.

LOUW L and KENDAL F (1986). *South Africa. The Solution.* Amagi Publications, Bisho, Ciskei. 237 pp.

LOVEJOY T (1988). In Hard choices ahead on biodiversity. (ed Roberts L) *Science* **241**, 1759 – 1761.

MAY R (1988). How many species are there on Earth? *Science* **241**, 1441 – 1449.

MERCER D C and PETERSEN J A (1988). Australia and the greenhouse effect: the science policy debate. In *Greenhouse*. (ed Pearman G I) E J Brill, Leiden. pp 708 – 724.

MYERS N (1988). Tropical forests and their species. In *Biodiversity*. (ed Wilson E O) National Academy Press, Washington DC. pp 28 – 35.

NORTON B G (1986a). On the inherent danger of undervaluing species. In *The preservation of species: the value of Biological Diversity*. (ed Norton B G) Princeton University Press, Princeton, New Jersey. pp 110 – 137.

NORTON B G (ed) (1986b). *The preservation of Species: The Value of Biological Diversity.* Princeton University Press, Princeton, New Jersey. 305 pp.

NORTON B (1988). Commodity, amenity, and morality. The limits of quantification in valuing biodiversity. In *Biodiversity*. (ed Wilson E O) National Academy Press, Washington DC. pp 200 – 205.

O'NEILL R V, DE ANGELIS D L, WAIDE J B and ALLEN T F H (1986). *A Hierarchical Concept of Ecosystems*. Princeton University Press, Princeton, New Jersey. 253 pp.

PARKER S P (ed) (1982). *Synopsis and classification of living organisms*. McGraw-Hill, New York. 2 vols.

PETERS R L (1988). The effect of global climatic change on natural communities. In *Biodiversity*. (ed Wilson E O) National Academy Press, Washington DC. pp 450 – 461.

PETERS R (ed) (1989). *Consequences of the Greenhouse effect for Biological Diversity*. Yale University Press. (In press).

PRESTCOTT-ALLEN C and PRESTCOTT-ALLEN R (1986). *The First Resource. Wild Species in the North American economy*. Yale University Press, New Haven. 529 pp.

SOULE M E (ed) (1986). *Conservation Biology: the Science of Scarcity and Diversity*. Sinauer Associates, Sunderland, Mass. 584 pp.

SOULE M E (ed) (1987). *Minimum Viable Population Size*. Cambridge University Press, Cambridge. 189 pp.

SOULE M E and SIMBERLOFF D (1986). What do genetics and ecology tell us about the design of nature reserves? *Biological Conservation* **35**, 19 – 40.

SOULE M E and WILCOX B A (eds) (1980). *Conservation Biology: An Evolutionary – Ecological Perspective*. Sinauer Associates, Sunderland, Mass.

SUNTER C (1987). *The World and South Africa in the 1990s*. Human and Rousseau Tafelberg, Cape Town. 111 pp.

USHER M B (ed) (1986). *Wildlife Conservation Evaluation.*, Chapman and Hall, London. 394 pp.

WILSON E O (1988a). The current state of biological diversity. In *Biodiversity*. (ed Wilson E O) National Academy Press, Washington DC. pp 3 – 18.

WILSON E O (ed) (1988b). *Biodiversity*. National Academic Press, Washington DC. 521 pp.

WOODWELL G (1989). How does the world work? In *Consequences of the Greenhouse Effect for Biological Diversity*. (ed Peters R) Yale University Press. (In press).

PART 1

The dynamic nature of biotic diversity

CHAPTER 1

Describing and conserving biotic diversity

W J Bond

INTRODUCTION

This chapter attempts to survey several topical issues relevant to the description of diversity, its geographical distribution, and the processes maintaining it. The chapter is by no means compre-hensive but may be useful in introducing several new topics relevant to the conservation of diversity and providing a status report on some older ones. The topics are sufficiently general to apply to most kinds of plants and animals and most types of habitat. This generality is at once a strength, since people from different specialities can share their insights and their experience with problems in different places and with different biota, and a weakness, since the intricacies of natural history may confound even the most elegant theory (Ricklefs 1987). The chapter is biased towards botanical examples, especially from the southern and south-western Cape, reflecting the author's research interests.

THE SPECIES CONCEPT AS A MEASURE OF DIVERSITY

The easiest and most common measure of biotic diversity when considering conservation priorities is a count of the number of species. However, one must ask how adequately species number measures diversity. Does specific status imply that one population of, say, plants differs from another in its ecological niche, or in its mix of secondary chemicals (implying new drug sources); or in the kinds of insects that feed on the plants; or in growth form, pollinators, dispersers or photosynthetic pathways? What, if anything, does the species concept measure and just how well does the number of species in an area or in a clade reflect diversity of form, physiology, biochemistry or ecological role? Should we be more concerned (as the logic of conserving species would seem to imply) about conserving *Erica* because there are over 500 species than about conserving yellowwoods (*Podocarpus* spp) because there are only four?

What are the benefits of biotic diversity?

If we consider for a moment the ideas underlying an interest in conserving biotic diversity, it should be easier to decide what we ought to be measuring in order to conserve most efficiently and effectively. Some of the arguments for conserving diversity are:

— the interdependence of nature, since if each species is part of an interdependent, holistic ecosystem, the loss of one part would lead to instability and eventual collapse of the whole;

— the somewhat contradictory notion that the more species in a community, the greater its resilience to perturbation (abandoned by the theoreticians, this notion persists in some literature), so that conserving species will maintain 'life support systems';

— that species are, or may in future be, useful for the production of substances — crops or animal products — useful to man (alternatively, diversity may be useful for selecting biological control agents to control crops and animals nobody wants any more);

— the scientific and cultural value of diversity which allows, among other things, application of the comparative method to understand evolution and adaptation; it also provides the inspiration and raw materials for making art, music and literature; and

— the aesthetic or collector's urge to preserve all rarities, which astute fund-raisers know has a powerful appeal to the public.

Does the taxonomic species measure biotic diversity?

From the ecological and utilitarian arguments for conserving diversity, one would hope for a taxonomy reflecting different ecological niches among its units. Thus a genus or family with many species should represent a group with diverse ecology. This is sometimes the case especially where speciation is associated with adaptive radiation such as in Darwin's finches (Grant 1986) or the Hawaiian honeycreepers (Carlquist 1980). Among southern African plant genera, good examples may include *Pelargonium* (Van der Walt and Vorster 1983) with its great diversity of growth forms and leaf morphologies, and *Cassia* which includes trees and shrubs, and perennial and annual herbs (Stebbins 1981). Frequently, however, the species in a genus or higher taxonomic grouping are ecologically monotonous. Speciation is not necessarily correlated with ecological differentiation. Indeed, if Paterson (1985) is right, much speciation is merely a by-product of the fickleness of fashion in recognizing suitable mates. The extraordinary diversity of land snails on Oahu, or of *Drosophila* over all the Hawaiian islands (Carlquist 1980), seem classic cases of such speciation in which many species making up very large genera are more or less ecologically equivalent. In the tiny Cape Floral Kingdom there are over 500 species of *Erica* of rather monotonous growth form. All are shrubs and all have small, evergreen, whorled, inrolled 'ericoid' leaves. It is difficult to imagine that profound changes to a fynbos ecosystem would take place if one of the many *Erica* species became extinct — there seem to be too many ecologically equivalent species ready to take its place — or if another significantly expanded its range. For organisms (such as small mammals) or processes (such as fire) dependent on vegetative characteristics, the species concept in *Erica* holds little information. A horticulturist, of course, would make a different assessment.

In contrast to the large genera with many very similar species is the case in which extraordinary variability is found within a single species. *Acacia karroo* covers a large geographical region and occurs in habitats as diverse as karroid shrublands, eastern Cape savanna and as a pioneer in gaps in subtropical coastal dune forest. Few other *Acacia* species cover as wide a climatic range in Africa. The species is extraordinarily variable among populations both morphologically and as measured by protein electrophoretic analysis (Brain 1986). Clearly conservation of this species, for example as a potential source of fuel wood, should entail conservation of populations from the

ST. PHILIP'S COLLEGE LIBRARY

whole of its wide range. *Themeda triandra, Eragrostis curvula* and *Sorghum* spp are other economically important examples of widespread species with diverse morphological and ecological attributes for which the species concept is useless in providing guidelines for conservation.

Perhaps the most bewildering cases are the cryptic, sibling species which appear morphologically identical to one another yet differ genetically, often in chromosome number, and do not interbreed. Some of the most notorious examples occur in small mammals. Thus sibling species have been identified in the South African rodents *Mastomys natalensis, Aethomys chrysophilus* and *Saccostomus campestris* (Gordon and Watson 1986). These species differ from their cryptic sibling pairs most obviously in chromosome numbers but also in haemoglobin electromorph patterns and, more intriguingly, in the morphology of spermatozoa and the genitalia. The differences between sibling pairs are not necessarily ecologically or economically trivial, for example, only one of the *Mastomys* sibling pairs carries plague (*M coucha*) (Isaacson et al 1983 in Gordon and Watson 1986). There are similar examples among the mosquito, *Anopheles gambiae* species complex, which influence antimalarial spraying campaigns (Paterson 1963).

This brief summary suggests that the species concept measures neither ecologically nor economically relevant diversity. Likewise, the number of species in a genus may be a very poor measure of ecologically or economically important diversity. Though it has become so familiar, we should perhaps not be too closely bound to the Linnaean classification system for guiding conservation priorities. After all, it was invented two centuries ago for a different purpose and without the least conception of the magnitude of the task of describing the diversity of life on earth (Raven et al 1971).

Genetic and biotic diversity

Those who argue that we should aim for conservation of genetic diversity rather than mere species diversity are faced with an even more thorny problem of how to measure and interpret diversity. There is a large and contentious literature on the relative merits of molecular and morphological data in systematics (Hillis 1987). In those cases in which both molecular and morphological data exist, the one kind of data may reveal more or less affinity between taxa than the other, and the data may be quite contradictory (Hillis 1987). It is not uncommon to find large morphological but negligible genetic change in closely related species. Chimpanzees and man, for example, are so close genetically (99% of measured polypeptide sequences are identical) that on molecular criteria alone, one might suggest that there is little point in attempting to conserve chimpanzees (King and Wilson 1975). Isozyme analysis fails to discriminate between different breeds of dog, a task simple for even the least discriminating layman (Hedrick 1983). Lewontin (1972) analysed genetic diversity in human populations using a sample of 17 blood group and allozyme loci. These data, the kind used by wildlife biologists for genetic information on conservation priorities, showed that 85% of the genetic diversity is within human populations, 7,5% among populations, and only 7,5% among major races. Politicians take note.

Similar examples occur among plants, including some of the most diverse genera. The genus *Bidens* in Hawaii includes prostrate and erect herbs, shrubs and small trees, encompassing greater morphological diversity than that found in the rest of the genus on five continents (Ganders and Nagata 1984; see also summary in Carr 1987). Despite this morphological and ecological diversity, all Hawaiian species of *Bidens* can hybridize, and isozyme studies indicate very high levels of genetic identity (approx 0,95) equivalent to within-population variability in mainland species (Helenurm and Ganders 1985).

Plants appear to differ from animals in that major morphological differences, such as those used to distinguish species or even genera, may be linked to mutations in one or two genes (Gottlieb

1984). In one case reviewed by Gottlieb (1984) *Rodrigia commutata* (Asteraceae) proved to be interfertile with *Crepis foetida*, with the morphological differences between genera resulting from allele substitutions at two loci. The crossing evidence was used to reduce the genus *Rodrigia* to a subspecies of *C foetida*! Large differences in fruit shape between taxa have been related to allometric changes triggered by a single gene (Sinnott 1935). Gottlieb suggests that only 'yield components' (length, width, weight and number) are determined by many genes and may therefore be expressed in protein electrophoresis. Major morphological differences of presence versus absence, or changed structure, shape or architectural orientation are frequently controlled by only one or two genes.

Although molecular genetic analysis may fail to correspond to major morphological differences, the opposite problem also occurs so that substantial differences at the molecular level are not necessarily related to differences in the behaviour, physiology or ecology of the whole organism. Very few studies have related molecular polymorphisms directly to the performance of populations in nature (Koen and Hilbish 1987). Most genetic variation at the level of the DNA molecule appears to be biologically neutral (Kimura 1983). Indeed, the assumption of neutrality underlies the extensive use of the molecular clock for dating purposes. The main value of the molecular methods is not as the ultimate method for describing diversity in order to set conservation priorities, but as a tool for studying population structure, inbreeding and gene flow. Since conserving genetic diversity is an almost impossible task, and the ecological and evolutionary importance of this diversity is difficult to demonstrate, the message for the conservation biologist seems clear: molecular methods for describing diversity cannot be the basis for conservation policy. Management objectives phrased in terms of 'maintaining genetic diversity' are too general to be workable.

To summarize, speciation is an arbitrary affair: the number of species in a genus or family may have little relevance for the ecologist, the chemist, the herbivore or possibly even for the conservationist. In the search for priorities in conservation, the number of species in a group is not an infallible, or even necessarily a moderately good, measure of biotic diversity. The taxonomists' view of diversity only sometimes coincides with the view of other disciplines and interests (Raven et al 1971). Diversity, like beauty, is in the eye of the beholder.

THE GEOGRAPHY OF BIODIVERSITY

Levels of diversity

The number of species on an island or in a nature reserve, a mountain range or a country is made up of the number of species within each habitat (alpha diversity), the number of habitats and the turnover of species between them (beta diversity), and the turnover within a habitat from one area to the next (gamma or delta diversity) (Whittaker 1972, 1977; Cody 1975). Whittaker (1972, 1977) used the term gamma diversity for the richness of an area or landscape, and delta diversity for the turnover of species between geographical areas in the same habitat. Cody, however, pointed out that landscape or regional diversity is some function of alpha, beta and geographic diversity (Figure 1.1) and suggested using gamma rather than delta diversity for geographic turnover. I follow Cody's practice here.

Cody has recently extended the use of these terms to define the status of rare species (Cody 1986). Thus an alpha rarity is a species that is rare within a community. A beta rarity occurs in a narrow portion of the habitat gradient — it is a habitat specialist. A gamma rarity is geographically restricted and narrowly endemic. There are eight permutations of alpha, beta and gamma rarity giving seven possible types of rarity and one that is common to all levels (Table 1.1). Rabinowitz used an essentially identical classification to describe 'seven forms of rarity' (Rabinowitz 1981;

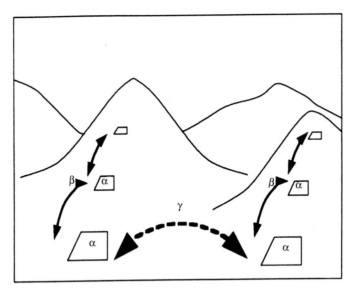

FIGURE 1.1 The three major components of species diversity. α = species richness within habitat, β = species turnover between habitats, γ = species turnover within habitats between geographic regions.

Rabinowitz et al 1986). The distribution of alpha, beta and gamma rarity is statistically independent in the British flora, the only flora in which species have been classified on all three levels (Rabinowitz et al 1986).

TABLE 1.1   The possible permutations of species distribution: alpha refers to within-community abundance; beta refers to habitat specificity (common = wide range, rare = habitat specialist); gamma refers to geographic range (common = widespread, rare = narrow endemic)

alpha	beta	gamma	alpha	beta	gamma
rare	common	common	common	common	common
rare	rare	common	common	common	rare
rare	common	rare	common	rare	common
rare	rare	rare	common	rare	rare

Rabinowitz's study of the British flora used subjective assessments of whether a particular species was alpha common in at least part of its range, beta common or rare (a habitat specialist), or gamma common or rare (geographically widespread or very local). The subjective classification, using published atlases, proved highly successful for gamma rarity, but there was less agreement on beta rarity, in part because of the absence of a standard list of habitats. Estimates of alpha rarity were also relatively poorly correlated between observers. Brown (1984) has provided evidence that although the abundance of a species may vary throughout its range, densities generally peak near the centre of distribution. This means that few species are likely to be alpha rarities throughout their range. Nevertheless, Rabinowitz's subjective method, given adequate monographs, seems a promising way of cataloguing rare species for conservation. More detailed quantitative methods of studying beta and gamma diversity have been developed by Whittaker, Cody and others (reviewed in Wilson and Shmida 1984).

The classification of species of interest according to the three criteria should help to define appropriate conservation action. Thus alpha rarities, especially where they are common in some parts of their range, will be most vulnerable to loss of genetic diversity at small population sizes (Franklin 1980), and to extinction arising from demographic or environmental stochasticity (Shaffer 1981; Soule 1987). Beta rarities can be conserved by setting aside land with the appropriate habitat, and preventing their deterioration (eg preventing drainage of wetlands). Finally, gamma rarities, perhaps the greatest problem for conservation, can be conserved in the wild only by creating many, usually small, reserves to incorporate local endemic species. Gamma rarities are likely to be the costliest to conserve and to manage, since the reserves may often be too small to warrant adequate attention. However, gamma rarities are perhaps the least critical to conserve, since they are most likely to be ecological homologues and poorly differentiated at morphological, ecological and genotypic levels. Gamma rarities are perhaps the best candidates for preservation through cultivation in botanical gardens or seed banks, particularly where their distribution does not coincide with other gamma rarities.

ISLAND BIOGEOGRAPHY THEORY AND AREA-RELATED DIVERSITY

One of the most general observations on the geography of diversity is that species number increases with area (Preston 1962; Williams 1964). On islands, species number increases not only with island area but also with proximity of the island to the mainland or other islands (Preston 1962; Williams 1964). Since many areas of natural habitat now occur as islands isolated from comparable habitats by pastures, cities or croplands, the study of true or habitat islands may offer insight into the effects of insularization. MacArthur and Wilson (1967) attempted to explain patterns of species diversity on islands as a dynamic equilibrium between the extinction rate of species already on the island and the immigration rate of species new to the island. Very few assumptions are needed to show that such an equilibrium is virtually guaranteed (Case and Cody 1983). The difficulty is in demonstrating whether the ecological processes of colonization and extinction are rapid enough to cause a significant shift in species numbers towards the equilibrium before historical events such as climatic changes or major catastrophes intervene to reshape the biota (Ricklefs 1987; Case and Cody 1987).

The equilibrium theory appeared to offer general predictions deduced from first principles for designing nature reserves (Wilson and Willis 1975; Diamond 1975; Diamond and May 1976). For example, it was suggested that single large reserves should support more species than several smaller ones (the single large or several small (SLOSS) debate, Diamond 1975). Basing their views on this, biologists have lobbied successfully for large reserves to maintain maximal species diversity (eg Kruger 1977 for Cape fynbos), although additional arguments such as economies of scale in management are also important. In recent years both the evidence for the equilibrium theory and its application in conservation biology have been strongly criticized (Simberloff and Abele 1976; Gilbert 1980; Boecklen and Gotelli 1984; Boecklen and Simberloff 1986). Burgman et al (1988) have written a particularly useful review of the theory from the perspective of conservation management. The argument that single large reserves are preferable to several smaller ones is among the more cherished tenets of island theory as applied to conservation that have fallen by the wayside. According to both protagonist and antagonist in the SLOSS debate (Soule and Simberloff 1986), the equilibrium theory has nothing to say on the relative merits of large or small reserves for capturing number of species. If small islands together cover a wider habitat range than a single large reserve, their combined species total should, as is often the case, be greater.

Does equilibrium theory still offer anything useful to conservation biologists? In my view the answer is most definitely yes. Unlike alternative theories attempting to explain the species-area curve (such as habitat heterogeneity and its relationship with area), the equilibrium theory explicitly predicts the loss of species due purely to insularization. It also predicts that the magnitude of the loss is directly related to the size of the habitat remnant and its isolation. Curiously, these fundamentally important predictions have barely been addressed in the current debate. A simple test is to compare species lists in isolated habitat patches with species lists from samples of the same area in more extensive remnants. 'Island effects' occur when the isolated patch has significantly fewer species than the 'mainland' remnant (Figure 1.2, Table 1.2).

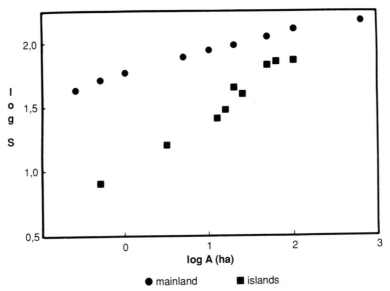

Island vs mainland species - area plots

● mainland   ■ islands

FIGURE 1.2   Species-area curves for fynbos islands surrounded by forest and an extensive 'mainland' tract of continuous fynbos, Knysna plateau, South Africa. The point at which lines drawn through the points intersect gives an estimate of the minimum reserve size (approx 590 ha) for species preservation (Bond et al 1988).

TABLE 1.2   Number of extinctions of vascular plant species on fynbos islands in the Knysna Forest. Extinctions are estimated from differences between 'mainland' and 'island' species lists. Initial island species numbers are assumed to be equal to mainland samples of the same area (estimated by regression in the case of Zeeliesrug) (Bond et al 1988).

| Island | Area (ha) | Species number | | Extinctions | |
		Initial	Present	No	% loss
Small	0,5	53	8	45	85
Forest Creek	3	70	16	54	77
Die Mark	13	88	26	62	70
Bakhuiskop	25	98	40	58	59
Zeeliesrug	70	115	71	44	38

Comparison of island and mainland biotas can sometimes reveal exactly which processes are responsible for insular extinctions (Table 1.3). In the example shown, fynbos islands isolated by Knysna Forest, the absence of short plants, and the scarcity of resprouting plants in the fynbos suggest that longer fire frequencies on islands are the main cause of species loss (Bond et al 1988). Extinctions from this cause could, at least in part, be reduced by careful fire management. Studies of this kind suggest that whether insularization causes extinction or not depends very much on the ability of the taxon of interest to colonize or persist on islands (Cody et al 1983; Case and Cody 1987). Plants that are good colonizers and good persisters tend to show limited island effects and limited island endemism (Simberloff and Gotelli 1984; Cody et al 1983; Bond et al 1988). Mammals, in contrast, are poor colonizers and poor persisters and show very strong island effects and island endemism (Case and Cody 1987; Brown 1971; Choate 1975 for the effects of Kariba).

TABLE 1.3   The distribution of species traits on fynbos islands and mainland in the Knysna Forest. Values are the number of species within each category. Traits that differ in frequency on islands and mainland suggest the causes of island extinctions. Mainland values were obtained from Groot Eiland and have been scaled to the species total for the islands. Small islands are those < 25 ha. X2 is chi square; NS = p < 0,05; * = p < 0,50; ** = p < 0,01.

Dispersal type				
	short (<10m)	medium (<50m)	long (>50m)	X2
mainland	51,7	29	35,3	NS
islands	54	31	31	0,76

Dispersal mode					
	wind	ants	ballistic	vertebrate	unknown	X2
mainland	58,2	22,3	6,9	4,3	22,3	NS
islands	58	27	5	5	19	2,12

Pollination syndrome (woody plants)								
	All islands				Small islands			
	wind	insect	bird	X2	wind	insect	bird	X2
mainland	5,6	47,4	5,0	NS	3,8	31,9	3,3	NS
islands	6	45	7	0,95	4	30	5	1

Growth form (1 = shrub; 2 = graminoid; 3 = forb; 4 = geophyte)									
	All islands					Small islands			
	1	2	3	4	X2	1	2	3	4	X2
mainland	65	18,1	16,3	13,6	*	39,7	11	9,9	8,4	NS
islands	59	28	18	8	8,45	39	15	10	5	2,84

Shrub height (short <1m; medium <2m; tall >2m)								
	All islands				Small islands			
	short	medium	tall	X2	short	medium	tall	X2
mainland	22,9	24,6	11,5	*	15,2	16,2	7,6	**
islands	14	26	19	8,43	7	16	16	13,71

Fire survival in shrubs						
	All islands			Small islands		
	seeder	sprouter	X2	seeder	sprouter	X2
mainland	42,8	16,1	*	28,3	10,7	*
islands	50	9	4,34	34	5	4,18

Where the geography is appropriate (ie 'islands' and 'mainland' exist) studies of island effects can provide useful information on the minimum size of reserve needed to maintain a desired fraction of mainland species, and the kinds of species most likely to be lost. If the origin of the 'islands' can be dated (such as islands in a man-made lake), it is possible to estimate rates of extinction. If rates of local extinction are slow (>200 years for a species), the whole issue falls away.

MINIMUM VIABLE POPULATION SIZE AND THE PREDICTION OF EXTINCTION

A common criticism of the application of the equilibrium theory in conservation is that it does not predict the fate of individual species (eg Burgman et al 1988). Frequently the conservation of a single species is the main management objective in a nature reserve. What factors govern species extinction and is there any relationship between population size and probability of extinction? A body of theory now exists which addresses this problem (Soule 1987). The aim is to predict the minimum number of individuals that will ensure (at an acceptable level of risk) that a population will persist in a viable state for a given period — the minimum viable population (MVP) (Gilpin and Soule 1986).

Deterministic extinction

The causes of extinction fall into two broad and overlapping categories labelled deterministic and stochastic (Shaffer 1981). Deterministic extinction is extinction for a reason — the result of some 'inexorable change or force from which there is no hope of escape' (Gilpin and Soule 1986). This is the most interesting and dramatic form of extinction. For example, tree euphorbias may be declining in the valley bushveld of the eastern Cape because there are no longer elephants to provide openings for their seedling growth (Midgley 1987 pers comm); orchids may be rare in Zululand game reserves because of past nagana spraying campaigns which killed their insect pollinators (Downing and Gibbs Russell 1981); myrmecochorous Proteaceae may be threatened because the invading Argentine ant fails to bury its seeds, which are then eaten by rodents (Bond and Slingsby 1984), and so on. Though southern African ecology has a long tradition of holistic thinking — of believing in the 'connectedness' of things — there is almost no theory and few guidelines on detecting, predicting or dealing with such 'holistic' problems. If ecological communities can be likened to a house, then the enormous effort expended by ecologists in the northern hemisphere on studying competition and other community interactions represents little more than an attempt to learn how to lay the bricks. Blueprints for the whole structure are still way beyond our present scientific capabilities in ecology. However, many field biologists and managers have an excellent 'feel' for the links that connect their systems, and are well placed to detect possible causes of 'deterministic extinction'. Expert systems may provide a useful tool for organizing, storing and extending this body of knowledge (Noble 1987). The challenge is to design methods of testing the predictions of the holistic thinkers without tinkering too much with the system they attempt to explain.

Stochastic extinction

Stochastic extinction is extinction due to unpredictable events, even in an environment that is favourable, on average, for growth and persistence. Shaffer (1981) has divided stochastic extinction into four different sources:

Demographic stochasticity — the variation among individuals in survival and reproductive success.

Environmental stochasticity — variation in births and deaths owing to changing environmental conditions affecting an entire population.

Natural catastrophes — unusual events such as floods, hurricanes and droughts, occurring at random intervals through time.

Genetic stochasticity — changes in gene frequencies due to founder effect, genetic drift, or inbreeding.

The last-mentioned is the source of the familiar 'magic numbers'of 50 to 500 individuals for MVP size (Franklin 1980). The effects of population size and population structure on genetic diversity can be made far more explicit provided information on the degree of genetic variability and the breeding structure of the species is supplied (eg Soule 1987). Although the 'magic numbers' do appear to be of approximately the right order of magnitude, there is little justification for using them when more is known about the biology of the species of interest (Soule and Simberloff 1986).

Demographic models completely independent of genetic considerations have also been developed for the MVP problem (Goodman 1987). They are based on the effects of demographic and environmental stochasticity on birth and death rates in small populations. Demographic stochasticity is familiar to all managers of very small populations (within the range of <10 to <50). It is the element of luck that produces a string of five heifers from the five cows in a new dairy, or the five consecutive heads in a coin-tossing exercise. A true-life example appears to be the case of the last Knysna elephant cow which gave birth to a daughter (the last bull is probably post-reproductive).* As population size increases, the effects of demographic stochasticity rapidly decline — in the same way as the number of heads and tails evens out to nearly 1:1, given enough throws (Richter-Dyn and Goel 1972; MacArthur and Wilson 1967).

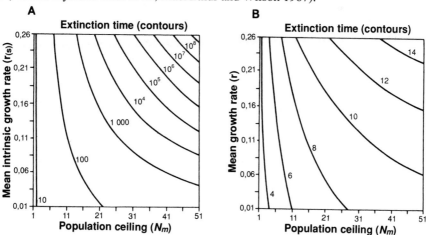

FIGURE 1.3 The effects of (a) demographic versus (b) environmental stochasticity on extinction. Contours represent mean population persistence time as a function of the population ceiling ($N(m)$) and mean population growth rate (r) (Goodman 1987).

However, the situation is far more serious when environmental stochasticity is also considered (Goodman 1987). Goodman developed a numerical model relating the persistence time of a population to its mean growth rate and the variance in growth rates:

$$T(N) = \sum_{x=1}^{N} \sum_{y=x}^{Nm} \frac{2}{y[yV(y) - r(z)]} \prod_{z=x}^{y-1} \frac{V(z)z + r(z)}{V(z)z - r(z)}$$

---

* The daughter turned out to be a son, and the mother has just given birth to a calf. (J Koen pers comm).

where $r(z)$ is the mean per capita population growth rate when the population size is z individuals, $V(z)$ is the variance in the per capita growth rate at population size z, N is the population size in the beginning, and $N(m)$ is the maximum possible population in the area of interest. Using this equation, Goodman showed that previous estimates of extinction greatly underestimate the population size needed if there is any appreciable environmental variance (Figure 1.3). Belovsky (1987) found a fair measure of agreement between the predictions of Goodman's model and the distribution of herbivorous mammals on isolated mountain tops in North America.

Simulating extinction in Proteaceae

The essential ideas of these models may be better understood with a southern African example. Several species of Proteaceae in the southern Cape fynbos are killed by fire but on their death release seeds which initiate a new generation of seedlings. Little or no effective seedling recruitment occurs between fires, so population growth is episodic and occurs only after fires. The per capita growth after a fire can thus be expressed simply as:

$$\lambda = \frac{Nt}{Nt-1} = \frac{number\ of\ seedlings}{number\ of\ parents}$$

Seedling censuses appear to be a reasonable measure of Nt because seedling mortality after the first few weeks of establishment is very low (Midgley 1988). Since 1979 seedling:parent ratios have been censused after many fires in the region for a number of *Protea* and *Leucadendron* species (Bond et al 1984; Midgley 1988). Means and variances of λ can thus be calculated (Table 1.4). The standard deviations are proportional to the mean in nearly all species which indicates that mean λ is distributed as a negative exponential (Figure 1.4). The probability of a population's becoming extinct was simulated by randomly generating λ from a negative exponential distribution with mean $\lambda = 1, 2, 3 \ldots 10$, and multiplying the number of plants before a burn by the simulated λ. The population was allowed to grow without density dependence (a realistic assumption cf Bond et al 1984) to a ceiling $N(m)$ representing the maximum density of individuals the site can support. Population growth was simulated until extinction $(N<1)$ or until the population had survived 100 fires.

TABLE 1.4 Mean *Leucadendron* and *Protea* population growth rate after fire, southern Cape. The data are taken from censuses of fires (n = number of fires) in which the number of pre-burn plants *(N(t–1))* and the number of seedlings *(N(t))* were counted. $\lambda = N(t)/N(t–1)$; sd = standard deviation. *Leucadendron* is dioecious so only female adults, and seedlings of both sexes, were counted.

Species	n	λ	sd
L album	11	2,15	3,073
L comosum	6	5,39	3,856
L eucalyptifolium	18	11,55	18,610
L rubrum	23	2,78	1,870
L uliginosum	13	4,24	5,610
P eximia	36	2,61	2,228
P lorifolia	62	3,77	4,437
P neriifolia	35	1,39	2,386
P punctata	34	2,74	3,257
P repens	56	1,82	2,790

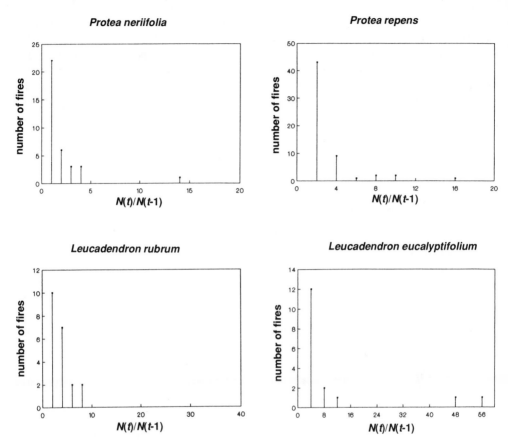

FIGURE 1.4   Frequency distribution of seedling:parent ratios ($N(t)/N(t\text{-}1)$) of some southern Cape Proteaceae. The distributions approximate a negative exponential.

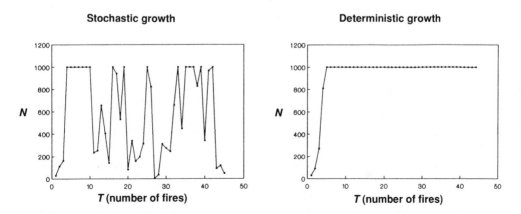

FIGURE 1.5   Stochastic versus deterministic population growth for $\lambda = 3$ (see text for growth model)

An example comparing a stochastic run with the deterministic model is shown in Figure 1.5. The probability of extinction was calculated as the number of times populations became extinct in 100 simulation runs at each λ. Results of the simulation are indicated in Figure 1.6. These indicate that if plants usually produce several seedlings (λ >8), their chances of extinction are very low unless available habitat patches are very small. However, where few seedlings are produced by each parent, (λ < 4), the probability of extinction is high, even where there is space for large populations *(N(m)* >100 000). According to this analysis several *Protea* species which are presently common and widespread (eg *Protea repens, P neriifolia*) would be at severe risk of extinction in isolated reserves.

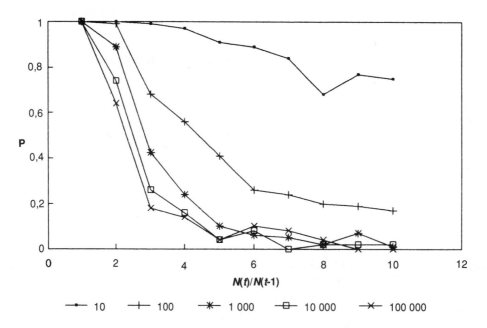

FIGURE 1.6 Probability of extinction for different mean λ *(N(t)/N(t-1))* calculated by simulating population growth over 100 fires. The different lines represent extinction probabilities for different population sizes.

Predictions of the simulation can be tested against actual distributions on isolated habitat patches. Ideally a large number of islands of known dates of origin and with similar N(m)s are needed to test observed patterns of species persistence against expected ones. Merely to illustrate the method I have plotted observed against expected distribution of *Protea neriifolia* and *Leucadendron eucalyptifolium* on fynbos islands isolated by Knysna Forest (Figure 1.7). These are the only Proteaceous species on the islands for which we have post-burn census information (Table 1.4). *P neriifolia* has a very low λ but a high variance in λ according to fire censuses (Table 1.4), whereas *L eucalyptifolium* has the highest mean and highest variance around λ of all southern Cape Proteaceae (Table 1.4), although these figures should be halved to approximate for dioecy. The simulation model predicts poor persistence of the *Protea* on all but the largest islands, and good persistence of *Leucadendron* across the entire range of island sizes. In fact *P neriifolia* is absent from all smaller islands but, contrary to prediction, does occur on larger islands. In

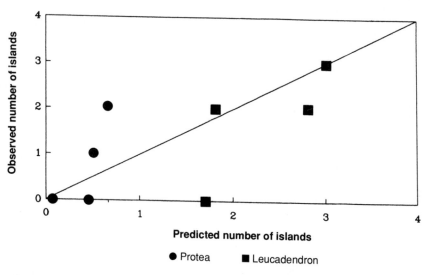

FIGURE 1.7 Predicted versus observed occurrence of *Protea neriifolia* and *Leucadendron eucalyptifolium* on fynbos islands in the Knysna Forest. Occurrence on groups of islands of different size was predicted from Figure 1.6 using the population size each class of islands could support, and mean λ of the two species (eg Table 1.4).

contrast to this *L eucalyptifolium* persists on all islands except the very smallest two. I conclude that the model, though crude and simple, provides at least a qualitative prediction of likely patterns of extinction in the Knysna fynbos islands.

The model could also be tested in the inverse way. *Protea mundii* occurs, as does *L eucalyptifolium*, on some of the smallest islands. There are no data on mean seedling recruitment of this species after fire. The MVP model, however, would predict substantially higher λ in this species than in *P neriifolia*. This prediction should be testable as fire census data accumulate.

In principle these models could be quite easily developed for several situations but there are a number of serious limitations. First they require a great deal of existing information (for example, on the shape of the frequency distribution of seedling:parent ratios), which limits their use to widespread or common species that are well studied and for which there is the least need for prediction of extinction likelihood. Secondly, they are difficult to test and the tests can be applied only to widespread species, preferably on many island-like remnants, although their most urgent area of application is in the conservation of rare species. Thirdly, nobody has yet succeeded in integrating the models with population genetics. Finally, they are almost completely without biology or, rather, the biology is hidden in the demographic statistics. Thus some consequences of small population size, such as predator evasion through rarity or the difficulty of finding mates at small population sizes may be crucial for survival but easily overlooked by the modeller.

CONCLUSIONS

We are still at the stage at which land can be acquired for conservation and choices can be made between different areas and different landuse options. I have expressed my doubts regarding the value of the species concept, and therefore species number, as a guide in determining where to spend the limited conservation budget, but I admit to being unable to offer any alternative. I do not believe we should accept uncritically the rationale that the importance of a genus, family or region with regard to conservation rises with the number of species it contains. I must still be persuaded of the value to humanity of the one million putative mite species before I can accept the urgency of conserving them merely because they have such inventive mate recognition systems, with such high species richness as a byproduct.

South African biologists have expended a great deal of effort on cataloguing and describing much of the biota of this very diverse corner of the earth. Although this task is still far from complete for some groups, I believe the time has come to move beyond inventory and to begin to explore the consequences of fragmentation and degradation of habitats, the invasion of alien species, and the many subtle and insidious effects of people and industry on the survival of our biota. Although new and more sophisticated scientific tools offer much promise for this task and for guiding conservation policy in this country, we should not lose sight of the rich field tradition of 'holistic' thinking in southern African ecology. A healthy dose of eccentric passion would probably do more than many pages of scientific argument to persuade people to conserve biotic diversity and re-examine congealed dogma.

Finally, as a periodic exercise in humility regarding our ability to predict the future and the needs of our descendants, you might like to contemplate the predicament of the Cretaceous botanist arguing the need for conserving the first angiosperm with an imaginary engineer who wishes to flood its riparian habitat. One can imagine the scorn of the politicians and the 'practical' men at the fuss over conserving this rare botanical peculiarity. Perhaps a giant sequoia or a huge *Araucaria* tree would justify the public expense, but surely these shrubby little weeds with their bizarre reproduction quite useless for anything have no redeeming features? The shrubby little weed, of course, gave us the modern biological world with its cacophony of birds, bats, butterflies, bees and all the other organisms dependent directly or indirectly on the astonishing profusion of flowering plants. What modern-day biologist would predict such a magnificent future from such an insignificant presence?

REFERENCES

BELOVSKY G E (1987). Extinction models and mammalian persistence. In *Minimum Viable Population Size.* (ed Soule M E). Cambridge University Press, Cambridge. pp 35 – 57.

BOECKLEN W J and GOTELLI N J (1984). Island biogeographic theory and conservation practice: species-area or specious area relationships? *Biological Conservation* 29, 63 – 80.

BOECKLEN W J and SIMBERLOFF D (1986). Area-based extinction models in conservation. In *Dynamics of Extinction* (ed Eliot D K) Wiley, New York pp 247 – 276.

BOND W J, MIDGLEY J and VLOK J (1988). When is an island not an island? Insular effects and their causes in fynbos islands. *Oecologia* 77, 515 – 521.

BOND W J and SLINGSBY P (1984). Collapse of an ant plant mutualism: the Argentine ant (*Iridomyrmex humilis*) and myrmecochorous Proteaceae. *Ecology* 65, 1031 – 1037.

BOND W J, VLOK J and VIVIERS M (1984). Variation in seedling recruitment of Cape Proteaceae after fire. *Journal of Ecology* 72, 209 – 221.

BRAIN P (1986). Leaf peroxidase types in *Acacia karoo.* Geographical distribution and influence of the environment. *South African Journal of Botany* 52, 47 – 52.

BROWN J H (1971). Mammals on mountaintops: nonequilibrium insular biogeography. *American Naturalist* 105, 467 – 478.

BROWN J H (1984). On the relationship between abundance and distribution of species. *American Naturalist* 124, 255 – 279.

BURGMAN M A, AKCAKAYA H R and LOEW S S (1988). The use of extinction models for species conservation. *Biological Conservation* 43, 9 – 25.

CARLQUIST S (1980). *Hawaii: A Natural History.* 2nd ed. Pacific Tropical Botanical Garden, Honolulu.

CARR G D (1987). Beggar's ticks and tarweeds: masters of adaptive radiation. *Trends in Ecology and Evolution* 2, 192 – 195.

CASE T J and CODY M L (1983). Synthesis: Pattern and Process in Island Biogeography. In *Island biogeography in the Sea of Cortez.* (eds Case T J and Cody M L) University of California Press, Berkeley, Ca. pp 307 – 341.

CASE T J and CODY M L (1987). Testing theories of island biogeography. *American Scientist* 75, 402 – 411.

CHOATE T S (1975). Island populations and their role in wildlife conservation, *Journal of the Southern African Wildlife Management Association* 5, 53 – 62.

CODY M L (1975). Towards a theory of continental species diversities; bird distributions over Mediterranean habitat gradients. In *Ecology and evolution of communities.* (eds Cody M L and Diamond J) Belknap Press, Cambridge, Mass. pp 214 – 257.

CODY M L (1986). Diversity, rarity, and conservation in Mediterranean climate regions. In *Conservation biology*. (ed Soule M E Sinauer, Sunderland Mass. pp 123 – 152.

CODY M L, MORAN R and THOMPSON H (1983). The plants. In *Island biogeography in the Sea of Cortez*. (eds Case T J Cody M L) University of California Press, Berkeley, Ca. pp 49 – 97.

DIAMOND J M (1975). The island dilemma: Lessons of modern biogeographic studies for the design of nature reserves. *Biological Conservation 7*, 129 – 146.

DIAMOND J M (1984). "Normal" extinctions of isolated populations. In *Extinctions* (ed Nitecki M H) University of Chicago Press, Chicago. pp 191 – 246.

DIAMOND J M and MAY R M (1976). Island biogeography and the design of natural reserves. In *Theoretical Ecology* (ed R May) Saunder, Philadelphia, Penn. pp 163 – 186.

DOWNING B H and GIBBS RUSSEL G E (1981). Phytogeographic and biotic relationships of a savanna in southern Africa: analysis of an angiosperm check-list from Acacia woodland in Zululand. *Journal of South African Botany 47*, 721 – 742.

FRANKLIN I R (1980). Evolutionary change in small populations. In *Conservation Biology: An evolutionary-Ecological Perspective*. (eds Soule M W and Wilcox B A Sinauer), Sunderland, Mass. pp 135 – 150.

GANDERS F R and NAGATA K M (1984). The role of hybridization in the evolution of *Bidens* in the Hawaiian Islands. In *Plant Biosystematics*. (ed Grant W F) Academic Press, New York, pp 179 – 194.

GILBERT F S (1980). The equilibrium theory of island biogeography: fact or fiction? *Journal of Biogeography 7*, 209 – 235.

GILPIN M E and SOULE M E (1986). Minimum viable populations: processes of species extinction. In *Conservation biology*. (ed Soule M E) Sinauer, Sunderland, Mass. pp 19 – 34.

GOODMAN D (1987). The demography of chance extinction. In *Viable Populations for Conservation*. (ed Soule M E) Cambridge University Press, Cambridge. pp 11 – 34.

GORDON D H and WATSON C R B (1986). Identification of cryptic species of rodents (Mastomys, Aethomys, Saccostomus) in the Kruger National Park. *South African Journal of Zoology 21*, 95 – 99.

GOTTLIEB L D (1984). Genetics and morphological evolution in plants. *American Naturalist 123*, 681 – 709.

GRANT P R (1986). *Ecology and Evolution of Darwin's finches*. Princeton University Press, Princeton, N.J.

HEDRICK P W (1983). *Genetics of Populations*. Van Nostrand Reinhold Company, Florence, Ky.

HELENURM K and GANDERS F R (1985). Adaptive radiation and genetic differentiation in Hawaiian Bidens. *Evolution 39*, 753 – 765.

HILLIS D M (1987). Molecular versus morphological approaches to systematics. *Annual Review of Ecology and Systematics 18* pp 23 – 42.

KOEHN R K and HILBISH T J (1987). The adaptive importance of genetic variation. *American Scientist 75*, 134 – 140.

KOEHN R K, ZERA A J and HALL J G (1983). Enzyme polymorphism and natural selection. In *Evolution of Genes and Proteins* (eds Nei M and Koehn R K) Sinauer, Sunderland, Mass. pp 115 – 136.

KIMURA M (1983). *The neutral theory of molecular evolution*. Cambridge University Press, London.

KING M and WILSON A (1975). Evolution at two levels: molecular similarities and biological differences between humans and chimpanzees. *Science 188*, 107 – 116.

KRUGER F J (1977). Ecological reserves in the Cape Fynbos: toward a strategy for conservation. *South African Journal of Science 73*, 81 – 85.

LEWONTIN R C (1972). The apportionment of human diversity. *Evolutionary Biology 6*, 381 – 398.

MACARTHUR R H and WILSON E O (1963). An equilibrium theory of insular zoogeography. *Evolution 17*, 373 – 387.

MACARTHUR R H and WILSON E O (1967). *The theory of island biogeography*. Princeton University Press, Princeton, NJ.

MIDGLEY J (1988). Mortality of Cape Proteaceae seedlings during their first summer. *South African Forestry Journal 145*, 9 – 12.

NOBLE I R (1987). The role of expert systems in vegetation science. *Vegetatio 69*, 115 – 121.

PATERSON H E (1963). The species, species control and antimalarial spraying campaigns. Implications of recent work on the *Anopheles gambiae* complex. *South African Journal of Medical Science 28*, 33 – 44.

PATERSON H E (1985). The recognition concept of species. In *Species and speciation, Transvaal Museum Monograph No 4* (ed Vrba E S). Pretoria pp 21 – 29.

PRESTON F W (1962). The canonical distribution of commonness and rarity. *Ecology 43*, 185 – 215.

RABINOWITZ D (1981). Seven forms of rarity. In *The biological aspects of rare plant conservation*. (ed Synge H) Wiley, Chichester.

RABINOWITZ D, CAIRNS S and DILLON T (1986). Seven forms of rarity and their frequency in the flora of the British Isles. In *Conservation Biology*. (ed Soule M E) Sinauer, Sunderland Mass. 182 – 204.

RAVEN P H, BERLIN B and BREEDLOVE D E (1971). The origins of taxonomy. *Science 174*, 1210 – 1213.

RICHTER-DYN N and GOEL N S (1972). On the extinction of a colonizing species. *Theoretical Population Biology 3*, 406 – 433.

RICKLEFS R E (1987). Community diversity: relative roles of local and regional processes. *Science 235*, 167 – 171.

SHAFFER M L (1981). Minimum population sizes for species conservation. *BioScience 31*, 131 – 134.

SHMIDA A and WILSON M V (1985). Biological determinants of species diversity. *Journal of Biogeography 12*, 1 – 20.

SIMBERLOFF D and ABELE L G (1976). Island biogeography theory and conservation practice. *Science 191*, 285 – 286.

SIMBERLOFF D and GOTELLI N (1984). Effects of insularisation on plant species richness in the prairie-forest ecotone. *Biological Conservation 29*, 27 – 46.

SINNOTT E W (1935). Evidence for the existence of genes controlling shape. *Genetics, 20*, 12 – 21.

SOULE M E (ed) (1986). *Conservation biology*. Sinauer, Sunderland, Mass.

SOULE M E (ed) (1987). *Viable populations for Conservation*. Cambridge.

SOULE M E and SIMBERLOFF D (1986). What do genetics and ecology tell us about the design of nature reserves? *Biological Conservation 35*, 19 – 40.

STEBBINS G L (1981). Why are there so many species of flowering plants? *BioScience 31*, 573 – 577.

VAN DER WALT J J A and Vorster P H (1983). Phytogeography of *Pelargonium. Bothalia 14*, 517 – 523.

WHITTAKER R H (1972). Evolution and measurement of species diversity. *Taxon 21*, 213 – 251.

WHITTAKER R H (1977). Evolution of species diversity in land-plant communities. *Evolutionary Biology 10*, 1 – 67.

WILCOX B A (1980). Insular ecology and conservation. In *Conservation Biology* (eds Soule M E and Wilcox B A) Sinauer, Sunderland, Mass. pp 95 – 118.

WILLIAMS C B (1964). *Patterns in the balance of Nature and related problems in quantitative ecology.* Academic Press, New York, NY.

WILSON E O and WILLIS E O (1975). Applied biogeography. In *Ecology and evolution of communities* (eds Cody M L and Diamond J M), Harvard University Press, Cambridge, Mass pp 522 – 534

WILSON M V and SHMIDA A (1984). Measuring beta diversity with presence-absence data. *Journal of Ecology* 72, 1055 – 1064.

WRIGHT S J and HUBBEL S P (1983). Stochastic extinction and reserve size: a focal species approach. *Oikos* 41, 466 – 476.

CHAPTER 2

Patterns of plant species diversity in southern Africa

R M Cowling, G E Gibbs Russell, M T Hoffman, C Hilton-Taylor

INTRODUCTION

The rising concern regarding the imminent extinction of many thousands of species (Soule and Wilcox 1980; Ehrlich and Ehrlich 1981) has rekindled interest in describing, understanding and predicting patterns of species richness (eg Soule 1986). This knowledge is essential for pinpointing areas vulnerable to species loss and for predicting the impact of that loss on community structure and functioning. This is especially true for southern Africa which, according to current estimates, has the world's highest plant species density at the subcontinental level (Gibbs Russell 1985a).

This paper summarizes patterns of plant species richness in southern Africa. As very few publications deal explicitly with this topic, a great deal of data, especially from outside the fynbos biome, had to be compiled and processed. Our approach has been to analyse different components of diversity (alpha, beta, gamma: see p 20) within each of the major southern African biomes (sensu Rutherford and Westfall 1986). The unfolding analysis of the patterns of diversity within biomes yielded predictions for regional richness at the biome and subcontinental levels. Although a grasp of the evolutionary processes which produce species is essential to comprehend diversity patterns (Ricklefs 1987), we have largely forgone discussion of the issue of speciation. Where there is an adequate theoretical and empirical framework, we attempt to discuss factors controlling diversity components. In the next section we outline briefly some of the conceptual and theoretical issues relevant to our analysis.

CONCEPTUAL AND THEORETICAL ISSUES

Attempts at understanding patterns of species richness and the factors controlling these are bedevilled by problems of scale, both spatial and temporal (Whittaker 1977; Bond 1983; Cowling 1983a; Rice and Westoby 1983; O'Neill et al 1986; Auerbach and Shmida 1987; Ricklefs 1987). It is therefore convenient to recognize components of diversity that are independently controlled (Whittaker 1972; Cody 1975, 1983; Bond this volume). These are alpha, beta and gamma diversity.

Alpha diversity refers to the number of species within a community regarded as homogeneous (Whittaker 1972, 1977). Theoretical population biologists (MacArthur 1972; Whittaker 1972; Cody 1975) attempt to explain and predict patterns of alpha diversity in terms of equilibrium models of niche relations, governed largely by biological interactions, especially competition (Shmida and Wilson 1985). More recently there has been a growing awareness that within-community richness is often governed by historical and stochastic factors operating in the short term (eg recurrent disturbance: see Auerbach and Shmida 1987 for a short review), or in the longer term, when evolutionary processes such as migration and speciation are involved (Whittaker 1977; Vuilleumier and Simberloff 1980; Cowling 1983a; Ricklefs 1987).

Beta diversity incorporates the concept of species turnover, or the rate at which species are replaced by others along habitat gradients (Whittaker 1972; Cody 1975). Beta diversity is also defined by Whittaker (1972) as species turnover among different communities in a landscape. As such, it has long been the interest of quantitative phytosociologists who have used a range of techniques to characterize communities floristically and environmentally (Shmida and Wilson 1985). Little attention has been devoted to developing a theory for factors controlling plant beta diversity (Whittaker 1972; Shmida and Wilson 1985). The evolution of habitat specificity, which is independent of predictable biological factors assumed to control alpha richness (Cody 1983), clearly plays a role in determining turnover along gradients. Other factors independent of alpha-level controls, invoked by Cody (1983) to explain beta diversity patterns of bird species, are habitat continuity and contiguity.

Gamma diversity describes the concept of species turnover among similar habitats along geographical gradients (Cody 1975, 1983). It is thus identical to Whittaker's (1977) delta diversity. Gamma diversity is independent of habitat differences (Cody 1983) and invokes the concept of ecologically equivalent species, ie species that occupy the same habitat in different geographical localities (Shmida and Wilson 1985). Very little is known regarding the controls of gamma diversity. Clearly, processes responsible for the evolution of ecologically equivalent vicariants must be involved. Traditionally these problems have provided the material for the (usually) narrative explanations of biogeographers.

Inventory diversity in areas larger than a homogeneous community (eg Whittaker's (1972) gamma diversity) is not independent of the diversity components described above. The size of the regional species pool is a function of the interaction between alpha, beta and gamma diversities which are variously controlled by predictable ecological and stochastic historical processes. In this paper we shall try to explain the richness of the southern African flora in terms of the relative importance of these three diversity components.

DATA AND METHODS

As this volume focuses on conserving biotic diversity, we concentrated on S, the number of species per unit area, rather than on any measure of equitability. Richness in this sense is biologically the most appropriate measurement of species diversity (Whittaker 1972, 1977).

We compared diversity within and among southern African biomes at the 1 m^2 and 1 000 m^2 levels. With a few exceptions, all data were from standard 20 x 50 m plots (127 in all) with nested 1,10 and 100 m^2 subsamples (Whittaker et al 1979; Shmida 1984). Richness at the 1 m^2 level describes point diversity (Whittaker 1977) where species numbers are presumably controlled by biological interactions such as competition (Grime 1979; Bond 1983). We regarded richness at the 1 000 m^2 level as a measure of alpha or within-habitat diversity (Whittaker 1977; also see Rice and Westoby 1983) although significant turnover or internal beta diversity (Whittaker 1977) may be associated with habitat heterogeneity at this scale.

We used inferential statistics to test for differences in mean species-richness among the biomes, although the data cannot be regarded as a fully randomized set of plots. For example, all savanna biome plots were either from the subtropical thickets of the south-eastern Cape or from the mesic savannas of the northern and eastern Transvaal. There were no plots from the extensive arid savannas in the western part of the subcontinent. With the exception of two plots from the humid coastal grasslands of Natal, all other grassland plots were from the south-eastern Cape. There were no plots from the extensive highveld grasslands which comprise the core of the biome. The fynbos biome was oversampled relative to the other biomes. We split the plots into renosterveld and separate south-western, southern and south-eastern fynbos categories.

We correlated richness data with an index of phytochorological diversity (PHD), where

$$PHD = \Sigma -p_i \log p_i \tag{1}$$

in which p_i is the number of species in a phytochorological group expressed as a fraction of the total number of species in the plot (Cowling 1983a).

Richness for larger areas in each biome was compiled from 55 published and unpublished check-lists (Appendix 2.1). These species-area data were fitted to the double logarithmic form of the power function (Preston 1962) by the method of least squares. Thus,

$$\log S = \log k + z \log A \tag{2}$$

in which S and A are species number and area respectively, and k and z are constants.

Richness data at the biome and subcontinental levels were compiled from the PRECIS (Pretoria National Herbarium Computerized Information System) specimen data base (Gibbs Russell 1985b). It is possible to generate check-lists of taxa derived from approximately 610 000 specimens for any combination of quarter-degree grids (see Gibbs Russell 1985a, 1985b, 1987 for further details on PRECIS). Because of the patchy distribution of forests in southern Africa, even relative to quarter-degree grids, we were unable to include forest biome data in the biome-level analysis.

Species turnover along gradients was measured using two simple techniques. As an index of similarity between adjacent communities (beta diversity) or geographic areas (gamma diversity), we used Sorenson's (1948) coefficient of community (CC) where

$$CC = 2S_s / (S_j + S_k) \tag{3}$$

in which S_s is the number of species shared by two samples, S_j being the number in the first and S_k the number in the second sample. CC values were computed for all pairwise comparisons of pooled sample floras from transect intervals along environmental gradients using both published and unpublished data (Appendix 2.1). Mean CC values for all sample comparisons gave a measure of species turnover diversity (Whittaker 1972). These data were graphically depicted by plotting along the gradient *CC* values of samples relative to the first sample. For each gradient, 'internal association', or the similarity of replicate samples from the first transect interval (Whittaker 1972), was arbitrarily set at 0,75. This is towards the lower threshold normally found in real communities (Whittaker 1972).

Beta and gamma diversity were also determined using Wilson and Shmida's (1984) measure. This measure is a modification of Cody's (1975, 1983, 1986a) measure of the rate of species loss and gain along a gradient. Diversity, *B*, is defined thus,

$$B = (g(H) + l(H))/2\bar{S} \qquad (4)$$

where *g(H)* is the number of species gained along the gradient, *H*; *l(H)* is the number of species lost along *H*; and $\bar{S}$ is the mean sample richness. The value of *B* equals exactly the number of community changes or *B*+1 distinct communities along a habitat gradient and hence conforms adequately with the notion of community turnover. Furthermore, *B* is independent of both alpha diversity and excessive sampling (Wilson and Shmida 1984).

ALPHA DIVERSITY

In this section we discuss alpha diversity at the 1 m^2 (point) and 1 000 m^2 scales for communities from different biomes. A brief characterization of the biomes is given in Table 2.1, and their distributions in southern Africa are mapped in Figure 2.1. Further details on the biomes are given in Huntley (1984) and Rutherford and Westfall (1986).

TABLE 2.1   Characteristics of southern African biomes. Data compiled from Huntley (1984) and Rutherford and Westfall (1986) .

Biome	% of southern Africa	Rainfall (mm yr^{-1})	Rainfall (season)	Dominant life forms	Structural characteristics
Fynbos	2,7	210–3 000	Winter (to all year)	Chamaephytes Phanerophytes Cryptophytes	Evergreen sclerophyllous shrubland
Savanna	54,0	235–900	Summer	Hemicryptophytes Phanerophytes	Wooded C4 grasslands
Grassland	13,3	400–2 000	Summer	Hemicryptophytes	Grassland, woody plants absent or rare
Nama-karoo	22,4	100–520	Summer (to all year)	Chamaephytes Hemicryptophytes	Dwarf and low open shrublands
Succulent karoo	4,3	20–290	Winter (to all year)	Chamaephytes	Dwarf and low open succulent shrublands
Desert	3,3	10–70	Summer	Therophytes	Open, ephemeral herbland
Forest	<1	525–2 000	Winter to summer	Phanerophytes	Closed evergreen and semi-deciduous forest

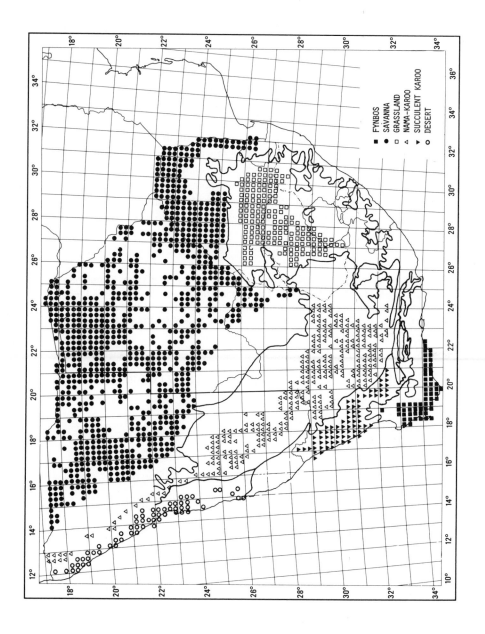

FIGURE 2.1   Biomes of southern Africa according to Rutherford and Westfall (1986). Symbols indicate quarter-degree grids searched in PRECIS (see text) for each biome, and comprise Gibbs Russell's (1987) core biome regions.

Patterns

Species richness in 1 m^2 plots (point diversity of Whittaker (1977)) is assumedly determined by biological interactions independent of the environmental heterogeneity responsible for internal beta diversity, and hence richness at the 1 000 m^2 scale (Bond 1983).

Point diversities in our plots ranged from 1,8 (Nama-karoo) to 33 (renosterveld) (Figure 2.2). Richness in 1 000 m^2 plots ranged from 21 (southern fynbos) to 143 (renosterveld). Fynbos and grassland communities had point diversities that were between 1,4 and 2,4 times higher than communities from the other biomes. At the 1 000 m^2 scale, the three highest mean values were renosterveld (86), grassland (82) and succulent karoo (74). Southern fynbos communities at this scale were marginally poorer in species than those from the south-west and south-east. The two biomes with the most species-poor communities were forest (51) and Nama-karoo (47). Although there were significant differences in mean species richness among biomes at both scales, overall heterogeneity among means was not great (Table 2.2). At both scales the species-poor Nama-karoo and species-rich renosterveld communities showed more significant differences than the communities from other biomes. Other than these, there were no pronounced differences in alpha diversity among southern African biomes.

TABLE 2.2   Results of a one-way analysis of variance, and Tukey's multiple-comparison test (Zar 1984) on plant species-richness data from southern African biomes at the 1m^2 and 1 000m^2 scales. Significant differences between means (P< 0,05) are shown by dissimilar letters or groups of letters in the columns for each scale.

	1m^2	1 000 m^2
F ratio	8,82	5,11
df	8	8
Significance	0,001	0,001
Fynbos, SW	a b	a b c
Fynbos, S	a b c	a b c
Fynbos, SE	a b	a b c
Renosterveld	b	c
Nama-karoo	d	a
Succulent karoo	c d	b c
Grassland	a b c	b c
Savanna	c d	a b c
Forest	a c d	a b

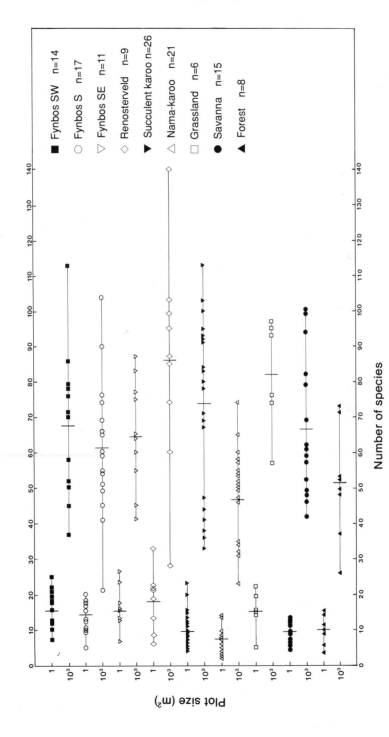

FIGURE 2.2   Plant species richness at the 1 m² and 1 000 m² scales for southern African biomes. Vertical bar indicates mean value. Data from Werger (1978a), Bond (1983), Cowling (1983a), and unpublished data of R M Cowling, C Hilton-Taylor and M T Hoffman.

Global patterns in species richness at scales at and below 1 000 m² have been discussed by Naveh and Whittaker (1979), Bond (1983), Cowling (1983a), Rice and Westoby (1983) and Gentry and Dodson (1987). It is clear from the foregoing that alpha diversity of southern African vegetation does not differ markedly from that in equivalent biomes elsewhere. Fynbos communities were as rich as south-western Australian kwongan but richer than other mediterranean shrublands, except for grazed communities in Israel. These last-mentioned communities are even richer than the richest renosterveld communities. Nama-karoo shrublands were somewhat poorer than comparable semi-arid communities in North America and Australia. The very high richness of succulent karoo communities (up to 113 species 1 000 m⁻²) was unparalleled for semi-arid shrublands. However, a clear assessment will require data from other winter rainfall semi-arid regions (eg Baja, Negev etc). Point diversity of southern African grasslands was similar to the mean of 18 m⁻² for North American grasslands (Peet et al 1983) but much lower than the very high values recorded for unproductive European grasslands (see refs in Bond 1983). At the 1 000 m² scale, southern African grasslands are surprisingly diverse and are on average richer than fynbos and kwongan communities. Point diversity in southern African savannas was lower than that recorded for the pine-wiregrass savannas of North Carolina, but at the 1 000 m² scale differences are not as marked (Walker and Peet 1983). At this scale, southern African savannas had marginally higher richness than tropical Australian savannas, but this appears to be in no way comparable to the extraordinary richness of the cerrado savannas in Brazil (230 spp 1 000 m⁻²) (Eiten 1978). Southern African forests, which include both temperate and subtropical types, were not particularly rich in species. However, all of our plots, except the richest which was from a Natal subtropical forest, came from the southern Cape temperate and transitional types which appear to be particularly species-poor. Richness in these types approximates values recorded for Australian temperate eucalypt forests. As Gentry and Dodson (1987) state so forcefully, in no southern African plant community does richness approach the value of between 173 and 365 species 1 000 m⁻² recorded for neotropical lowland rain forests.

Maintenance

Much attention has been given to the development of theoretical models to explain and predict patterns and maintenance of plant alpha diversity. Ecologists usually distinguish between equilibrium and non-equilibrium models (Chesson and Case 1986; Crawley 1986). The former attribute coexistence to niche differentiation; species coexist only if differences in resource utilization are large enough to prevent competitive exclusion (eg Tilman 1982, 1986; Cody 1986b). Non-equilibrium models invoke a wide array of plant processes or disturbances which prevent competitive exclusion (eg Connell 1978; Grime 1979; Huston 1979). Theoretical and empirical studies have shown that some intermediate level of disturbance, approximately equivalent to the regime prevailing over evolutionary time (Walker and Peet 1983), will maximize species diversity. An overriding determinant of alpha diversity, often ignored in the literature, would be the size of the regional species pool (Bond 1983; Cowling 1983a; Ricklefs 1987).

Most studies on the patterns and maintenance of plant alpha diversity in southern African vegetation come from the fynbos (Campbell and van der Meulen 1980; Bond 1983; Cowling 1983a, 1987; Cody 1986a, 1986b). Cody (1986a, 1986b) presents data on the varied leaf morphology of proteoid shrubs which co-occur in several fynbos communities. This suggests that structural niche differentiation may play an important role in the maintenance of alpha diversity. Tilman (1982, 1986) has developed equilibrium models which predict high alpha richness on the nutrient-poor fynbos soils, because a wider range of nutritional trophic niches is available in these habitats. Bond (1983) and Cowling (1983a) found a negative correlation between species richness and measures of soil fertility in a wide range of fynbos communities in the southern and south-eastern Cape respectively.

Bond (1983) showed that point diversity of southern Cape fynbos communities was related to biomass in a 'humpbacked' or quadratic relationship: highest richness was recorded in mesic sites with intermediate biomass. Campbell and van der Meulen (1980) related the decline in alpha richness in 36-year-old proteoid fynbos relative to more recently burnt vegetation (also see Kruger and Bigalke 1984), to the suppression of the understorey species by a dense overstorey (cf Specht and Morgan 1981). Fire and the effects of understorey-overstorey interactions could thus play an important role in determining species richness in multi-layered proteoid fynbos communities. As predicted by Bond (1983), greatest diversity should be expected at intermediate fire frequencies and biomass levels (Huston 1979). Variable post-fire recruitment of overstorey proteoid shrubs (Bond et al 1984) is also an important determinant of alpha richness (Esler and Cowling 1989).

We have shown that at the alpha level fynbos communities are not especially diverse. What is unusual, and a feature fynbos shares with tropical rain forests (Hubbell and Foster 1986a), is high within-guild richness. Many of these apparently trophically equivalent species (Shmida and Ellner 1984), usually within the ericoid growth form (Campbell 1985) (eg 15 species of *Erica* in 50 m^2 plots (Bond 1981)), have similar regeneration niche components (Grubb 1977) in that recruitment from small soil-stored seed banks is confined to post-fire conditions (S M Pierce pers comm). Inspired by the non-equilibrium coexistence models of Chesson and others (Chesson and Warner 1981; Comins and Noble 1985; Chesson 1986; Chesson and Huntly 1989), Cowling (1987) argued that the coexistence of trophically equivalent fynbos species could be a function of differential post-fire recruitment. The relative abundance of each species may vary after each fire, depending on the size of the pre-fire viable seed bank, post-fire moisture conditions and seed and seedling predation. Many of these factors vary stochastically so that each fire is unique in its effect on the recruitment potential of any given co-occurring species. No single species may dominate for any length of time, and coexistence is maintained.

Bond (1983) stressed the importance of the size of the regional species pool in determining alpha richness of fynbos communities. He explained the low diversity of fynbos 'islands' enclosed by southern Cape forests in terms of their isolation from 'mainland' mountain communities (Bond et al 1988), and ascribed the richness of the transitional waboomveld (communities dominated by *Protea nitida*) to the availability of both renosterveld and fynbos species pools. Shmida and Wilson (1985) associate the richness of transitional regions with a 'mass effect' ie the establishment of species in sites where they are not self-maintaining in the long term, as a result of the continual rain of propagules from adjoining favourable sites. The high beta diversity within most fynbos landscapes (see later) results in a large regional species pool, and a proportionally high area of inter-community contact zones. Species may establish in adjacent communities at low densities or, depending on the degree of habitat specialization and post-fire population changes of other trophic equivalents, their populations may expand. The greater the size of the species pool available for immigration to any particular community, the richer that community will be. This model is similar to that proposed by Hubbell and Foster (1986a, b) for the maintenance of diversity in neotropical lowland rain forests.

Since the size of the regional species pool is strongly determined by historical processes, it follows that these could be important determinants of alpha diversity. In the south-eastern Cape pure fynbos communities (ie with low phytochorological diversity indices) are confined to the mesic upper slopes of the Folded Belt (Cowling 1983b). They are poorer in species than the phytochorologically mixed grassy fynbos communities of drier, more fertile sites (Figure 2.3; also see Cowling and Campbell 1983, 1984). Grassy fynbos comprises a rich admixture of species from grassland, renosterveld and karroid shrublands, all of which are juxtaposed in the phytochorologically complex south-eastern Cape (Cowling 1983b). The comparatively depauperate pure fynbos communities are also poorer in species than the communities in similar habitats with the same dominants on the 'mainland', to the west (Cowling 1983a). The lower richness of these south-eastern Cape communities may be a function of smaller species pools on the pure fynbos 'islands'. The extent to which this is a true island effect (MacArthur and Wilson 1967) or a consequence of the time that has elapsed for the migration into these uplands of mesic fynbos species since climatic amelioration some 12 000 BP, remains unclear (Cowling 1983a). These hypotheses could be tested by comparing beta diversity patterns along comparable gradients on mainland and island sites. There is tremendous scope, for those brave enough to face the floristic complexity, for examining the relationship between the size of regional species pools and local richness in different parts of the fynbos.

Much less is known about the maintenance of alpha diversity in communities from other biomes. In their studies on the community structure of savanna vegetation at Nylsvley, Whittaker et al (1984) observed 1,3 to 2,5 times more species in 1 000 m^2 plots in oligotrophic (*Burkea*) than in eutrophic (*Acacia*) communities (cf Huston 1979; Tilman 1982). Whittaker et al (1984) interpreted savanna diversity in terms of an essentially equilibrium model invoking a complex array of microsites for species with different regeneration and trophic niches (also see Smith and Goodman 1987; Yeaton 1988a). A more dynamic view of tropical African savannas offered by Pellew

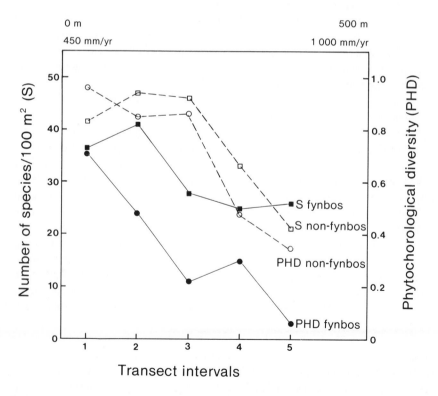

FIGURE 2.3   Relationship between species richness and phytochorological diversity (PHD) at five correspond-ing fynbos and non-fynbos sites along an altitudinal gradient in the Gamtoos River Valley, SE Cape. For fynbos sites: $r = 0,79$, $P < 0,05$; non-fynbos: $r = 0,93$, $P < 0,01$. See Methods for computation of PHD. Details on the gradient are given in Appendix 2.2.

(1983), Bell and Jackmann (1984) and others, focuses on the role of fire, herbivory and the physical impact of large ungulates on the balance between grass and woody components. Recently, Yeaton (1988a) discussed the role of fire and ringbarking by porcupines in preventing mesic *Burkea* savanna from developing into a closed-canopy woodland, and presumably losing many shade-intolerant species. Other studies (eg Smith and Goodman 1986) suggest that interspecific competition among tree species is an important determinant of savanna structure, particularly in xeric *Acacia*-dominated communities. We have not found any published papers that have investigated experimentally the implications of these processes for the maintenance of species diversity.

The subtropical thickets of the south-eastern Cape differ from savannas in that they are mostly closed shrublands or low forests and are not fire-prone (Cowling 1984; Everard 1987). Their inclusion in the savanna biome is problematic (Rutherford and Westfall 1986). The high diversity of these communities could be a function of the wealth of growth forms (including evergreen and deciduous trees, arborescent rosette and stem succulents, vines, grasses and geophytes) occupying different trophic niches and recruited from several phytochoria (Cowling 1983a). However, the thickets are invariably dominated by evergreen shrubs and trees recruited from adjacent subtropical forests (eg *Cassine, Euclea, Sideroxylon*). Superficially, these species seem to have similar requirements for regeneration and growth. Most produce bird-dispersed fruits asynchoronously, both within and among species (Pierce and Cowling 1984). This variable reproduction, in association with the temporally and spatially patchy occurrence of sites suitable for seedling establishment (mainly large herbivore damage in precolonial times), could explain the high richness of these communities (Cowling 1983a). In contrast, thickets in the south-western Cape are often species poor, with one or few dominant species (Cowling 1983a). This situation is probably a result of a low regional species pool, caused by insufficient time for the westward diffusion of subtropical taxa into the south-western Cape since Holocene warming.

We have seen no data on patterns and maintenance of species richness in grasslands, despite many decades of intensive research in this biome (Tainton 1981). Up to 18 grass species may coexist in 1 000 m^2 of humid grassland in Natal (Cowling 1983a). How do so many species within the same guild coexist? Differential responses to grazing, fire and local disturbances (Yeaton et al 1986) could affect competitive hierarchies. There is evidence for phenological separation and resource partitioning among mesic grassland species (Tainton 1981 and refs therein). In semi-arid grasslands long-term competitive interactions may be dampened by year to year changes in species composition and dominance resulting from variable rainfall patterns (Dye and Walker 1987).

There appears to be a latitudinal gradient in the richness of southern African forest communities. The most diverse forests occur in north-eastern subtropical regions and the poorest in the temperate south-western Cape (Cowling 1983a). Even within the Cape region, forest alpha diversity decreases dramatically from the Knysna region to the Cape Peninsula (McKenzie 1978). Again, these patterns could reflect the south-westward diffusion of the forest flora after climatic amelioration in the early Holocene (Cowling 1983b).

The roles of competition and disturbance in determining the richness of forest communities in southern Africa have not been studied. In this respect there could be fundamental differences between the temperate Afromontane and the subtropical forest types. The latter have many more vine species which could be responsible for a high frequency of gap disturbances (Putz 1984) and higher species richness (Hubbell and Foster 1986a).

Nama-karoo communities comprise a matrix of long-lived dwarf shrubs and a group of shorter-lived interstitial species, particularly grasses. The communities in the harsh and unpredictable Nama-karoo environment are comparatively poor in species and growth forms. Competition may be an important determinant of shrub community structure but spatial patterning and the richness of these communities have still to be studied. Overgrazing, which may result in the elimination of certain growth forms and species in these semi-arid shrublands, resulted in a reduction in richness of a community in the south-eastern Cape (Hoffman 1988).

Succulent karoo communities include a spectacular array of growth forms, particularly within the succulents (Hilton-Taylor 1987; Van Jaarsveld 1987). Annuals and geophytes coexist with short-, medium- and long-lived shrubs, mostly within the Mesembryanthemaceae, in this relatively benign region with its predictable but low winter rainfall (Jurgens 1986; Hoffman and Cowling 1987). As a result of increased population turnover among shorter-lived mesembs (*Aridaria, Malephora, Psilocaulon*), competition may be less important in structuring these communities than it is in the Nama-karoo. A study of the maintenance of species richness in these very diverse communities is an exciting prospect.

There are no published data on the diversity of southern African desert communities. Namib plant communities appear exceptionally species-poor (approximately $0 - 10$ species per 1 000 m^2) (Cowling and Hoffman pers obs) but this would exclude ephemerals which appear only after exceptional rains (Walter and Breckle 1986). In this hyperarid region, richness is related directly to moisture levels. For example, dune slope grasses in the Namib increased from a single species in the west (approximately 10 mm yr^{-1}) to four species in the east (approx 85 mm yr^{-1}) (Yeaton 1988b).

BETA DIVERSITY

Despite the importance of beta diversity in determining regional species richness (Whittaker 1977; Shmida and Wilson 1985; Cody 1986a) there are, with the exception of early papers by Whittaker and colleagues (Whittaker 1972), few empirical studies on plant patterns. Theoretical and empirical insights on the controls of beta diversity are also limited. Whittaker (1977) has explained the evolution of beta diversity in terms of displacement and divergence associated with the arrival of additional species along habitat gradients. Shmida and Wilson (1985) suggest that niche differentiation increases alpha rather than beta diversity. They argue, quite logically, that habitat diversity is the ultimate determinant of beta diversity. Mass effects may increase or decrease beta diversity depending on the spatial arrangement of contiguous habitats (see examples in Shmida and Wilson 1985). Cody (1975, 1983) has shown that bird beta diversity patterns are correlated with the area, contiguity and spatial arrangements of different habitat types. For example, turnover would be high between two relatively large and contiguous habitats, and low between a large and a small habitat. Species uniquely adapted to the large habitat would establish, possibly as a result of mass effects, in the smaller habitat, resulting in a low turnover between them. This model assumes that the small, isolated habitat would have fewer habitat specialists and, overall, fewer species. Regional and historical processes associated with the size of habitat species pools and the evolution of habitat specialists are also important in understanding beta diversity patterns.

Because of the lack of studies on plant beta diversity in southern Africa, we present previously unpublished beta diversity data for a number of community gradients or coenoclines (Figure 2.4). Comparisons of turnover are problematic owing to differences in the underlying habitat gradients. Details on the coenoclines and habitat gradients are given in Appendix 2.2.

Figure 2.4a confirms the impression of high turnover in succulent karoo communities. The coenocline comprises four communities along a gradient of decreasing soil depth over a horizontal distance of 100 m in southern Namaqualand. Beta diversity was associated with a high turnover of Mesembryanthemaceae species and growth forms which indicates extraordinary habitat specialization within this group (Jurgens 1986).

The Sundays River Valley coenocline (Figure 2.4b) includes four communities located at 50 km intervals along a complex gradient of increasing aridity, from the coast (mesic) inland. Vegetation ranges from coastal subtropical succulent thicket (savanna biome) to inland karroid shrubland (Nama-karoo biome), and thus the gradient transcends a biome boundary (Hoffman 1989). Considering the nature of the underlying gradient (Appendix 2.2), beta diversity is not particularly high, which indicates a gradual transition between the savanna and Nama-karoo (Werger 1978a). Turnover was marginally higher along the overgrazed coenocline, in spite of the grazing-induced local extinction of some habitat specialists (Hoffman 1989). However, the loss due to overgrazing of widespread palatable species, together with the establishment of a habitat-specific ephemeral disturbance flora ('opslag'), could explain these patterns.

Matched fynbos and non-fynbos sites were sampled along a topographic-moisture gradient (0 – 500 m altitude) in the Gamtoos River Valley (Figure 2.4c). The fynbos coenocline ranged from grassy fynbos at the lower, drier sites to pure proteoid fynbos at the mesic, high-altitude site. The non-fynbos coenocline ranged from xeric succulent thicket (savanna biome) to mesic Afromontane forest (forest biome), and thus crossed a biome boundary (Cowling and Campbell 1983). Beta diversity was very similar for both coenoclines, confirming the impression that fynbos beta diversity in the southern and south-eastern Cape mountains is not exceptionally high, particularly when compared with patterns in the south-western fynbos. This is especially true of the wet coastal Outeniqua and Tsitsikamma Mountains where relatively uniform *Leucadendron eucalyptifolium*-dominated communities occur from the foothills (400 m) to mid-elevations (1 000 m) on south-facing slopes (Bond 1981; Campbell 1985). There appears to be greater turnover on the inland mountains (eg Swartberg; cf Bond 1981), owing possibly to more pronounced between- habitat differences in soil moisture in these drier regions.

Studies on the Agulhas Plain in the south-western Cape (Figure 2.4d) reflect fully the high beta diversity attributed to fynbos (Whittaker 1977; Kruger and Taylor 1979; Linder 1985; Cody 1986a). Five communities sampled within a 5-km transect on a topographically and climatically uniform coastal plain, and representing a gradient from least to most fertile soils, had 3,9 complete changes in community composition, which is equivalent to 4,9 communities. Somewhat lower, but nonetheless comparable turnover patterns along a similar habitat gradient, were recorded for analogous vegetation types (Beard and Pate 1984) in a climatically and edaphically similar region in south-western Australia (Milewski and Cowling 1985).

What is responsible for the high habitat specificity and hence the high turnover between these fynbos communities? Preliminary data show low growth rates and high mortality of reciprocally transplanted seedlings of Agulhas Plain species confined to calcareous (calcicole) and acidic (calcifuge) substrata (Mustart and Cowling unpublished; Van der Leij and Cowling unpublished). Even small differences in the relative growth rates of seedlings may translate into a substantial competitive effect in the field (Givnish 1986). The physiological mechanisms, for the calcicole-calcifuge effect, could include the effect of calcium per se (Jeffries and Willis 1964), a pH effect on the solubility of heavy metals (Rorison 1960) or a pH effect on nitrification (Gigon and Rorison 1972). A further factor could be the high degree of host specificity of symbiotic microbes (Cowling 1987) which enhance nutrient uptake in the infertile fynbos soils (Lamont 1982). Mycorrhizal fungi play a vital role in the nutrition of fynbos Ericaceae (Straker and Mitchell 1985). Mycosymbionts are particularly sensitive to pH and other soil characteristics (Lapeyerie and Bruchet 1986; Porter et al 1987) so it is reasonable to expect a high degree of edaphic specificity. There is evidence of host specificity among ericiod endomycorrhizae (Straker et al 1988). Of interest here is the study by Berliner et al (1986) which suggested that the absence of *Cistus incanus* from basaltic soils in Israel was related to the failure of ectomycorrhizal development in these soils. Bacteria of the genus *Rhizobium* also show marked host specificity

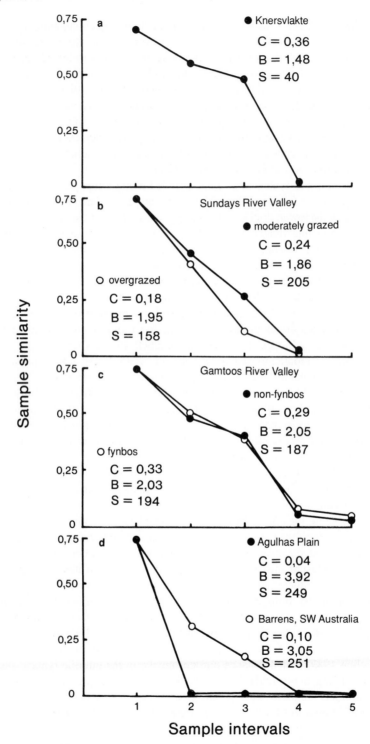

Sample similarity

Sample intervals

with fynbos nitrogen-fixing legumes, including the large genus *Aspalathus* (Staphorst and Strijdom 1975; Deschodt and Strijdom 1976, also see Spaink et al 1987). The role of microbial host-specificity in determining habitat specificity and the evolution of edaphic endemics (cf Thompson 1987) is an exciting research prospect for fynbos biologists.

Turning to Cody's (1983) habitat-area hypothesis, we find no evidence to suggest that beta diversity between two large contiguous communities is higher than between adjacent communities occupying large and small habitats. The *CC* values between *Protea compacta* proteoid fynbos on acid sand, which occupies a large area on the Agulhas Plain (Cowling et al 1988), and *P obtusifolia* (limestone) and *P susannae* (weakly acid sand), which occupy relatively small areas in the region, were 0,07 and 0,08 respectively. Similarly, the *CC* value between *P compacta* and *P repens* fynbos, which is confined to isolated silcrete and ferricrete outcrops and is one of the most restricted communities on the Agulhas Plain, was 0,05. The corresponding value between the *P compacta* community and mesotrophic asteraceous fynbos, another widespread community on the Agulhas Plain, was 0,08. There appears to be a similarly large turnover between widespread midslope proteoid fynbos communities and restricted summit heathlands in the mountains of the fynbos biome (W J Bond pers comm). In this respect, the beta diversity of sedentary and often highly habitat-specific organisms such as plants is determined by factors fundamentally different from those controlling bird species turnover.

Our data show moderately to exceptionally high beta diversity for a range of communities from the Cape coastal region. We know nothing of a the turnover patterns in other parts of the subcontinent. Yeaton et al (1986) and Yeaton (1988a) hint at considerable turnover in the grass and tree layers of *Burkea* savanna along a moisture gradient at Nylsvley. Undeniably, more data are required from biomes other than the fynbos.

GAMMA DIVERSITY

Gamma diversity, which here is the same as Whittaker's (1977) delta diversity, is controlled largely by historical processes associated with the speciation of ecological equivalents in similar habitats along geographical gradients (Cody 1986a). In a region of high gamma diversity, the slope of a species area curve with nested quadrats would be steeper than in a region of lower gamma diversity, owing to the contribution of ecologically equivalent species (Shmida and Wilson 1985). However, most geographical gradients incorporate some degree of environmental and thus habitat change, so it is usually difficult to separate the beta and gamma components of diversity (Cody 1986a).

FIGURE 2.4   Species turnover along environmental gradients (beta diversity) in southern Africa. Plotted values are Sorenson's similarity coefficients (CC) to first gradient sample. Internal association of first sample is standardized for all gradients at 0,75. C = mean CC for all possible pairwise combinations of samples; B = Wilson and Shmida's (1984) beta diversity measure; S = number of species in the composite sample. See Methods for details. a) Gradient of increasing soil depth spanning approximately 100 m horizontal distance in the succulent karoo. b) Gradient of increasing aridity spanning approximately 250 km from sub-humid subtropical succulent thicket to semi-arid karroid shrubland in the Sundays River Valley, SE Cape. Overgrazed and moderately grazed vegetation were sampled at each site. c) Complex topographic-moisture gradient from sea-level to 500 m in the Gamtoos River Valley, SE Cape, incorporating altitudinally matched fynbos and non-fynbos sites. Fynbos sites were largely forms of proteoid fynbos; non-fynbos sites ranged from subtropical succulent thicket to Afromontane forest. d) Sites located on different soil types representing a crude gradient of increasing soil fertility. At all but the most fertile sites, vegetation was fynbos (at the Agulhas Plain) and kwongan (the Barrens). Details on all the gradients are given in Appendix 2.2.

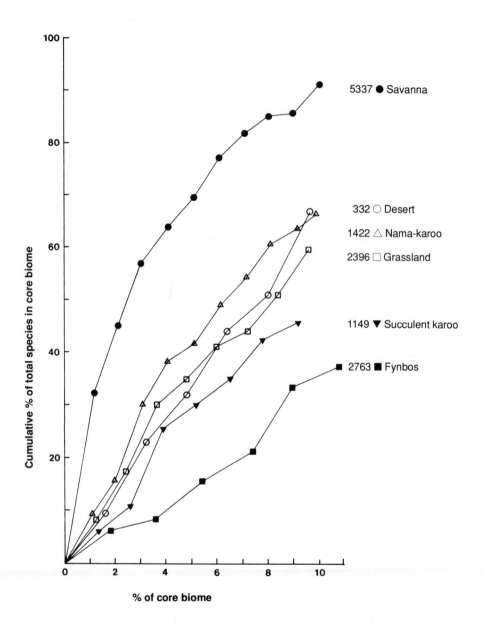

FIGURE 2.5   Relationship between area and cumulative species numbers in core regions of southern African biomes (see Figure 2.1). Data compiled from a random selection of quarter-degree grids with a collecting intensity of >100 specimens, searched in PRECIS to cover approximately 10% of each core biome. Number of quarter-degree grids searched in savanna biome = 91; Nama-karoo = 29; grassland = 17; succulent karoo = 8; desert = 6; fynbos = 6.

The south-western fynbos exhibits exceptionally high gamma diversity. Kruger and Taylor (1979) showed that the percentage difference (complemented Sorenson's coefficient) among four sites within 75 km of one another, was between 50 and 70%. The floras of Swartboschkloof and Biesievlei, which lie on opposite sides of the Jonkershoek valley at similar elevations, differ by 46% (Kruger and Taylor 1979; also see Linder 1985). Cody (1986a) has shown that turnover of *Protea* and *Leucadendron* species along geographical gradients is greater in the western than in the southern parts of the biome, which he attributes to steeper environmental gradients in the former region. There is little turnover among dominant species at least between proteoid fynbos communities in similar habitats in the Outeniqua (Bond 1981) and Elandsberg Mountains (Cowling 1984) some 300 km apart in the southern and south-eastern Cape respectively. On a more limited scale, turnover among three *Protea compacta* sites occupying similar habitats and occurring within 15 km of one another on the Agulhas Plain in the south-western Cape, was greater (CC = 0,21; B = 1,32) than similarly dispersed *P neriifolia* sites on the Humansdorp lowlands of the south-eastern Cape (CC = 0,43; B = 0,89).

Figure 2.5 shows for each biome the relationship between the cumulative species number and the area, both expressed as a percentage of the total for each core biome (see Table 2.4). Using PRECIS data, approximately 10% of each biome was sampled. Shallow curves, in which a low proportion of the species pool is sampled with increasing area, could indicate high gamma diversity. A simple alternative is that the pattern is a consequence of the larger biomes' (eg savanna) receiving effectively larger samples than the smaller ones (eg fynbos). Ideally, comparisons should be confined to similar sized biomes (see Table 2.4). These biomes often showed markedly different curves indicating that the pattern could have biological relevance. After 10% of the core region of the fynbos had been sampled, 37% of the biome species pool was counted compared with 91% for the savanna. The other biomes showed intermediate but similar values (c 60 – 70%) except for the succulent karoo which, at 46%, was closer to the fynbos.

BIOME-LEVEL RICHNESS

Most of the biomes of southern Africa are floristically distinct indicating considerable compositional turnover at this scale (Gibbs Russell 1987). With the exception of savanna and grassland, all pairwise comparisons of similarity yield Sorenson's coefficient of less than 30% (Figure 2.6). Fynbos showed strongest similarities to the succulent karoo and these two biomes share many centres of diversity for large genera, strengthening the earlier claim of Bews (1925) for a distinct southern, winter-rainfall flora (Gibbs Russell 1987). The Nama-karoo is the least distinctive biome, showing a similarity of more than 10% with all biomes and the strongest relationships with the succulent karoo and desert. The floristic relationships among the biomes are discussed more fully by Gibbs Russell (1987).

Before proceeding with a discussion of biome-level richness, we refer briefly to the species-area curve (Figure 2.7) which was compiled from inventories of species richness for different-sized areas within each of the biomes (Appendix 2.1). Sites with transitional floras, ie outside the core biome regions (Figure 2.1), are indicated as such. Species-area relationships were computed separately for the full data set and for fynbos, savanna, grassland and succulent karoo sites (Table 2.3). We found that in each case the double logarithmic transformation of the power curve (Preston 1962) gave a better fit than untransformed or semi-log plots. With the exception of the biome core regions and the Cape Peninsula-Cape of Good Hope Nature Reserve sites, all sites were independent and never contiguous (cf Preston 1962). Removal of these larger sites did not alter the slopes significantly (F test; P< 0,05; Sokal and Rohlf 1987).

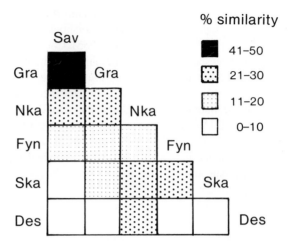

FIGURE 2.6   Matrix of Sorenson's coefficients of similarity for southern African biomes. Check-lists compiled from PRECIS for core biomes (see Data and Methods) were used to calculate similarity coefficients. Sav = savanna biome; Gra = grassland biome; Nka = Nama-karoo biome; Fyn = fynbos biome; Ska = succulent karoo biome; Des = desert biome. Data from Gibbs Russell (1987).

There has been much debate on the biological significance of the slope (z) and intercept (k) of the power function as applied to species-area curves (eg Connor and McCoy 1979). The magnitude of z is a measure of the rate of accumulation of species with increasing area. It is logically and statistically impossible to apportion beta diversity to the slope parameter (Connor and McCoy 1979). Only in the traditional nested form of the species-area curve is information about the similarity in species composition of different samples retained (Shmida and Wilson 1985). We have no intrinsic problems with the biological interpretation of k, whether in absolute or comparative terms, since all our areas were measured in the same units (km^2) and k (or y when x=1) was never a distant extrapolation from the actual data (Gould 1979). For all data sets area ranges overlapped extensively (Figure 2.7) and no slopes were significantly different. We could therefore factor out area by comparing the ratio of k values for curves with similar, constant slopes

TABLE 2.3   Regression coefficients for log-log species-area curves ($\log S = \log k + z \log A$). Observed species numbers are taken from Tables 2.4 and 2.5. Predicted species numbers are from regression.

Region	n	k	z	R^2	No species Observed	Predicted
All biomes	55	302,0	0,180	0,560	21 000	
Fynbos	13	457,1	0,210	0,791	7 316	4 751
Savanna	10	195,0	0,200	0,758	5 788	3 312
Grassland	6	165,9	0,248	0,800	3 788	3 917
Succulent karoo	6	199,5	0,204	0,846	2 125	2 150

(Gould 1979). Thus, for certain biomes there is an invariant ratio in the number of species regardless of area.

With the exception of the recent work by Gibbs Russell (1987), all biome-level treatments of species richness have focused on the fynbos (eg Levyns 1938; Weimarck 1941; Goldblatt 1978; Oliver et al 1983; Bond and Goldblatt 1984). Owing to the high levels of alpha, beta and gamma diversity, we would expect the fynbos to be extremely species-rich (also see Linder 1985). Indeed, the biome is remarkably rich, with an estimated 7 316 species in the core region (Table 2.4). The biome includes the Cape Floristic Kingdom, with exceptionally high levels of regional and local endemism (Goldblatt 1978; Gentry 1986). It is by far the richest of the world's mediterranean regions and one of its most speciose parts (Goldblatt 1978; Raven and Axelrod 1978; Kruger and Taylor 1979).

TABLE 2.4   Number of species in core regions of southern African biomes. The quarter-degree grids searched in PRECIS (see Methods) are shown in Figure 2.1. Each quarter-degree grid covers approximately 666 km² . Data from Gibbs Russell (1987).

	No Species	Core area $(10^3 km^2)$	% of total biome[1]	Species/km²
Fynbos	7 316	36,6	52,4	0,19
Savanna	5 788	632,0	45,3	0,01
Grassland	3 788	111,9	32,6	0,03
Nama-karoo	2 147	198,5	34,3	0,01
Succulent karoo	2 125	50,6	45,5	0,04
Desert	497	41,3	48,8	0,01

1 Sensu Rutherford and Westfall (1986)

The z value of 0,21 for the fynbos data (Table 2.3) was lower than that recorded by Kruger and Taylor (1979) for a subset of the sites in Figure 2.7. However, its magnitude is best appreciated by comparing it with the value of 0,16 ($k = 2,61$) recorded by Johnson and Raven (1970) over a similar range of areas in the environmentally similar southern Californian region. Fynbos had the highest value of k (Table 2.3) and intercept ratios of 2,3 with savanna and grassland, all of which had homogeneous slopes. Thus for any given area, fynbos will have 2,3 times more species than the other two biomes. All fynbos sites in Figure 2.7 are on or well above the regression line for the combined data with the exception of site 5, which is located in the depauperate Outeniqua Mountains (Bond et al 1988). Most other comparably rich sites included transitional floras whose enrichment could be a consequence of regional mass effects (Shmida and Wilson 1985). The other fynbos sites are in the south-western Cape which, as we have shown, has the highest beta and gamma diversity in the biome, and hence, the highest concentration of species (Figure 2.8). Not surprisingly, it is this region that has the greatest number of local endemics (Weimarck 1941; Cowling unpublished) and rare and endangered taxa (Hilton-Taylor and Le Roux this volume).

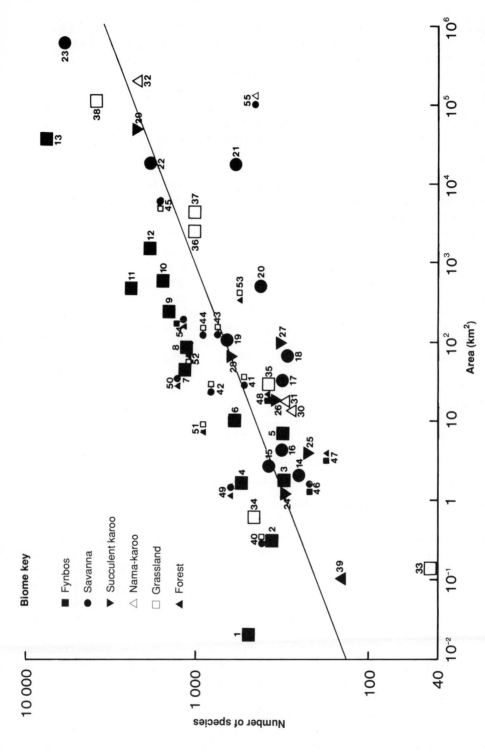

FIGURE 2.7   Species-area relationship for floras from a range of sites in southern Africa. Combined symbols indicate regions with floras transitional between biomes. List of sites is given in Appendix 2.1.

The southern and eastern portions of the fynbos biome are probably not exceptionally rich in species (Levyns 1938; Weimarck 1941).

Although savanna communities are species-rich, the moderately low beta and gamma diversity limit the size of the flora. The core region includes 5 788 species (Table 2.4) of which 43% are endemic (Gibbs Russell 1987). The slope of the species-area curve is similar to that of fynbos but the intercept is appreciably lower (Table 2.3). Dry and topographically uniform regions such as the Kruger National Park (Site 21) and the savanna-Nama-karoo transition in the southern Kalahari (Site 55) are relatively depauperate (Figure 2.7). However, by African standards, the savanna is very rich. For example Exell (1971) records just 6 000 species for the combined and largely savanna areas of Botswana, Malawi, Mozambique, Zambia and Zimbabwe, covering more than four times the area of the core savanna region. Thus, the southern African savanna flora ought not to be viewed as a depauperate intrusion of the Sudano-Zambezian flora (White 1983). It is an important centre of diversity and speciation for many genera (Gibbs Russell 1987).

On the basis of the high alpha and moderate gamma diversity in the grasslands, we would expect the biome to be relatively species-rich. A core region richness of 3 788 species (Table 2.4) is surprisingly high, particularly since it precludes the species-rich Afromontane grasslands of the eastern escarpment (eg Killick 1963; site 51 in Figure 2.7). The grassland sites in Figure 2.7 were drawn largely from the interior Highveld, included by White (1983) in the 'depauperate' Kalahari-Highveld Transition Zone. This flora cannot be regarded as a southward extension of the savanna herbaceous layer. Floristically, the grassland biome is distinct and comprises a centre of diversity for many large genera including many that are shared with the fynbos (Gibbs Russell 1987), probably via the Afromontane grasslands (Cowling 1983b, Linder 1983).

Floristic analyses of the arid biomes are hindered by a dearth of specimens in the PRECIS data base (Gibbs Russell et al 1984). From an evaluation of the diversity component values recorded from the succulent karoo, we would expect more than the 2 125 species recorded for the core biome region. A Le Roux (pers comm) estimates that the Namaqualand flora alone includes some 3 500 species. By any account the succulent karoo has a very rich flora for a semi-arid region, especially when compared with the Sonoran Desert flora of 2 441 species in an area six times that of the core biome (Raven and Axelrod 1978), and a Sinai Desert flora of 1 130 species in an area 1,2 times its size (Danin 1983). Of particular interest in the succulent karoo is the enormous richness of succulents. These include many families (Hilton-Taylor 1987; Van Jaarsveld 1987) but especially the Mesembryanthemaceae, the richest family in the southern African flora (Gibbs Russell 1985a), which is centred in the biome (Figure 2.9). The region shares with the fynbos an astonishing richness of both species and genera of petaloid monocots (Goldblatt 1978). Patterns of endemism are complex (Hilton-Taylor unpublished) as evidenced by the high density of threatened and rare taxa (Hilton-Taylor and Le Roux this volume).

The Nama-karoo emerges as having a relatively depauperate flora for its size (2 147 species in the core biome) (Table 2.4). For a given area most Nama-karoo core and transitional sites were poorer than those from the other biomes (Figure 2.7). We have no data on beta or gamma diversity but these are expected to be low. Combining these with the generally low alpha diversity, we would thus predict a low richness for the biome. However, the Nama-karoo is poorly collected (Gibbs Russell et al 1984), and patterns of diversity and endemism remain largely unstudied. On a global scale the Nama-karoo has a rich flora for a semi-arid region, certainly as rich as the Sonoran Desert, 1,5 times its size.

The poorly collected desert biome core region has only 497 species and is by far the most species-poor of the southern African biomes (Table 2.4). Only 16% of the species are endemic but these include interesting and sometimes ancient taxa. These often show continental-scale disjunctions (Goldblatt 1978) which suggests pre-Pliocene aridity in the south-western part of the

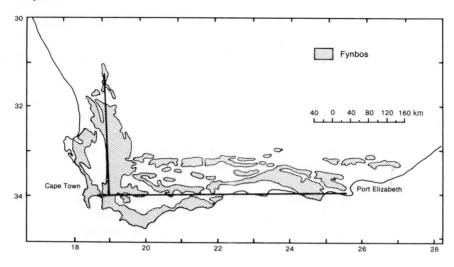

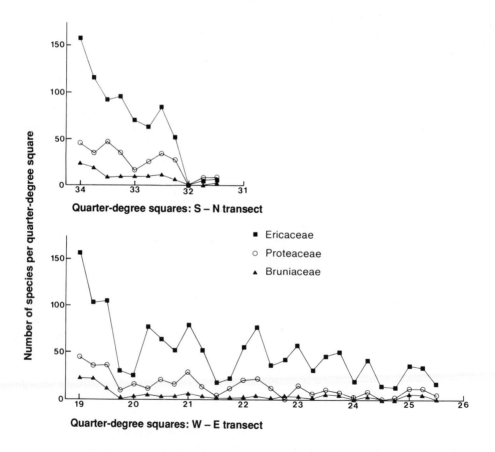

FIGURE 2.8   Number of species for three families in quarter-degree grids along north-south and west-east transects in the fynbos biome. Data from Oliver et al (1983).

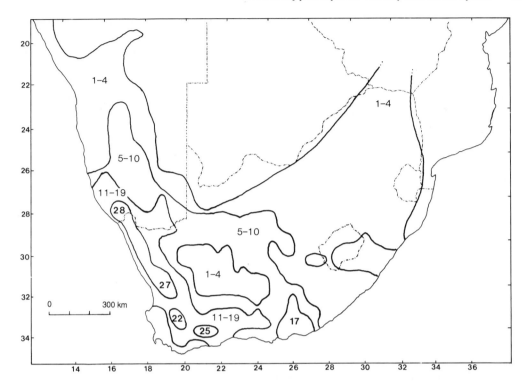

FIGURE 2.9   Patterns of generic diversity in the Mesembryanthemaceae, currently the largest family in the
southern African flora, based on 107 genera. Contours are isoflores joining half-degree grids
with equal numbers of genera. Redrawn from Jurgens (1986).

subcontinent (cf Ward et al 1983). A detailed analysis of the desert flora in terms of diversity
patterns and biogeographic relationships is long overdue.

We have no accurate data on the size of the forest flora. The temperate Afromontane forests are
probably species-poor. Geldenhuys and MacDevette (this volume) list 465 species for the southern
Cape Afromontane forests which comprise the most extensive forest region (approximately
65 000 ha) in southern Africa. Moll and White (1978) report 650 forest trees, shrubs and climbers
for the Tongaland-Pondoland region, which is virtually confined to the subcontinent and includes
all the forest patches on the subtropical east coast.

SUBCONTINENTAL-LEVEL RICHNESS

Among areas of comparable size southern Africa has the richest flora in the world, including both
temperate and tropical regions (Gibbs Russell 1985a) (Table 2.5). This is hardly surprising when
one considers that the subcontinent contains seven floristically distinct biomes, five of which are
rich in species, and one of the world's floral kingdoms. However, even if the diverse Cape flora
were excluded from the total, species density in southern Africa would still be the highest in the
world. The high level of species endemism (80%) is another distinctive feature of the southern
African flora (Goldblatt 1978).

The southern African landscape is not uniformly rich in species. Richer-than-average areas
occur in the south-west Cape, the mesic phytogeographically mixed parts of the south-east Cape,
the eastern seaboard and the north eastern escarpment (Figure 2.7). Relatively depauperate areas

include parts of the southern Cape coastal plain and the arid savanna, karroid and desert regions in the central and western parts of the subcontinent. For example, Namibia comprises 32% of southern Africa yet only 15% of the region's species occur there. Corresponding figures for Natal are 3,5% (area) and 23% (spp), and for the Cape Floristic Region, 3,5% (area) and 41% (spp).

TABLE 2.5   Number of plant species for regions of the world. Data compiled from Gibbs Russell (1985a) and Cody (1986a).

Region	No species	Area $(10^6$ km^2)	Species/area[1]
Southern Africa	21 000	2,57	8,1
Southern Africa (excluding Capensis)	15 100	2,48	6,1
Brazil	40 000	8,46	4,7
India, Pakistan Bangladesh and Burma	20 000	4,89	4,1
Australia	25 000	7,71	3,2
Europe	14 000	5,68	2,5
USA	20 000	9,36	2,1
West tropical Africa	7 300	4,50	1,6
Tropical Africa	30 000	20,00	1,5
Eastern North America	4 425	3,24	1,4
Sudan	3 200	2,51	1,3

1 Species-area ratios are calculated as 10^3 plant species per 106 km^2.

The subcontinent includes five phytochoria (three endemic) (Werger 1978b), four centres of endemism (two endemic) (White 1983) and some 70 Veld Types (Acocks 1953). The diversity of these floristic units indicates a great deal of turnover of regional species pools.

Why is the southern African flora so rich? Richness at the subcontinental scale is likely to be controlled by a bewildering array of ecological and historical factors which affect the components of diversity to different degrees in the various biomes. Over the years there have been several brave attempts at unravelling these mysteries (eg Weimarck 1941; Adamson 1958; Levyns 1964; Goldblatt 1978; Gibbs Russell 1985a). Much of the region's diversity can be attributed to contemporary ecological conditions. The subcontinent's transitional location relative to the subtropical summer rainfall and temperate winter rainfall climates, combined with a varied topography, results in steep ecological gradients along which many species can be packed. Yet more species are packed into southern African landscapes than in topographically and climatically similar landscapes in other parts of the world. For example, the California Floristic Province, which has a greater habitat diversity than the Cape Region (Raven and Axelrod 1978), has 4 452 species in 324 000 km^2 compared with the Cape's 8 550 species in 89 000 km^2 (Goldblatt 1978). Similarly, the Sonoran Desert, which occupies much more rugged country, and is an order of magnitude larger than the succulent karoo, has approximately the same number of species. Clearly we need to invoke historical processes, with all their attendant idiosyncrasies which frustrate ecologists, to explain these differences. Most models that attempt to explain the speciation of the southern African flora are geographic, attributing population fragmentation and speciation to the considerable movement and mixing of vegetation types associated with Quaternary climatic fluctuations (eg Goldblatt 1978). Surely southern Africa is not unique in its potential for geographic speciation? It would appear that speciation is a ceaseless and self-augmenting process (Whittaker 1977) and that there are no theoretical limits to the number of species able to coexist in a community (Walker and Valentine 1984). Why is this process so pronounced in southern Africa? Why are certain lineages such as the Ericaceae in the fynbos and the Mesembryanthemaceae in the succulent karoo prone to rampant speciation? Is it possible to generalize in answering these questions? Earlier in this paper we promised to eschew discussion on speciation. We hold true to that promise but affirm a caveat of thematic importance in this paper: regional historical processes must be considered when formulating hypotheses on the patterns and maintenance of diversity, irrespective of scale or diversity component.

CONCLUSIONS

For conservation biologists, the richness of the southern African flora is a dubious distinction. If the long-term goal is to maintain this diversity, and this usually seems to be the case, more distributional and inventory data are urgently required to ensure that reserve location has a sound biogeographical basis (Kruger 1977; Siegfried this volume). It is almost certainly true that in the short term the control of speciation rates is beyond our grasp. However, we are beginning to gain insights into extinction processes for a wide range of organisms (Wilcox 1986). In the fynbos there is a growing understanding of the effects of fire on local species extinction (eg Bond et al 1984). The development of models to predict the impact of processes on species diversity can be developed without necessarily understanding speciation *per se*.

The acknowledgement that regional historical processes determine species diversity patterns does not imply ecological chaos. Indeed there are very few studies in southern African vegetation that have searched explicitly for the effects of ecological processes such as competition and predation on species diversity. However, studies must be designed so as to facilitate the effective partitioning of variance associated with ecological and historical factors (eg Endler 1982; Milne and Forman 1986; Currie and Paquin 1987).

Finally, this review has identified the enormous gaps in our knowledge regarding all aspects of plant species diversity in southern Africa. Some of our deliberations do have testable implications, but we shall not be detailing these here. We hope, rather, that the patterns discussed will stimulate novel ideas and fresh research.

ACKNOWLEDGEMENTS

We thank William Bond and Fred Kruger for commenting on an earlier draft. Shirley Pierce provided useful discussion. CH-T and MTH were funded by FRD:NPER and GEG-R by the Botanical Research Institute. RMC acknowledges the support of FRD:MSP and the University of Cape Town.

APPENDIX 2.1

List of sites and sources used in determining species-area relationships in Figure 2.7

No[1]	Site	Source
1	Swartboschkloof	Van der Merwe (1966)
2	Biesievlei	Rycroft (1953)
3	Assegaaibos	CDNEC[2] (pers comm)
4	Jakkalsrivier	Kruger (1974)
5	Outeniquas	J Midgley & J Vlok (pers comm)
6	Stellenbosch Flats	Duthie (1929)
7	Jonkershoek	F J Kruger (pers comm)
8	Cape of Good Hope	Taylor (1985)
9	Cape Hangklip	Boucher (1977)
10	De Hoop	CDNEC (pers comm)
11	Cape Peninsula	Adamson and Salter (1950)
12	Agulhas Plain	R M Cowling (pers comm)
13	Fynbos biome core region	Gibbs Russell (1987)
14	Blaaukranz	P Lloyd (pers comm)
15	Thomas Baines	Jessop and Jacot Guillarmod (1969)
16	Springs	Olivier (1981)
17	Waterberg, Namibia	Rutherford (1972)
18	Andries Vosloo	A R Palmer (pers comm)
19	Ndumu	Pooley (1978)
20	Umfolozi	Downing and Gibbs Russell (1981)
21	Springbok Flats	Galpin (1926)
22	Kruger National Park	Van der Schijff (1969)
23	Savanna biome core region	Gibbs Russell (1987)
24	Niewoudtville	Snijman and Perry (1987)
25	Rocher Pan	CDNEC (pers comm)
26	Vrolikheid	CDNEC (pers comm)
27	Gamka	CDNEC (pers comm)
28	Hester Malan	CDNEC (pers comm)
29	Succulent karoo biome core region	Gibbs Russell (1987)
30	Oviston	CDNEC (pers comm)
31	Karoo Reserve, Graaff Reinet	A R Palmer (pers comm)
32	Nama-karoo biome core region	Gibbs Russell (1987)
33	Norscott Koppies	BRI[3] (pers comm)
34	Cathedral Peak	Granger (1976)
35	Ventersdorp Klipveld	Louw (1970)
36	Bloemfontein and Brandfort	Mostert (1958)
37	Potchefstroom	Louw (1951)
38	Grassland biome core regions	Gibbs Russell (1987)
39	Mapelana	Venter (1976)
40	Magaliesberg	Van Vuuren (1961)
41	Jack Scott Reserve	Coetzee (1972)
42	Doornkop	Du Plessis (1972)
43	Suikerbosrand	Bredenkamp and Lamprechts (1979)
44	Loskop Dam	Theron (1973)
45	Estcourt and Weenen Counties	West (1951)
56	Seekoei Reserve	CDNEC (pers comm)
47	Cape Recife	Olivier (1983)
48	Goukamma	Van der Merwe (1976)
49	Mtunzini	Venter (1972)
50	Umtamvuna	A E van Wyk (pers comm)
51	Cathedral Peak	Killick (1963)
52	Westfalia	Scheepers (1978)
53	Blouberg	R J Scholes (pers comm)
54	Zuurberg	Van Wyk et al (1988)
55	Southern Kalahari	Leistner (1967)

1 Site numbers correspond to those in Figure 2.7
2 Cape Department of Nature and Environmental Conservation
3 Botanical Research Institute

APPENDIX 2.2

Characterization of environmental gradients for beta diversity analyses (See Figure 2.4)

Gradient	General description	Vegetation and environment at gradient stations 1	2	3	4	5	References [1]
Knersvlakte, Namaqualand	Gradient of decreasing soil depth (and decreasing soil moisture storage capacity), located along a horizontal distance of 100 m from upslope station 1) to downslope (vertical distance approximately 50 m). Winter rainfall of approximately 150 mm yr⁻¹. Vegetation is succulent karroid shrubland dominated by Mesembryanthemaceae.	*Sphalmanthus, Aridaria, Psilocaulon.* Soil depth = 0,5 m.	*Ruschia, Drosanthemum.* Soil depth = 0,3 m.	*Drosanthemum malephora.* Soil depth = 0,2 m.	*Cephalophyllum, Argyroderma, Dactylopsis, Oophytum.* Soil depth = 0,1 – 0,15 m; quartz pebbles.		Jurgens (1986)
Sundays River Valley, SE Cape	Gradient of increasing aridity, spanning 250 km. Stations located in major vegetation types at approximately 50-km intervals. At each site both moderately grazed and overgrazed vegetation was sampled.	Subtropical succulent thicket. *Capparis, Portulacaria, Maytenus, Euphorbia.* Equinoctial rainfall of 450 mm yr⁻¹	Subtropical succulent thicket *Brachylaena. Maytenus, Euphorbia, Crassula.* Equinoctial rainfall of 380 mm yr⁻¹.	Subtropical succulent thicket/karroid shrubland. *Euphorbia, Rhigozum, Grewia, Diospyros.* Autumn rainfall of 290 mm yr⁻¹.	Karroid shrubland. *Pentzia, Selago. Eragrostis, Aristida.* Autumn rainfall of 230 mm yr⁻¹.		Hoffman (1989)
Gamtoos River Valley, SE Cape	Matching fynbos and non-fynbos sites located at approximately 100-m intervals between altitudes of 0 m and 500 m. Rainfall (equinoctial peaks) ranged from 400-1 000 mm yr⁻¹. Fynbos soils were shallow, infertile and sandy; non-fynbos soils were deep, fertile and loamy.	Fynbos: renosterveld/grassy fynbos. *Elytropappus, Themeda, Erica* — Non-fynbos: subtropical succulent thicket. *Euclea, Portulacaria, Pappea.*	Fynbos: grassy proteoid fynbos *Protea, Restio, Trachypogon.* — Non-fynbos: subtropical succulent thicket. *Euclea, Sideroxylon, Aloe.*	Fynbos: grassy and pure proteoid fynbos. *Protea Leucadendron, Hypodiscus.* — Non-fynbos: subtropical thicket. *Sideroxylon, Olea, Euclea.*	Fynbos: proteoid fynbos. *Leucadendron, Thamnus, Hypodiscus.* — Non-fynbos: afro-montane forest. *Gonioma, Rapanea, Podocarpus.*	Fynbos: proteoid fynbos. *Leucadendron, Erica, Hypodiscus.* — Non-fynbos: afromontane forest. *Ocotea, Olea, Curtisia.*	Cowling and Campbell (1983)
Agulhas Plain, SW Cape	Sites located along a crude fertility gradient over a distance of 5 km on a relatively level coastal plain. Winter rainfall of approximately 500 mm yr⁻¹.	Proteoid fynbos. *Erica, Willdennowia, Protea.* Deep, acid sand.	Proteoid fynbos. *Euchaetes, Protea, Mastersiella.* Deep, weakly acid sand.	Proteoid fynbos. *Leucadendron, Phylica, Agathosma.* Shallow, alkaline sand on limestone.	Dune asteraceous fynbos. *Passerina Euclea, Ischyrolepis.* Deep alkaline sand.	Subtropical forest. *Stipa, Sideroxylon, Celtis.* Deep organic-rich alkaline sand.	Thwaites and Cowling (1988), Cowling et al (1988)
Barrens, SW Australia	Sites located along a crude fertility gradient, and except for site 4, over a distance of 8 km on a relatively level coastal plain. Winter rainfall approximately 500 mm yr⁻¹.	Scrub-heath. *Banksia, Melaleuca, Hibbertia.* Deep, acid sand.	Scrub-heath. *Adenanthos, Leucopogon, Anarthria.* Shallow, acid sand on quartzite.	Heath. *Amphipogon, Beaufortia, Mesomalaena.* Shallow, acid sand over ferricrete.	Scrub-heath *Dryandra. Eucalyptus, Melaleuca.* Deep organic-rich alkaline sand.	Thicket *Acacia, Tetragonia, Melaleuca.* Deep organic-rich alkaline sand.	Milewski and Cowling (1985)

1 References give additional details on gradients.

REFERENCES

ACOCKS J P H (1953). Veld Types of South Africa. *Memoirs of the Botanical Survey of South Africa* **28**,1 – 192.

ADAMSON R S (1958). The Cape as an ancient flora. *Reports of the British Association for the Advancement of Science* **15**, 118 – 127.

ADAMSON R S and SALTER T M (1950). *The flora of the Cape Peninsula*. Juta, Cape Town.

AUERBACH M and SHMIDA A (1987). Spatial scale and the determinants of plant species richness. *Trends in Ecology and Evolution* **2**, 238 – 242.

BEARD J S and PATE J S (eds)(1984). *Kwongan, plant life of the sand plain*. University of Western Australia Press, Nedlands.

BELL R H V and JACHMANN H (1984). Influence of fire on the use of *Brachystegia* woodland by elephants. *African Journal of Ecology* **22**, 157 – 163.

BERLINER R, JACOBY B and ZAMSKI E (1986). Absence of *Cistus incanus* from basaltic soils in Israel : effect of mycorrhizae. *Ecology* **67**, 1283 – 1288.

BEWS J W (1925). *Plant forms and their evolution in South Africa*. Longmans, London.

BOND P and GOLDBLATT P (1984). Plants of the Cape Flora. A descriptive catalogue. *Journal of South African Botany*. Supplementary volume 13, 1 – 455.

BOND W J (1981). Vegetation gradients in the southern Cape mountains. MSc thesis, University of Cape Town, 228 pp.

BOND W J (1983). An alpha diversity and the richness of the Cape Flora: a study in southern Cape fynbos. In *Mediterranean-type ecosystems: The role of nutrients*. (eds Kruger F J, Mitchell D T and Jarvis J U M) Springer, Berlin. pp 225 – 243.

BOND W J, MIDGLEY J J and VLOK J (1988). When is an island not an island? Insular effects and their causes in fynbos shrublands. *Oecologia* (in press).

BOND W J, VLOK J and VIVIERS M (1984). Variation in seedling recruitment of Cape Proteaceae after fire. *Journal of Ecology* **72**, 209 – 221.

BOUCHER C (1977). A provisional checklist of flowering plants and ferns in the Cape Hangklip area. *Journal of South African Botany* **43**, 57 – 150.

BREDENKAMP G J and LAMBRECHTS A V W (1979). A check list of ferns and flowering plants of the Suikerbosrand Nature Reserve. *Journal of South African Botany* **45**, 25 – 47.

CAMPBELL B M (1985).A classification of the mountain vegetation of the Fynbos Biome. *Memoirs of the Botanical Survey of South Africa* **50**, 1 – 115.

CAMPBELL B M and VAN DER MEULEN F (1980). Patterns of plant species diversity in fynbos vegetation, South Africa. *Vegetatio* **43**, 43 – 467.

CHESSON P L (1986). Environmental variability and the coexistence of species. In *Community ecology*. (eds Diamond J and Case T J) Harper and Row, New York, NY. pp 263 – 287.

CHESSON P L and CASE T J (1986). Overview: nonequilibrium community theories: chance, variability, history and coexistence. In *Community ecology*. (eds Diamond J and Case T J) Harper and Row, New York, NY. pp 229 – 239.

CHESSON P L and HUNTLY P (1989). Community consequences of life history traits in variable environments. *American Naturalist* (in press).

CHESSON P L and WARNER R R (1981). Environmental variability promotes coexistence in lottery competitive systems. *American Naturalist* **117**, 923 – 943.

CODY M L (1975). Towards a theory of continental diversities: bird distribution over mediterranean habitat gradients. In *Ecology and evolution of communities*. (eds Cody M L and Diamond J) Harvard University Press, Cambridge, Mass. pp 214 – 257.

CODY M L (1983). Continental diversity patterns and convergent evolution in bird communities. In *Mediterranean-type ecosystems: the role of nutrients*. (eds Kruger F J, Mitchell D T and Jarvis J U M) Springer, Berlin.

CODY M L (1986a). Diversity, rarity and conservation in mediterranean-climate regions. In *Conservation biology. The science of scarcity and diversity*. (ed Soule M E) Sinauer, Sunderland, Mass. pp 122 – 152.

CODY M L (1986b). Structural niches in plant communities. In *Community ecology*. (eds Diamond J and Case T J) Harper and Row, New York, NY. pp 381 – 405.

COETZEE B J (1972). 'n Plant sosiologiese studie van die Jack-Scott Natuurreservaat. Unpublished MSc thesis. University of Pretoria, 124 pp.

COMINS H N and NOBLE I R (1985). Dispersal, variability and transient niches: species coexistence in a uniformly variable environment. *American Naturalist* **126**, 706 – 723.

CONNELL J H (1978). Diversity in tropical rain forests and coral reefs. *Science* **199**, 1302 – 1309.

CONNOR E F and McCOY E O (1979). The statistics and biology of the species-area relationship. *American Naturalist* **113**, 791 – 833.

COWLING R M (1983a). Diversity relations in Cape shrublands and other vegetation in the south eastern Cape, South Africa. *Vegetatio* **45**, 103 – 127.

COWLING R M (1983b). Phytochorology and vegetation history in the south eastern Cape, South Africa. *Journal of Biogeography* **10**, 393 – 419.

COWLING R M (1984). A syntaxonomic and synecological study in the Humansdorp region of the fynbos biome. *Bothalia* **15**, 175 – 227.

COWLING R M (1987). Fire and its role in coexistence and speciation in Gondwanan shrublands. *South African Journal of Science* **83**, 106 – 112.

COWLING R M and CAMPBELL B M (1983). A comparison of fynbos and non-fynbos coenoclines in the lower Gamtoos River Valley, south eastern Cape, South Africa. *Vegetatio* **53**, 161 – 178.

COWLING R M and CAMPBELL B M (1984). Beta diversity along fynbos and non-fynbos coenoclines in the lower Gamtoos River Valley, south-eastern Cape. *South African Journal of Botany* **50**, 187 – 189.

COWLING R M, CAMPBELL B M, MUSTART P, MCDONALD D, JARMAN M L and MOLL E J (1988). Vegetation classification in a floristically complex area: The Agulhas Plain. *South African Journal of Botany* **54**, 290 – 300.

CRAWLEY M J (1986). The structure of plant communities. In *Plant Ecology* (ed Crawley M J) Blackwell, Oxford. pp 1 – 50.

CURRIE D J and PAQUIN V (1987). Large-scale biogeographical patterns in the species richness of trees. *Nature* **239**, 326 – 327.

DANIN A (1983). *Desert vegetation of Israel and Sinai*. Cana Publishing House, Jerusalem.

DESCHODT C C and STRIJDOM B W (1976). Effective nodulation of *Aspalathus linearis* ssp *linearis* by Rhizobia from other *Aspalathus* species. *Phytophylactica* **8**, 103 – 104.

DOWNING B H and GIBBS RUSSELL G E (1981). Phytogeographical and biotic relationships of a savanna in southern Africa: analysis of an angiosperm checklist from *Acacia* woodlands in Zululand. *Journal of South African Botany* **47**, 721 – 742.

DU PLESSIS C J (1972). 'n Floristies-ekologiese studie van die plaas Doornkop in die distrik Middelburg, Transvaal. Unpublished MSc thesis, University of Pretoria, 188 pp.

DUTHIE A V (1929). Vegetation and flora of the Stellenbosch Flats. *Annals of the University of Stellenbosch* **7**, 1 – 59.

DYE P J and WALKER B H (1987). Patterns of shoot growth in a semi-arid grassland in Zimbabwe. *Journal of Applied Ecology* **24**, 633 – 644.

EHRLICH P R and EHRLICH A H (1981). *Extinction: the causes and consequences of the disappearance of species*. Random House, New York, NY.

EITEN G (1972). The cerrado vegetation of Brazil. *Botanical Review* **38**, 201 – 341.

ENDLER J A (1982). Problems in distinguishing historical from ecological factors in biogeography. *American Zoologist* **22**, 441 – 452.

ESLER K J and COWLING R M (1989). The effects of density on the reproductive output of *Protea lepidocarpodendron*. *South African Journal of Botany* (submitted).

EVERARD D A (1987). A classification of the subtropical transitional thicket in the eastern Cape, based on syntaxonomic and structural attributes. *South African Journal of Botany* **53**, 329 – 340.

EXELL A W (1971). Flora Zambeziaca. *Mitteilungen Botanik Staatsammeling Munchen* **10**, 69 – 70.

GALPIN E E (1926). A botanical survey of the Springbok Flats (Transvaal). *Memoirs of the Botanical Survey of South Africa* **12**, 58 – 69.

GENTRY A H (1986). Endemism in tropical versus temperate plant communities. In *Conservation biology. The science of scarcity and diversity*. (ed Soule M E) Sinauer, Sunderland, Mass. pp 153 – 181.

GENTRY A H and DODSON C (1987). Contribution of nontrees to species richness of a tropical rain forest. *Biotropica* **19**, 149 – 156.

GIBBS RUSSELL G E (1985a). Analysis of the size and composition of the southern African flora. *Bothalia* **15**, 613 – 629.

GIBBS RUSSELL G E (1985b). PRECIS, the National Herbarium's computerized information system. *South African Journal of Science* **81**, 662 – 65.

GIBBS RUSSELL G E (1987). Preliminary floristic analysis of the major biomes in southern Africa. *Bothalia* **17**, 213 – 227.

GIBBS RUSSELL G E, RETIEF E and SMOOK L (1984). Intensity of plant collecting in southern Africa. *Bothalia* **15**, 131 – 138.

GIGON A and RORISON J H (1972). The response of some ecologically distinct plant species to nitrate and ammonium nutrition. *Journal of Ecology* **63**, 393 – 422.

GIVNISH T J (1986). Introduction. In *On the economy of plant form and function*. (ed Givnish T J) Cambridge University Press, Cambridge. pp 1 – 19.

GOLDBLATT P (1978). An analysis of the flora of southern Africa: Its characteristics, relationships and origins. *Annals of the Missouri Botanical Garden* **65**, 369 – 436.

GOULD S J (1979). An allometric interpretation of species-area curves: the meaning of the coefficient. *American Naturalist* **114**, 335 – 343.

GRANGER J E (1976). The vegetation changes, some related factors and changes in the water balance following 20 years of fire exclusion in Catchment IX, Cathedral Peak. Unpublished PhD thesis. University of Natal, Pietermaritzburg, 612 pp.

GRIME J P (1979). *Plant strategies and vegetation processes*. Wiley, Chichester.

GRUBB P J (1977). The maintenance of species richness in plant communities: the importance of the regeneration niche. *Biological Review* **52**, 107 – 145.

HILTON-TAYLOR C (1987). Phytogeography and origins of the Karoo flora. In The Karoo Biome: a preliminary synthesis. Part 2 – vegetation and history. (eds Cowling R M and Roux P W) *South African National Scientific Programmes Report* **142**. CSIR. Pretoria. pp 70 – 95.

HOFFMAN M T (1989). Vegetation studies and the impact of grazing in the semi-arid Eastern Cape. PhD thesis, University of Cape Town.

HOFFMAN M T and COWLING R M (1987). Plant physiognomy, phenology and demography. In The Karoo Biome: a preliminary synthesis. Part 2 – vegetation and history. (eds Cowling RM and Roux R W) *South African National Scientific Programmes Report* **142**. CSIR, Pretoria. pp 1 – 34.

HUBBELL S P and FOSTER R B (1986a). Commonness and rarity in the neotropical forest: implications for tropical tree conservation. In *Conservation biology. The science of scarcity and diversity*. (ed Soule M E) Sinauer, New York, NY. pp 205 – 236.

HUBBELL S P and FOSTER R B (1986b). Canopy gaps and the dynamics of a neotropical forest. In *Plant ecology*. (ed Crawley M J) Blackwells, Oxford. pp 77 – 96.

HUNTLEY B J (1984). Characteristics of South African biomes. In *Ecological effects of fire in South African ecosystems*. (eds Booysen P de V and Tainton N M) Springer, Berlin. pp 2 – 17

HUSTON M (1979). A general hypothesis of species diversity. *American Naturalist* **113**, 81 – 101.

JEFFRIES R L and WILLIS A J (1964). Studies on the calcicole-calcifuge habit. II. The influence of calcium on the growth and establishment of four species in soil and sand cultures. *Journal of Ecology* **52**, 691 – 707.

JESSOP J P and JACOT GUILLARMOD A (1969). The vegetation of the Thomas Baines Nature Reserve. *Journal of South African Botany* **35**, 367 – 392.

JOHNSON M P and RAVEN P H (1970). Natural regulation of plant species diversity. *Evolutionary Biology* **4**, 127 – 162.

JURGENS N (1986). Untersuchungen zur okologie sukkulenter pflanzen des sudlichen Afrika. *Mitteilungen aus dem Institut fur Allgemeine Botanik Hamburg* **21**, 129 – 365.

KILLICK D J B (1963). An account of the plant ecology of the Cathedral Peak area of the Natal Drakensberg. *Memoirs of the Botanical Survey of South Africa* **34**, 1 – 143.

KRUGER F J (1974). The physiography and plant communities of Jakkalsrivier catchment. MSc thesis, University of Stellenbosch, 148 pp.

KRUGER F J (1977). Ecological reserves in the Cape fynbos. Towards a strategy for conservation. *South African Journal of Science* **73**, 81 – 85.

KRUGER F J and BIGALKE R C (1984). Fire in fynbos. In *Ecological effects of fire in South African ecosystems.* (eds Booysen P de V and Tainton N M) Springer, Berlin. pp 67 – 114.

KRUGER F J and TAYLOR H C (1979). Plant species diversity in Cape fynbos. Gamma and delta diversity. *Vegetatio* **41**, 85 – 93.

LAMONT B B (1982). Mechanisms for enhancing nutrient-uptake in plants with particular reference to mediterranean South Africa and Western Australia. *Botanical Review* **48**, 597 – 689.

LAPEYRIE F F and BRUCHET G (1986). Calcium accumulation by two strains, calcicole and calcifuge, of the mycorrhizal fungus *Paxillus involutus.* *New Phytologist* **103**, 133 – 141.

LEISTNER O A (1967). The plant ecology of the Southern Kalahari. *Memoirs of the Botanical Survey of South Africa* **38**, 1 – 133.

LEVYNS M R (1938). Some evidence bearing on the past history of the Cape Flora. *Transactions of the Royal Society of South Africa* **26**, 401 – 424.

LEVYNS M R (1964). Migrations and origins of the Cape Flora. *Transactions of the Royal Society of South Africa* **37**, 85 – 107.

LINDER H P (1983). The historical biogeography of the Disinae (Orchidaceae) *Bothalia* **14**, 565 – 570.

LINDER H P (1985). Gene flow, speciation and species diversity patterns in a species-rich area: the Cape Flora. In *Species and speciation.* (ed Vrba E S) Transvaal Museum, Pretoria. pp 53 – 57.

LOUW W J (1951). An ecological account of the vegetation of the Potchefstroom area. *Memoirs of the Botanical Survey of South Africa* **24**, 58 – 66.

LOUW W J (1970). Klipveld studies: I. Checklist of the vegetation. *Journal of South African Botany* **36**, 199 – 206.

MACARTHUR R H (1972). *Geographical ecology.* Harper and Row, New York, NY.

MACARTHUR R H and WILSON E O (1967). *The theory of island biogeography.* Princeton University Press, Princeton, NJ.

MCKENZIE B (1978). A quantitative and qualitative study of the indigenous forests of the south western Cape. MSc thesis, University of Cape Town.

MILEWSKI A V and COWLING R M (1985). Anomalies in the plant and animal communities in similar environments at the Barrens, Western Australia and the Caledon Coast, South Africa. *Proceedings of the Australian Ecological Society* **14**, 199 – 212.

MILNE B T and FORMAN R T T (1986). Peninsulas in Maine: woody plant diversity, distance and environmental patterns. *Ecology* **67**, 967 – 974.

MOLL E J and WHITE F (1978). The Indian Ocean Coastal Belt. In *Biogeography and ecology of southern Africa.* (ed Werger M J A) Junk, The Hague. pp 561 – 598.

MOSTERT J W C (1958). Studies of the vegetation of parts of the Bloemfontein and Brandfort Districts. *Memoirs of the Botanical Survey of South Africa* **31**, 1 – 200.

NAVEH Z and WHITTAKER R H (1979). Structural and floristic diversity of shrublands and woodlands in northern Israel and other mediterranean areas. *Vegetatio* **41**, 171 – 190.

OLIVER E G H, LINDER H P and ROURKE J P (1983). Geographical distribution of present-day Cape taxa and their phytogeographical significance. *Bothalia* **14**, 427 – 440.

OLIVIER M C (1981). An annotated checklist of the Spermatophyta at the Springs Reserve, Uitenhage. *Journal of South African Botany* **47**, 813 – 828.

OLIVIER M C (1983). An annotated checklist of the Angiospermae of the Cape Recife Nature Reserve. *Journal of South African Botany* **49**, 161 – 174.

O'NEILL R V, DE ANGELIS D L, WIDE J B and ALLEN T F H (1986). *A hierarchical concept of ecosystems.* Princeton University Press, Princeton, NJ.

PEET R K, GLENN-LEWIN D C and WOLF J W (1983). Prediction of man's impact on plant species diversity. In *Man's impact on vegetation.* (eds Holzner W, Werger M J A and Ikusima I) Junk, The Hague. pp 41 – 54.

PELLEW R A P (1983). The impact of elephant, giraffe and fire upon the *Acacia tortillis* woodlands of the Serengeti. *African Journal of Ecology* **21**, 41 – 74.

PIERCE S M and COWLING R M (1984). The phenology of fynbos, renosterveld and subtropical thicket in the south eastern Cape. *South African Journal of Botany* **4**, 1 – 17.

POOLEY E F (1978). A checklist of the plants of Ndumu Game Reserve, north eastern Zululand. *Journal of South African Botany* **44**, 1 – 54.

PORTER W M, ROBSON A D and ABBOTT L K (1987). Factors controlling the distribution of vesicular-arbuscular mycorrhizal fungi in relation to soil pH. *Journal of Applied Ecology* **24**, 663 – 672.

PRESTON F W (1962). The canonical distribution of commonness and rarity. *Ecology* **43**, 181 – 215, 410 – 432.

PUTZ F E (1984). The natural history of Vines on Barro Colorado Island, Panama. *Ecology* **65**, 1713 – 1724.

RAVEN P H and AXELROD D I (1978). *Origin and relationships of the California flora.* University of California Press, Los Angeles, Ca.

RICE B and WESTOBY M (1983). Plant species richness at the 0.1 ha scale in Australian vegetation compared to other continents. *Vegetatio* **52**, 129 – 140.

RICKLEFS R E (1987). Community diversity: relative roles of local and regional processes. *Science* **235**, 167 – 171.

RORISON I H (1960). Some experimental aspects of the calcicole-calcifuge problem. The effects of competition and mineral nutrition upon seedling growth in the field. *Journal of Ecology* **48**, 585 – 599.

RUTHERFORD M C (1972). Notes on the flora and vegetation of the Omuverume Plateau-Mountain, Waterberg, South West Africa. *Dinteria* **8**, 3 – 55.

RUTHERFORD M C and WESTFALL R H (1986). Biomes of southern Africa — an objective categorization. *Memoirs of the Botanical Survey of South Africa* **54**, 1 – 98.

RYCROFT H B (1953). A quantitative ecological study of the vegetation at Biesievlei, Jonkershoek. PhD thesis, University of Cape Town, 298 pp.

SCHEEPERS J C (1978). Vegetation of Westfalia Estate on the north-eastern Transvaal Escarpment. *Memoirs of the Botanical Survey of South Africa* **42**, 1 – 230.

SHMIDA A (1984). Whittaker's plant diversity sampling method. *Israel Journal of Botany* **33**, 41 – 46.

SHMIDA A and ELLNER S (1984). Coexistence of plant species with similar niches. *Vegetatio* **58**, 29 – 55.

SHMIDA A and WILSON M V (1985). Biological determinants of species diversity. *Journal of Biogeography* **12**, 1 20.

SMITH T M and GOODMAN P S (1986). The effects of competition on the structure and dynamics of *Acacia* savannas in southern Africa. *Journal ofEcology* **74**, 1031 – 1044.

SMITH T M and GOODMAN P S (1987). Successional dynamics in an *Acacia nilotica — Euclea divinorum* savannah in southern Africa. *Journal of Ecology* **75**, 603 – 610.

SNIJMAN D and PERRY P (1987). A floristic analysis of the Niewoudtville wild flower reserve. *South African Journal of Botany* **53**, 445 – 454.

SOKAL R K and ROHLF F J (1987). *Introduction to biostatistics*. (2nd edn). Freeman, New York, NY.

SORENSON T (1948). A method of establishing groups of equal amplitude in plant sociology based on similarity in species content. *Biol Skr K Danske Vidensk Selsk* **5**, 1 – 34.

SOULE M E (ed)(1986). *Conservation biology. The science of scarcity and diversity*. Sinauer, Sunderland, Mass.

SOULE M E and WILCOX B A (eds)(1980). *Conservation biology: an evolutionary approach*. Sinauer, Sunderland, Mass.

SPAINK H P, WIJFFELMAN C A, PEES E, OKKER R J H and LUGTENBERG B J J (1987). *Rhizobium* nodulation gene nodD as a determinant of host specificity. *Nature* **238**, 337 – 340.

SPECHT R L and MORGAN D G (1981). The balance between the foliage projective covers of overstorey and understorey strata in Australian vegetation. *Australian Journal of Ecology* **6**, 193 – 202.

STAPHORST J L and STRIJDOM B W (1975). Specificity in the Rhizobium symbiosis of *Aspalathus linearis* (Burm fil) R Dahlgr ssp *linearis*. *Phytophylactica* **7**, 95 – 96.

STRAKER C J, CLEYET-MAREL J C, GIANINAZZI-PEARSON V, GIANINAZZI S and BOUSQUET N (1988).Electrophoretic and immunological studies on acid phosphatases from ericoid endomycorrhizal fungi. *New Phytologist* (in press).

STRAKER C J and MITCHELL D T (1985). The characterization of polyphosphates in the endomycorhizas of the Ericaceae. *New Phytologist* **99**, 431 – 440.

TAINTON N M (ed) (1981). *Veld and pasture management in South Africa*. Shuter and Shooter, Pietermaritzburg.

TAYLOR H C (1985). An analysis of the flowering plants and ferns of the Cape of Good Hope Nature Reserve. *South African Journal of Botany* **51**, 1 – 13.

THERON G K (1973). 'n Ekologiese studie van die plante groei van die Loskopdam Natuurreservaat. Unpublished DSc thesis, University of Pretoria, 430 pp.

THOMPSON J S (1987). Symbiont-induced speciation. *Biological Journal of the Linnean Society* **32**, 385 – 393.

THWAITES R N and COWLING R M (1988). Landscape-vegetation relationships on the Agulhas Plain, South Africa. *Catena* **15**, 333 – 345.

TILMAN D (1982). *Resource competition and community structure*. Princeton University Press, Princeton, NJ.

TILMAN D (1986). Resources, competition and the dynamics of plant communities. In *Plant ecology*.(ed Crawley M J) Blackwell, Oxford. pp 51 – 76.

VAN DER MERWE C V (1976). Die plantekologiese aspek en bestuurprobleme van die Goukamma-Natuurreservaat. MSc thesis, University of Stellenbosch.

VAN DER MERWE P (1966). Die flora van Swartbosch-Kloof, Stellenbosch en die herstel van die soorte na 'n brand. *Annals of the University of Stellenbosch* **41**, series A, 691 – 736.

VAN DER SCHIJFF H P (1969). A checklist of the vascular plants of the Kruger National Park. *Publications of the University of Pretoria* **53**. 100 pp.

VAN JAARSVELD E (1987). The succulent riches of South Africa and Namibia. *Aloe* **24**, 45 – 92.

VAN VUUREN D R JANSE (1961). 'n Ekologiese studie van die plantegroei van 'n noordelike and suidelike kloof van die Magaliesberge. Unpublished MSc thesis, University of Pretoria, 254 pp.

VAN WYK B E, VAN WYK C M and NOVELLIE P A (1988). Flora of the Zuurberg National Park 2. An annotated checklist of ferns and seed plants. *Bothalia* (in press).

VENTER H J T (1972). An annotated checklist of the vascular flora of the Ubisana Valley, Mtunzini, Zululand. *Journal of South African Botany* **38**, 215 – 235.

VENTER H J T (1976). An ecological study of the dune forest at Mapelana, Cape St Lucia, Zululand. *Journal of South African Botany* **42**, 211 – 230.

VUILLEUMIER F and SIMBERLOFF D (1980). Ecology versus history as determinants of patchy and insular distribution in high Andean birds. *Evolutionary Biology* **13**, 235 – 379.

WALKER J and PEET R K (1983). Composition and species diversity of pine-wiregrass savannas of the Green Swamp, North Carolina. *Vegetatio* **55**, 163 – 179.

WALKER T D and VALENTINE J W (1984). Equilibrium models of evolutionary species diversity and the number of empty niches. *American Naturalist* **124**, 887 – 889.

WALTER H and BRECKLE S W (1986). *Ecological systems of the geobiosphere. 2. Tropical and subtropical zonobiomes*. Springer, Berlin.

WARD J D, SEELY M K and LANCASTER N (1983). On the antiquity of the Namib. *South African Journal of Science* **79**, 175 – 183.

WEIMARCK H (1941). Phytogeographical groups, centres and intervals within the Cape Flora. *Acta Univ Lund* **37**, 5 – 143.

WERGER M J A (1972). Species-area relationships and plot size with some examples from South African vegetation. *Bothalia* **10**, 583 – 594.

WERGER M J A (1978a). The Karoo Namib Region. In *Biogeography and ecology of southern Africa*. (ed Werger M J A) Junk, The Hague. pp 147 – 170.

WERGER M J A (1978b). Biogeographical division of Southern Africa. In *Biogeography and Ecology of Southern Africa*.(ed Werger M J A) Junk, The Hague. pp 233 – 299.

WEST O (1951). The vegetation of Weenen County, Natal. *Memoirs of the Botanical Survey of South Africa* **23**, 1 – 153.

WHITTAKER R H (1972). Evolution and measurement of species diversity. *Taxon* **21**, 213 – 251.

WHITTAKER R H (1977). Evolution of species diversity in land communities. *Evolutionary Biology* **10**, 1 – 67.

WHITTAKER R H, MORRIS J W and GOODMAN D (1984). Pattern analysis in savanna — woodlands at Nylsvley, South Africa. *Memoirs of the Botanical Survey of South Africa* **49**, 1 – 51.

WHITTAKER RH, NIERING W A and CRISP M D (1979). Structure, pattern and diversity of a malleee community in New South Wales. *Vegetatio* **39**, 65 – 76.

WHITE F (1983). *Vegetation of Africa.* UNESCO, Paris.

WILCOX B A (1986). Extinction models and conservation. *Trends in Ecology and Evolution* **1**, 46 – 48.

WILSON M V and SHMIDA A (1984). Measuring beta diversity with presence-absence data. *Journal of Ecology* **72**, 1055 – 1064.

YEATON R I (1988a). Porcupines, fires and the dynamics of the tree layer of the *Burkea africana* savanna. *Journal of Ecology* (in press).

YEATON R I (1988b). The structure and function of Namib dune grasslands: characteristics of the environmental gradients and species distributions. *Journal of Ecology* (in press).

YEATON R I, FROST S K and FROST P G H (1986). Direct gradient analysis of grasses in a Savanna. *South African Journal of Science* **82**, 482 – 487.

ZAR J H (1984). *Biostatistical analysis.* (2nd ed). Prentice-Hall, New Jersey.

CHAPTER 3

Man's role in changing the face of southern Africa

I A W Macdonald

INTRODUCTION

This chapter aims to provide a quantitative account of human impacts on natural ecosystems in southern Africa. Natural ecosystems are considered to be those still dominated by communities of indigenous plants and animals. The review thus does not attempt to document ongoing human impacts in those ecosystems transformed by man, for example, agricultural croplands, urban areas and plantations. The initial process of ecosystem transformation is, however, dealt with in detail, since it is possibly the most important human impact within the subcontinent. Only two of the major man-induced modifications of the remaining natural ecosystems are dealt with in the review. Of necessity, the chapter concentrates on those human actions that affect large areas within the subcontinent. This should not be taken to mean that other human impacts are insignificant, many being of extreme importance at local scales.

Although, ideally, this account should cover the various ecological processes occurring in the subcontinent, and describe the way each of these is influenced by man, this is not possible. Quantification of ecological processes on a regional scale has generally not yet been carried out, let alone quantification of the impact man has had on them. Instead the approach adopted has been, first, to try to categorize those human actions likely to have affected ecological processes; second, to attempt to discover the areas affected by each category of action, either in the subcontinent or in portions thereof; third, wherever possible, to document the historical spread of an activity; and, finally, to address the implications for ecological processes and the maintenance of biotic diversity.

The review relates mainly to the Republic of South Africa and included territories (hereafter called 'South Africa') for which the most comprehensive data are available, but also refers to relevant information from neighbouring states. The approximately four per cent of Africa's surface area that constitutes South Africa should provide a useful study area for the ecological effects of modern man's activities on African ecosystems. This country is by far the most industrialized of any on the continent, with its electrical generating capacity being 60% of that of the whole continent.

The paucity of 'control areas'

Surprisingly, critical analyses of human influences on the structure and function of natural ecosystems are not readily conducted even in this, the most highly 'developed' region in Africa. This is not because these influences are thought to be slight, but rather because the appropriate 'experimental controls' — natural ecosystems unaffected by man — are lacking. In most continents there are either large tracts of land where man's influence is so small it can be considered negligible or, more commonly, there are 'historical controls' — knowledge about periods when man's influence was either totally lacking (as in certain oceanic islands and the island continent of Australia) or very slight.

   For southern Africa protected areas cover 72 700 km^2 or 5,8% of the total area (Siegfried, this volume). The proportional coverage of the country's different ecosystems by protected areas is skewed towards those savannas that still have large herds of native ungulates, and montane ecosystems (Edwards 1974; Huntley 1978; Cowan 1987). Of the 70 Veld Types (Acocks 1975) occurring in the region only 10 have more than five per cent of their area protected (Cowan 1987). As could be predicted, many of the ecosystem types most heavily utilized by man have exceedingly low coverage in the nature conservation estate, for example, the highveld grasslands of the Transvaal and the coastal renosterveld of the fynbos biome (Huntley 1978; Moll and Bossi 1984; Clinning 1986; Cowan 1987). Even in ecosystem types with large protected areas, their value as natural controls is limited by the extensive human management that has been applied within them. This has been necessary in order to mitigate the effects of the inclusion of incomplete ecosystems, fencing, water-regime alterations and other unnatural influences (Macdonald 1976, 1988; Pienaar 1983; Brooks and Macdonald 1983; Walker et al 1987).

The effects of prehistoric man

The possibility of using historical controls is complicated by the long duration of human occupation of the subcontinent in the prehistoric period. Stone-Age populations are thought to have occupied the subcontinent for the last one to one and a half million years (Deacon 1986). Initially their impacts were probably slight but since about 150 000 BP, they have employed fire (Deacon 1986). This is likely to have led to profound changes in natural fire regimes, particularly in ecosystems not characterized by high frequencies of lightning-strike ignitions such as those of the fynbos biome (Deacon 1983; Kruger and Bigalke 1984).

   It is being postulated more and more often that prehistoric man, particularly early pastoralist (since approximately 2 000 BP) and Iron-Age (since approximately 1 700 BP) populations, had considerable impacts on the vegetation (West 1971; Feely 1979; Hall 1979; Coetzee et al 1983; Avery 1987; Bredenkamp and Van Vuuren 1987), soils (Marker and Evers 1976) and fauna (Klein 1974; Avery 1987) of the subcontinent. However, owing to the incompleteness of the archaeological and palaeoecological record it is not possible to quantify these prehistoric human influences. In the absence of data the effects of early human activities can easily be exaggerated. For example, detailed archaeological studies in the Transkei have clearly shown the gross assumption that grassland vegetation in this area is man-induced (Acocks 1975) to be false (Feely 1987). The end result of the above situation is that the only conclusions that can be drawn concerning the effects of modern man's activities are that they have changed ecosystems from some pre-existing state. It is generally impossible to say how this state relates to the pristine condition. As the introduction of a system of mixed agriculture 'implies a quantum increase in the human impact on natural ecosystems' (Deacon 1983) it is convenient to consider here only those impacts that have resulted from the overland settlement of the subcontinent from the north-east by Iron-Age farmers in about AD 250 (Feely 1987 gives a useful map showing the spatio-temporal distribution of early farming settlement) and settlement by seaborne European colonists from the south and south-east starting in AD 1652.

THE MAJOR LAND TRANSFORMATIONS

Changes to natural ecosystems in which the structure and species composition are completely or almost completely altered, are termed transformations (Poore 1978). It is these changes that most deleteriously affect the conservation of native species and communities, and have the greatest potential to alter ecosystem function.

Cultivation

Currently 129 156 km^2 (12,26%) is estimated to be under cultivation throughout the 'white'-controlled portion of South Africa (Table 3.1; Schoeman and Scotney 1987). In a study of a 47 800-km^2 area in the south-western Cape centred on Cape Town and having the Piketberg, Ceres, Montagu and Heidelberg magisterial districts as its outer limits, Gasson (in press) has estimated that 13 705 km^2 (28,7%) was cultivated by 1976. In one of the most heavily cultivated vegetation types in the fynbos biome, the west-coast renosterveld, the extent of cultivation in 1979 was estimated to be 73% (McDowell 1988). In the six 'national states' within South Africa, 9 225 km^2 (13,51%) was estimated to be under cultivation in 1987 (Table 3.1; Department of Development Aid pers comm). For Zimbabwe, Whitlow (1980a) has estimated that 16,4% of the area available for agriculture (the total area minus the protected areas) had been cultivated by the mid-1970s.

TABLE 3.1   The estimated extents of current cultivation during 1987 in the agricultural regions and national states of South Africa (Schoeman and Scotney 1987; Department of Development Aid unpublished data)

Agricultural regions and national states	Total area km^2	Area of current Cultivation km^2	Cultivation as % of total
Agricultural regions			
Transvaal	152 490	22 020	14,44
Natal	58 321	9 335	16,01
Highveld	115 859	56 782	49,01
Free State	204 823	16 836	8,22
Winter rainfall	138 000	18 000	13,04
Eastern Cape	54 000	4 263	7,89
Karoo	290 600	1 920	0,66
National states			
Gazankulu	6 790	630	9,28
KaNgwane	3 850	470	12,21
KwaNdebele	3 140	905	28,82
KwaZulu	31 750	3 830	12,06
Lebowa	22 120	3 330	15,05
Qwaqwa	620	60	9,68

HISTORICAL INCREASES IN THE AREA CULTIVATED

Although the extent of cultivation by Iron-Age people in eastern southern Africa has never been quantified, it is thought to have been considerable and to have markedly influenced current patterns in the wooded vegetation types of this area (Swynnerton 1918; Feely 1979, 1987). Statistics on the areas cultivated generally date from the beginning of the colonial era and show that there was an exponential rate of increase towards the end of the 19th century (Deacon 1986), since which time the area has increased steadily (Figure 3.1).

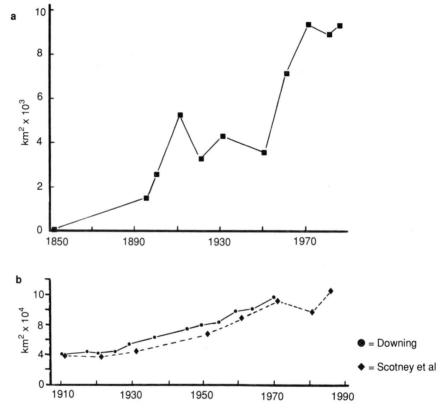

FIGURE 3.1   The change in area of cultivated land in (a) Natal (1852 – 1987) and (b) South Africa (1911 – 1987) (Downing 1978; Scotney et al 1988)

In the subsistence agricultural areas of Zimbabwe the area under cultivation doubled between 1962 and 1977 to an area of 22 000 km^2, or 24,7% of the total (Whitlow 1980b).

THE CONTRIBUTION OF 'OLD LANDS'

The amount of land that was cultivated in the past and has now been allowed to revert to 'natural' vegetation is uncertain. Using published statistics for the extent of transformation and cultivation of the coastal renosterveld vegetation type of the south-western Cape (Boucher 1981; Moll and Bossi 1984; McDowell 1988), I estimate that old lands constitute between 12 and 21% of the total area transformed by cultivation. In the more arid north-western subsection of the winter-rainfall region these old lands made up 39% of the total area cultivated in 1983 (Anonymous 1985).

In many of the landuse studies carried out for different portions of southern Africa there is no indication whether abandoned fields have been included within the area classed as cultivated. For example, the statistics on cultivated areas in the national states (Table 3.1) appear to exclude them. This extremely important component of the land transformation process is thus difficult to quantify on a regional scale.

THE AREAS MOST EXTENSIVELY CULTIVATED

In addition to lowland areas of the fynbos biome it appears that grasslands of the Transvaal highveld (Clinning 1986), the mesic coastal lowlands and the mistbelt region of Natal (Bourquin 1987) have been extensively transformed by cultivation. In all three of these areas natural ecosystems are now to be found only as small, isolated fragments.

CULTIVATION AS A SELECTIVE TRANSFORMATION

The fragments that remain are not necessarily representative of the original spectrum of communities present in a particular area or vegetation type. For example, McDowell (1988) has shown that of the three per cent of the original area of 7 280 km^2 of west-coast renosterveld that remained in 1980, only 26% occurs on slopes of less than five per cent, whereas this statistic for the original area was 74%. The 'kopje' vegetation has been selectively left by farming, with 63% of the remnant area having slopes in excess of 10%, compared with only 10% of the original area. This selective transformation has resulted in a high level of endangerment to plant taxa restricted to bottomland communities within the coastal renosterveld (McDowell 1988).

THE CULTIVATION OF MARGINAL LANDS

Unfortunately, cultivation is not only a major factor in reducing the extent of ecosystems that could be considered legitimate arable land. Large tracts of semi-arid ecosystems (which cover the largest area of the subcontinent) have been injudiciously cultivated in the past. In the Namaqualand agricultural region (68 719 km^2), for example, it was stated that 2 449 km^2 had been cultivated by 1983. Of this area only 180 km^2 was under irrigation (Anonymous 1985).

For the 290 600-km^2 karoo agricultural region, Schoeman and Scotney (1987) state that 1 920 km^2 is currently cultivated, whereas only 200 km^2 is considered suitable for rain-fed crop production. Given that 1 410 km^2 is currently under irrigation in the region, with only 730 km^2 under regular irrigation (J L Schoeman pers comm), it would appear that 310 km^2 (or 16% of the area cultivated) is being unwisely cultivated under dry-land conditions. There are considerable areas in the karoo, where cultivation was practised in the historic past, which are no longer cultivated. These areas have not yet reverted to the plant communities characteristic of uncultivated sites (W R J Dean pers comm).

THE EFFECT OF CULTIVATION ON ECOSYSTEM PROCESSES

Probably the most important effect of cultivation on ecological processes in the subcontinent is the effect on soil erosion. Currently the estimated mean annual soil loss throughout South Africa is between 2,5 and 3,0 t ha^{-1} yr^{-1} (Roux 1986; Schoeman and Scotney 1987). The contribution of cultivated areas to this overall loss rate has not been estimated, but it is known that water erosion on pineapple fields in the eastern Cape can exceed 60 t ha^{-1} yr^{-1}, and wind erosion from vulnerable soils in the western Orange Free State has given rise to losses of 40 t ha^{-1} yr^{-1} (Schoeman and Scotney 1987). In Zimbabwe, poor cultivation practices and the cultivation of steeply sloping ground have been shown to give rise to erosion rates several times to orders of magnitude higher than rates in comparable areas having well-managed natural vegetation (Elwell 1972).

By comparing ocean-basin sedimentation rates with recent estimates of the sediment yield of rivers along the eastern seaboard of the subcontinent, Martin (1987) has concluded that mean erosion rates in southern Africa over the last century are between 12 and 22 times higher than the geological average. Independent estimates of the mean rate of natural soil formation under African conditions are generally less than 0,1 t ha^{-1} yr^{-1} (Stocking 1984), so the current loss rate for South Africa is approximately 25 to 30 times this figure. In Zimbabwe, estimates of soil formation rates on granite range from 0,15 to 0,4 t ha^{-1} yr^{-1}, whereas measured soil loss rates from tobacco lands are at least one order of magnitude higher (Elwell 1975).

Afforestation and alien tree invasions

Southern Africa is not well endowed with indigenous forests which lend themselves readily to commercial timber exploitation. South Africa, for example, has only approximately 3 000 km^2 of closed forests (Geldenhuys and MacDevette, this volume). These cover only some 1,9% of its area. As a result, large areas have been planted to fast-growing, introduced trees, mainly *Pinus* and *Eucalyptus* species, to meet the region's timber needs. In addition to the intentional afforestation of large tracts of country for commercial purposes there are three other ways in which introduced tree species have contributed to the transformation of southern African landscapes. These are through the invasive spread of certain of the species, through the intentional planting of woodlots and shelterbeds, and through their widespread use in gardens, both rural and urban.

PLANTATION FORESTRY

Currently an area of 13 960 km^2 is afforested with these introduced species throughout southern Africa (Scharfetter 1987). Although, on a regional scale, these plantations account for only a very small percentage of the total area, for smaller portions of the area on the eastern seaboard and along the eastern escarpment the percentage becomes substantial (Table 3.2). It is in these areas that afforestation has had deleterious effects on certain localized species of fauna (eg, the blue swallow *Hirundo atrocaerulea*) (Snell 1979; Allan et al 1987) and flora (eg, 30% of the rare and endangered plants of the Transvaal occur in habitats subject to afforestation, even though plantations cover only two per cent of the province) (Raal 1986).

TABLE 3.2   The extent of afforested areas in southern Africa (Scharfetter1987)

Country	Total area (km^2)	Area of afforestation (km^2)	% afforested
Republic of South Africa	1 123 226	11 160	0,99
Transkei	45 000	610	1,36
Ciskei	9 000	60	0,67
Venda	6 000	60	1,0
Lesotho	30 344	0	0
Swaziland	17 366	1 070	6,16
Zimbabwe	389 361	1 000	0,26
Bophuthatswana	40 000	0	0
Botswana	575 000	0	0
Namibia	823 168	0	0
Southern Africa	3 058 465	13 960	0,46

THE EFFECT OF AFFORESTATION ON ECOLOGICAL PROCESSES

The most important ecological effect of afforestation in the region is that it invariably reduces natural runoff from the afforested catchment. The reduction brought about by mature plantations is often in the order of 50% in high-rainfall fynbos and grassland areas, rising to 100% in the drier portions of the area in which plantations are commercially viable (Versfeld and Van Wilgen 1986). Accordingly, it has been necessary to impose strict controls on afforestation in South Africa to protect the country's already extremely limited water supply. By 1980 it was estimated that afforestation had reduced runoff throughout South Africa by 1 284 million $m^3 yr^{-1}$. This constitutes 2,4% of the total mean annual runoff of the country's rivers (Department of Water Affairs 1986). The biological effects of the reduction in stream flow brought about by afforestation have apparently not yet been researched in southern Africa (O'Keeffe 1986a). This lacuna in available knowledge has apparently been overlooked in several recent reviews of river conservation in the region (Davies and Day 1986; O'Keeffe 1986a, b).

THE INVASIVE SPREAD OF ALIEN TREES AND SHRUBS

In the fynbos biome the area of natural vegetation densely infested (quasi-transformed) by *Hakea* and *Pinus* species in 1984 was estimated to be 1 254 km^2. The analogous estimate for *Acacia* and other thicket-forming alien plant species was 2 896 km^2 (Macdonald et al 1985). Although these infestations overlap to a limited extent (the former being mainly in mountain fynbos, the latter mainly in lowland vegetation types), their combined area is considerably greater than the 990 km^2 covered by *Eucalyptus* and *Pinus* plantations throughout the Cape Province in 1980 (Forestry Department 1980). The area of natural vegetation modified by sparser invasions of these two categories of alien trees covered 4 798 km^2 and 5 373 km^2 respectively.

In Natal the invasion of some 8 000 km^2 of the mesic coastal lowlands by the scrambling shrub *Chromolaena odorata* by 1982 (Liggitt 1983) makes this the province's most important alien plant species (Macdonald and Jarman 1985). There are, however, a number of other tree and shrub species having significant infestations in Natal, in particular along the banks of streams and rivers (Macdonald and Jarman 1985). The same applies to the Transvaal where roadside surveys carried out in the central highveld during the late 1970s showed that 33% of the length of watercourses had been invaded by introduced woody plants (Wells et al 1980). Twelve per cent of their point samples had more than four species present.

In the more arid areas of southern Africa the alien plant situation is generally less serious. However, several cactus species have invaded large areas of the drier savanna and karoo areas mainly in the eastern Cape (Zimmerman et al 1986). The most widespread of these, *Opuntia ficus-indica*, reached its peak in the 1930s when an area of some 9 000 km^2 was densely infested and much larger areas were lightly infested. The smaller *O aurantiaca* is currently estimated to occupy some 8 300 km^2. The successful release of biological control agents has served to limit the density and distribution of most of these cactuses (Zimmerman et al 1986). Currently, the most serious invasions of alien trees and shrubs in the arid karoo, the southern Kalahari and the Namib Desert are those of *Prosopis* spp, *Nicotiana glauca* and *Datura* spp into the riverbeds and riparian fringes of ephemeral watercourses (Brown and Gubb 1986).

THE ECOLOGICAL IMPACTS OF PLANT INVASIONS

The most important impacts of introduced plants on biotic diversity in the region have been their effects on the endemic flora of the fynbos biome. These invasions currently pose the most serious threat to the survival of the numerous endangered plant species occurring in this biome (Hall and Ashton 1983; Hall and Veldhuis 1985). In the fynbos-like shrublands of the eastern highlands of

Zimbabwe, alien tree invasions are known to have been responsible for the documented local extinction of a population of blue swallows during the last decade (Snell 1988).

The generally high level of invasion of riparian habitats throughout South Africa has led several researchers to conclude that the survival of the indigenous flora and fauna of these habitats is now in jeopardy on a national scale (Wells et al 1980, 1983; Macdonald 1983, 1988; Henderson and Wells 1986). The ecological process most affected by this invasion is thought to be stream flow, which is likely to be significantly reduced (Henderson and Wells 1986; Versfeld and Van Wilgen 1986). Another important impact of these invasions is accelerated erosion of stream banks (Versfeld and Van Wilgen 1986).

EFFECTS OF URBAN AND RURAL TREE PLANTINGS

The extent of these plantings of introduced trees has never been quantified. However, it is now extremely difficult to find a landscape anywhere in South Africa outside the larger conservation areas in which copses of intentionally planted introduced trees are not visible. In the formerly treeless grassland and karoo areas of the country, these plantings (together with subsequent invasions) have made possible range expansions of numerous animal species dependent on trees. Examples are several hawks of the genus *Accipiter* (Allan and Tarboton 1985; Macdonald 1986), the hadeda ibis *Bostrychia hagedash* (Macdonald et al 1986), and the cricket-like insect *Libanasidus vittatus* (Toms 1985).

On the Transvaal highveld the extensive planting of trees in gardens has shifted the avifauna away from its original grassland species composition to one dominated by woodland species (Dean 1969; Fraser 1987).

Urbanization

Although Iron-Age villages have had lasting effects on soil properties and resultant vegetation patterns (Marker and Evers 1976; Feely 1979; Hall 1979), these effects have been rather localized within the better-watered eastern parts of the country. It is since the advent of European colonization that the impact of towns and cities has become significant on a regional basis.

The total area transformed by urbanization throughout southern Africa has never been estimated. In the mid-1970s two independent estimates were carried out for South Africa: Edwards (1974) gives the area as being 21 270 km^2 (1,7%) of a total of 1 221 110 km^2, whereas Serfontein (1976) gives 16 260 km^2 (1,3%) of a total area of 1 114 830 km^2. For his 47 800-km^2 south-western Cape study area, Gasson (in press) estimated that urban areas covered 1,6% by 1976. It would appear that a decade ago less than 1,8% of South Africa, the most 'developed' country in the subcontinent, had been transformed by urbanization. However, urbanization has proceeded at an accelerated rate throughout the region in recent years. In 1982 urban areas were estimated to cover 37 366 km^2 (3,5%) of the total area of 1 068 721 km^2 within South Africa (Department of Water Affairs 1986). However, this figure includes areas proclaimed for urban development but not yet developed, so the real figure for the area already transformed may be closer to 2,5% (L G Pienaar, pers comm).

Although it is difficult to draw a distinction between smallholdings and agricultural land, it appears that from the viewpoint of conserving wild species this form of landuse (or land misuse, depending on one's viewpoint) can best be considered as a special category of urban transformation. Smallholdings were estimated to cover 32 028 km^2 (3,0%) of South Africa by 1982 (Department of Water Affairs 1986).

THE ECOLOGICAL EFFECTS OF URBAN TRANSFORMATIONS

Apart from the above-mentioned effects of planting introduced trees in suburban gardens, urban areas have had several ecological impacts, their importance being out of proportion to the relatively small percentage of the region the urban areas occupy. One of these has been the effect on the rivers of the subcontinent. First, the water requirements of these urban areas (urban and industrial usage comprised 21% of the total demand for stored water in 1980 in South Africa: Department of Water Affairs 1986) has made necessary the impoundment of virtually every perennial river in their vicinity. So great have their water requirements become that in certain instances it has been necessary to transfer water between catchments in order to meet their needs (Department of Water Affairs 1986). The resultant mixing of the biotas of formerly separated catchments has serious implications for their long-term conservation (Bruton and Van As 1986). Second, the effluents from these urban areas have resulted in the eutrophication of several artificial bodies of water downstream from major cities, for example, the Hartbeespoort Dam in the Transvaal (Allanson and Gieskes 1961) and Lake McIlwaine in Zimbabwe (Thornton 1982).

Because urban sprawl generally involves the complete transformation of relatively large areas of contiguous ground, it can lead to the endangerment or even extinction of species with very localized ranges. Such a situation exists in the Greater Cape Town metropolitan area on the Cape Peninsula and the adjacent Cape Flats where five species are already known to have become extinct and a further 169 have an uncertain future (Hall and Ashton 1983). Of the 58 species considered extinct, endangered or vulnerable, 50% are already directly threatened by urban sprawl, and 28% will be if proposed developments are implemented. This alarming situation is not surprising given the extraordinary species richness and local endemism of the fynbos flora. For example, the 3,43 km^2 Bergvliet farm held 595 species of indigenous plants in 1919, since which time the entire area has been transformed for suburban development (Rourke et al 1981).

Certain of the urban areas in South Africa have already reached sizes at which they have begun to affect local climatic conditions, for example, through the formation of urban heat islands (Preston-Whyte and Tyson 1988). Air pollution is already a significant problem in some of South Africa's major urban areas. However, as a result of the implementation of controlling legislation, pollutant levels now generally fall within acceptable limits (Fuggle and Rabie 1983). Air pollution in southern Africa has been viewed mainly as a human health problem and it is only very recently that atmospheric pollutant levels in the eastern Transvaal highveld have been shown to be high enough to pose a potential hazard to natural ecosystems (Tyson et al 1988).

The impoundment of rivers

Although almost all of man's activities tend to increase rainfall runoff from the land (eg, cultivation (Elwell 1972), urbanization, the construction of roads, and the over-utilization of rangelands (Elwell 1972; Gifford and Hawkins 1978)), the paradoxical situation of no measurable change or even a decrease in mean annual river flow has obtained in southern Africa over the last few decades. Prior to this there was no detectable trend in river flows over the historic period, so one can assume that any landuse practices increasing this flow were balanced out by those that decreased it (Department of Water Affairs 1986). The recently observed decreases in river flow are mainly thought to have arisen from the increasing impoundment of rivers in the region.

THE EXTENT OF THE AREA INUNDATED

The Department of Water Affairs (1986) lists a total of 517 major reservoirs as having been constructed in South Africa by July 1986. This figure excludes privately owned farm dams, the number and extent of which have not yet been estimated for the whole country. Using a regression approach to estimate the size of the 191 reservoirs for which this statistic was not given in their list, I have calculated that the total surface area at full supply level for these 517 major reservoirs is 2 786 km^2. These impoundments can therefore be considered to have totally transformed approximately 0,22% of the surface area of South Africa (1 253 570 km^2, including all the enclosed states). The contribution of farm dams on a regional basis is uncertain, but in the well-watered south-western Cape, Gasson (in press) has estimated the total cover of surface water to be 1,3%. This statistic includes farm dams and major reservoirs.

In Natal the number of farm dams has doubled over the period from 1956 to 1986, and their proliferation throughout South Africa is now thought to pose a threat to the efficient storage and utilization of the country's scarce water resources (Department of Water Affairs 1986).

THE MEGA-IMPOUNDMENTS ON THE ZAMBEZI RIVER

On a subcontinental basis the most significant impoundments are the Kariba (5 250 km^2) and Cahora Bassa (2 739 km^2) dams on the Zambezi River (Bowmaker 1970). These two impoundments have inundated two thirds of the Middle Zambezi Valley (550 km), and constitute the single most radical ecosystem transformation in the subcontinent. If the proposal to construct a further impoundment at the Mupata Gorge is implemented, this will result in the total destruction of the Middle Zambezi ecosystem (Ferrar and Pitman 1980). This is probably the single most significant conservation/development conflict in the subcontinent.

The wisdom behind the construction of these huge dams has been seriously questioned (Balon 1978). Kariba Dam has manifested many of the classical side effects of such impoundments. The inundation of the valley has led to the disruption of the normal seasonal movement patterns of large mammals (Jarman 1968, 1972). Below the dam, the altered discharge pattern has had a deleterious effect on the vegetation and fauna (Attwell 1970), while the discharge of 'hungry water' has led to accelerated bank erosion (Guy 1981).

THE ECOLOGICAL IMPACTS OF IMPOUNDMENTS

Within South Africa, it is only in the fynbos biome with its highly localized plant species that the inundation of river valleys by impoundments poses a major threat to biotic diversity; for example, the proposed Palmiet River Phase II impoundment will flood the last two populations of the protea shrublet *Spatalla prolifera*. Two other impoundments have already submerged the other known populations of this river valley species (Hall and Veldhuis 1985).

The major regional impacts of dams, given the small percentage of most ecosystems actually inundated, arise from their downstream effects. These are both quantitative (eg, reduced river flows and flood peaks) and qualitative (eg, altered seasonality of flows and sediment loads). These impacts have manifested themselves in classic fashion in South Africa's largest river system, the Orange–Vaal (Cambray et al 1986).

Two further examples of important aquatic ecosystems where impoundment and extraction of water in the catchment have led to quantitative reductions in water flows are Lake St Lucia in Natal (Hutchison 1976) and the Nyl floodplain in the Transvaal (Tarboton 1987). In the latter case it is the cumulative effect of numerous small farm dams that is causing the reduction. Altered seasonality and scale of flooding of the Pongola floodplain has been a major impact of the Josini Dam in Natal (Heeg and Breen 1982). Although the 1980 decision to build the Klipfontein Dam in the upper catchment of the White Mfolozi River will alter the flow of the last major river system which was then still intact in South Africa (Porter 1981), the Mfolozi still provides a uniquely 'pristine' river in a subcontinent where lotic systems are rapidly becoming endangered (Davies and Day 1986).

Other transformations

MINING

The development of southern Africa has to a large extent been a function of the mining of its non-renewable geological resources. This mining has not been accomplished without some major impacts on southern African landscapes and ecosystems. In the Hluhluwe Valley in the Iron Age the mining of surface iron deposits and its subsequent smelting are thought to have given rise to some of the vegetation patterns that characterize the area to this day (Hall 1979). However, the major mining transformations have occurred from the last few decades of the last century to the present date.

The most obvious of these impacts has been the creation of the enormous artificial hillocks, called mine dumps, of treated spoil from the gold mines on the Witwatersrand. These have not only radically altered the appearance of what was formerly a grassland 'steppe' area but have proven exceptionally difficult to revegetate. As a consequence, in the early years of their creation they gave rise to some of the worst large-particle air pollution rates in the world. In recent years leachates from these mine dumps have been implicated in the deteriorating quality of water flowing into the Vaal River (Department of Water Affairs 1986).

Other important areas affected by mining transformations have been the surface diamond mining areas in the western Transvaal (Morris 1976) and those along the west coast (Heydorn and Tinley 1980). In the latter area approximately 22 km^2 has been subjected to overburden stripping in the approximately 51-km coastal strip between the Orange River Mouth and Kleinzee. A further 48 km^2 has been radically disturbed as a result of wind erosion from mining and prospecting sites (Talkenberg 1982). In total, between three and five per cent of the proclaimed mining area has thus been transformed by mining activities in this coastal strip. Mining activities have also been one of the few major transformations occurring within the Namib Desert, an ecosystem otherwise remarkably unaffected by man's influence (Seely and Ward 1988).

Several satellite images have shown that wind erosion is apparently being accelerated as a result of surface mining activities north and south of the Orange River Mouth. A crude estimate of the seaward transportation of dust from this area during one dust storm in 1979 was about 50 x 10^6 tonnes. This approximates the mean annual waterborne input of sediments by the Orange River itself (Shannon and Anderson 1982).

No assessment has yet been published of the ecological impacts of the creation of enormous open mines on what were the kimberlite pipes of the northern Cape and Botswana. It is known that their presence can be detected geo-botanically so it is likely that there were certain specialized ecotypes, if not species, of plants growing on these pipes in the Kimberley area prior to the total removal of their surface deposits which had occurred by the turn of the century. Unfortunately there were no comprehensive botanical collections made in this area prior to this date (Wilman 1946).

Mining of surface coal deposits in the eastern Transvaal and northern Natal has also transformed extensive areas. This transformation, in addition to that of cultivation, is rapidly reducing the area available to endemic grassland species such as the lizard *Cordylus giganteus* (Petersen et al 1985). Similarly, the mining of the forested dunes of the northern Natal coastline for heavy metals threatens one of South Africa's most species-rich ecosystems. Fortunately this mining activity was preceded by surveys which identified critically important areas (Weisser 1982) that are to be excluded from these mining operations (Tinley 1985). Nevertheless, it is likely that these dune mining operations will substantially reduce the diversity of these forests for many years to come.

TRANSPORTATION NETWORKS

The subcontinent currently boasts an exceptionally fine transport system. The historical development of this system is best illustrated by the road network. The formal national road system was initiated in 1910. By 1935 there were 120 000 km of unsurfaced roads. This had grown to 5 840 km of surfaced roads and 136 000 km of unsurfaced roads by 1948. By 1984 (excluding the recently independent TBVC states) there had been an almost tenfold increase in the length of surfaced roads running between towns (50 034 km), whereas unsurfaced rural roads measured 134 296 km. In addition, municipalities were estimated to have 35 400 km of surfaced and 9 000 km of unsurfaced roads under their jurisdiction by this date (Heyns 1986).

The other transport networks, railways and airports, have shown similar increases since the turn of the century with the major expansion of railways occurring earlier than that of roads, and the expansion of the airport system occurring later.

The area transformed by these networks within South Africa has been estimated as having been 11 550 km^2 (0,94% of 1 224 050 km^2) by the mid-1970s (Serfontein 1976), but almost double this area, 21 352 km^2 (2% of 1 068 721 km^2) by 1982 (Department of Water Affairs 1986). The accuracy of these area estimates is questionable and a more detailed investigation of this statistic for roads in one portion of South Africa is presented below.

THE AREA OF ROADS IN THE FYNBOS BIOME

The extent of the area covered by roads has been estimated for a 93 018-km^2 region in the southern Cape Province approximating to the fynbos biome (Dawson 1986). This area includes substantial portions of adjacent and interspersed karoo vegetation, since the total extent of the fynbos biome has been calculated to be 77 172 km^2 (Macdonald and Richardson 1986). The estimated total area included within road reserves is 740 km^2, of which 232 km^2 is under tarmacadam or gravelled road surface (Dawson 1986). These areas represent 0,80% and 0,25% of the total fynbos region, respectively. In addition, within the Cape Town Municipal area there are approximately 2 000 km of roads which have an approximate average road reserve width of 11 m (P J Pretorius, pers comm). In the municipal area virtually this whole road reserve can be considered totally transformed. This would constitute an additional 22 km^2 of roads. Including all the other municipal road systems in the fynbos biome it is unlikely that the area included within road reserves will exceed 0,83%, or the area under road surfaces exceed 0,27%.

It would thus appear unlikely that transformation due to road building is having major regional impacts on environmental characteristics such as runoff and albedo. However, in local situations such as at estuary mouths (Begg 1979) and in coastal sand dune areas (Tinley 1985), road building can have important effects on geomorphological processes such as the migration of estuary mouths and the transportation of sand via headland bypass dunes. Dark tarmacadam roads presumably also play an important role in the urban heat island phenomenon (Preston-Whyte and Tyson 1988).

SOME ECOLOGICAL EFFECTS OF ROADS

Roads have other effects on the conservation of biotic diversity disproportionate to their small areal extent. They can bring about unnaturally high levels of mortality in wildlife species likely to be killed by road traffic. In the Cape Province, several studies (Table 3.3) have shown that mortality is highly selective with regard to species affected. Of the birds, the spotted eagle-owl *Bubo africanus* is possibly the most likely to be killed by vehicles (Siegfried 1966). However, there is no indication that vehicle kills currently pose a significant threat to the survival of even this highly susceptible species.

TABLE 3.3   Numbers of road kills recorded per 1 000 km of road surveyed in some Cape studies

Area		Cape[1]	S W Cape[2]	Stellenbosch[3] area	Cape[4] (mainly northern)
Period		1962 – 1964	1960 – 1961	1963 – 1965	1976 – 1977
No of km surveyed		989 535	66 799	10 573	18 088
REPTILES					
Molesnake	*Pseudaspis cana*	0,025			
Cape cobra	*Naja nivea*	0,015	N/R	N/R	N/R
Puffadder	*Bitis arietens*	0,013			
BIRDS					
Cape turtle dove	*Streptopelia capicola*	0,121	0,659	1,324	N/R
Laughing dove	*Streptopelia senegalensis*	0,078	0,629	0,284	N/R
Eagle owl species mainly	*Bubo africanus*	0,195	0,299	0,095	0,166
Nightjar species mainly	*Caprimulgus pectoralis*	0,042	0,015	0,284	N/R
European swallow	*Hirundo rustica*	0,123	0,180	0	N/R
Cape thrush	*Turdus olivaceus*	0,006	0,210	0,473	N/R
Cape robin	*Cossypha caffra*	0,023	0,195	0,284	N/R
Fiscal shrike	*Lanius collaris*	0,060	0,569	0,567	N/R
Bokmakierie	*Telophorus zeylonus*	0,034	0,644	0,662	N/R
Cape sparrow	*Passer melanurus*	0,267	0,868	0,567	N/R
Cape weaver	*Ploceus capensis*	0,082	0,569	0,284	N/R
Cape canary	*Serinus canicollis*	0,070	0,509	0,662	N/R
MAMMALS					
Bat-eared fox	*Otocyon megalotis*	0,122			0,221
Striped polecat	*Ictonyx striatus*	1,035	N/R	N/R	2,156
Cape grey mongoose	*Herpestes pulverulentus*	0,180			N/R
Yellow mongoose	*Cynictis penicillata*	0,135			N/R
Suricate	*Suricata suricatta*	0,087	N/R	N/R	N/R
Dassie	*Procavia capensis*	0,043			N/R
Steenbok	*Raphicerus campestris*	0,061			N/R
Hare species	*Lepus* spp	1,055			N/R

Source: 1   Siegfried 1965
          2   Broekhuysen 1965
          3   Siegfried 1966
          4   Macdonald, unpublished data

Road-kill frequencies of certain of the mammal species are even higher than those recorded for bird species (Table 3.3). Although it appears that the frequency of kills reflects to a large extent the relative abundance of the different species (with common animals being the most commonly killed), the mortality rate does appear to be biased towards certain susceptible species.

The interaction of fencing along roadsides and an increase in traffic densities could make this effect of roads more pronounced in the future.

ROADS IN PROTECTED AREAS

Roads constitute the only ecosystem transformation the general populace readily accepts, sometimes even welcomes, within protected areas. However, roads in protected areas have tremendous and wide-ranging ecological effects (eg, Pienaar 1968; Coetzee et al 1979; Milton and Macdonald 1988). The quarrying of material to maintain gravel roads — which, ironically, appear to be more aesthetically acceptable to South African reserve visitors than tarmacadam roads, which are likely to be less ecologically damaging — constitutes a continuous mining of the protected area's non-renewable resources (Brooks et al 1980). The indirect effects of quarrying are also sometimes serious, for example, the role of quarries as anthrax reservoirs in the Etosha National Park (Ebedes 1977). The widespread creation of poorly maintained four-wheel drive tracks in southern African protected areas is often an important factor in accelerated soil erosion, and in associated changes in soil hydrology and resultant vegetation patterns (Macdonald 1979; Macdonald et al 1989).

The overall extent of transformations

If the various classes of transformation considered previously are taken as being additive, it appears that some 248 674 km^2 of South Africa has been transformed (Figure 3.2). Cultivated agricultural fields constitute more than half this area. Urban areas, smallholdings, and areas transformed for transport networks cover the next largest portions of the area transformed. Afforestation and mining plus other non-agricultural intensive transformations each cover just over four per cent of the transformed area. Dense alien plant infestations and impoundments are shown here to be the two least extensive classes of transformation in South Africa. It should be borne in mind that only dense alien woody plant invasions of the fynbos biome are classed as transformations in this review. Dense infestations do exist in the other biomes but are either not dense enough to bring about transformations, as distinct from modifications, or the real extent of their dense infestations has not been estimated. Similarly, the extent of impoundments is underestimated since small farm dams are excluded from this figure.

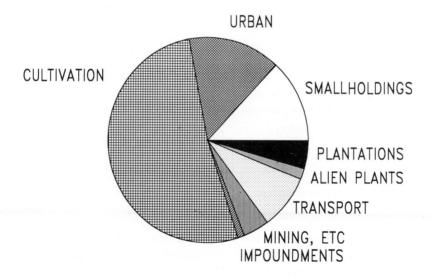

FIGURE 3.2   The extent of the areas transformed by different usages in the white-controlled portion of South Africa using statistics from 1982 to 1987. (Total transformed area = 248 674 km^2.)

    The proportion of South Africa that transformed areas comprise is about 22,2% (Figure 3.3). An uncertain proportion of the area of the national states (KwaZulu, KwaNdebele, etc) has also been transformed. As these areas fall mainly in the better-watered eastern half of South Africa, and are densely populated, it is likely that at least 50% of their surface area has been transformed, or at least so radically modified that it approximates a state of transformation. In similar subsistence landuse areas in Zimbabwe, cultivation alone had transformed 24,7% of the total area by 1977 (Whitlow 1980b). As the extent of the area cultivated has increased rapidly in recent years and other forms of transformation are occurring, it is estimated that the total area affected will now be closer to 50%. If this estimate is accepted, it would bring the total area transformed within South Africa to 278 968 km^2 (24,9% of the total area).

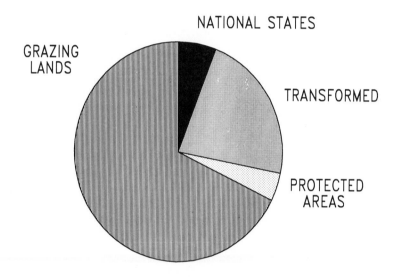

FIGURE 3.3   An approximate breakdown of South Africa's surface area using statistics from 1982 – 1987. (Total area = 1 119 894 km^2.)

THE OVERALL EXTENT OF TRANSFORMATION OF VEGETATION TYPES

Various authorities have estimated the total extent of transformations for Acocks's (1975) Veld Types which occur within particular geographical or ecological subdivisions of South Africa (Table 3.4). As these estimates have been carried out only in the more intensively transformed eastern and southern portions of the country, they should not be construed as representative of the extent of transformation of Veld Types throughout the country. For the rest of the subcontinent it is only in the high-rainfall miombo vegetation types of the Zimbabwean plateau that comparably high percentage transformations are likely to be found (Whitlow 1980a, b) (excluding the Middle Zambezi Valley discussed previously).

    In order to place these estimates in some sort of global perspective they have been compared with estimated extents of transformation for agriculture of the world's vegetation types (World Resources Institute 1986). Although this comparison is not perfect it was thought to be worthwhile both to see how the major biomes compared and to gain some idea of how South Africa rated relative to the world as a whole (Figure 3.4).

TABLE 3.4   Some estimates of the extent of transformation of individual Veld Types in different portions of southern Africa (Sources: Natal (Bourquin 1987; Colvin 1984); Transvaal (Johnston 1979; Batchelor 1985; Fynbos biome (Moll and Bossi 1984))

No	Veld Type name (Acocks 1975)	Total area in RSA (Edwards 1974) (km²)	Natal (pre-1979) % of total area in province	Natal area in province (km²)	Natal % of provinces area transformed	Transvaal** (pre-1979) % of total area in province	Transvaal area in province (km²)	Transvaal % of provinces area transformed	Fynbos biome (1981) % of total area in biome	Fynbos area in biome (km²)	Fynbos % of biomes area transformed
1	Coastal forest and thornveld	20088	75	14984	†1a = 16% 1b = 93%						
3	Pondoland coastal plateau sourveld	730	14	104	94						
4	Knysna forest	3844*							100	3844	24
5	'Ngongoni veld	11130	76	8481	87						
6	Zululand thornveld	3383	95	3226	54						
8	Northeastern mountain sourveld	9541	6	624	61	5	157	?			
9	Lowveld sour bushveld	11120				67	5030	66			
10	Lowveld	18685	61	11342	39	48	5765	76			
11	Arid lowveld	17724	4	624	40	58	8190	32			
12	Springbok flats turf thornveld	3409				77	11471	29			
13	Other turf thornveld	7973				60	2884	92			
14	Arid sweet bushveld	18936				53	3909	60			
15	Mopani veld	20719				87	13186	27			
16	Kalahari thornveld & shrub bushveld	139718				86	15655	8			
18	Mixed bushveld	43170				5	5655	55			
19	Sourish mixed bushveld	33340				66	24208	55			
20	Sour bushveld	18306				46	15916	64			
23	Valley bushveld	24264	34	8325	81	92	10514	28			
34	Strandveld of west coast	6308								4453	24
43	Mountain Renosterveld	11172								4754	27
44	Highland and Dohne sourvelds	39535	33	12986	60						
45	Natal mistbelt	3694	100	3694	89						
46	Coastal Renosterveld	15285*							100	15285	85
47	Coastal Macchia	8770*							100	8770	47
48	Cymbopogon — Themeda veld (sandy)	40381				89	12853	72			
50	Dry Cymbopogon — Themeda veld	43376				22	10342	67			
52	Themeda veld (turf highveld)	10911				91	9845	79			
53	Themeda veld to Cymbopogon — Themeda veld transition	11703				1	79	59			
54	Themeda veld to highland sourveld transition	2749				64	1861	64			
55	Themeda veld to bankenveld transition	682				100	600	82			
56	Highveld sourveld to Cymbopogon — transition	9851	10	1004	92	77	11258	52			
57	Northeastern sandy highveld	14678	8	1145	19	83	24932	65			
58	Themeda — Festuca alpine veld	8222	21	1694	32	100	1545	41			
61	Bankenveld	23568				62	4891	44			
62	Bankenveld to sour sandveld transition	1154				32	1723	47			
63	Piet Relief sourveld	7594	19	1457	39						
64	Northern tall grassveld	4475	100	4475	78						
65	Southern tall grassveld	14251	81	11551	78						
66	Natal sour sandveld	5463	100	5463	90						
67	Pietersburg plateau grassveld	2413				61	1272	81			
68	Eastern Province grassveld	600									
69	Macchia	18345*							100	18345	11
70	False Macchia	18965*							100	18965	3

* Area estimates ex Moll and Bossi (1984) where these are greater than total given by Edwards (1974)

** Only the white-controlled area of the Transvaal was assessed

Note: The only statistic available for the eastern Cape is for valley bushveld which, in 1981, had 31% of its total area of 13 000 km² transformed (G D La Cock pers comm).

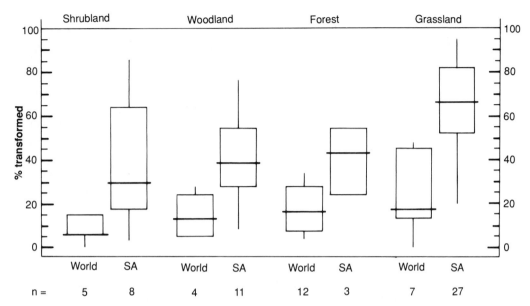

FIGURE 3.4   Estimates of the percentage of vegetation types in the major biomes (a) converted by cultivation on a global scale (World Resources Institute 1986), and (b) transformed within portions of South Africa (Table 3.4). Modal values, quartile ranges and total ranges are given for each sample.

We can conclude that South Africa shows the same relative ordering of its biomes as regards the extent of their transformation as the rest of the world; shrublands are on the average the least transformed, woodland the next, forest the next, and grassland the most. Shrublands have a much wider dispersion of transformation extents in South Africa than in the rest of the world, with certain of the fynbos biome's shrublands being among our most transformed vegetation types. Even if cultivation is only approximately half the extent of transformations globally (as it is in South Africa), it would appear that the modal values for comparable vegetation types in the higher-rainfall eastern and southern portions of South Africa would still be considerably higher than the global average. This is despite the fact that the country as a whole has a mean annual precipitation of only 497 mm which is well below the global average of 860 mm (Department of Water Affairs 1986). (In general more arid regions and vegetation types show smaller extents of transformation — see Table 3.4 and World Resources Institute 1986).

THE MAJOR MODIFICATIONS

Only two classes of modifications are covered here. The first class is the modifications brought about by the replacement of wild ungulate communities with a few domesticated species, and the second class is the modifications of native predator communities.

Modifications of large mammalian herbivore communities

With the advent of Iron-Age pastoralist communities in the region in about AD 250 the long established balance between man, the hunter-gatherer, and his favoured prey animals was disturbed. When European colonists arrived, bringing with them firearms, the balance was completely upset and several species or subspecies of large mammals became extinct or reduced to tiny remnant populations (Sidney 1965; Du Plessis 1969; Skead 1980, 1987). These changes took place mainly prior to the advent of scientific observers in the subcontinent, and the composition of these pristine ungulate communities — their densities, feeding and migratory patterns — is, to this day, largely conjectural (Child and Le Riche 1969; Mentis 1970; Liversidge 1978). For example, it is noteworthy that the species composition of the allegedly pristine ungulate community of the Umfolozi Game Reserve area, as documented by the nagana campaign removal records (Mentis 1970), was highly biased towards species either not favoured as food by the Zulu people or those not readily hunted using 'drives' (see Brooks (1978) for an account of the responses of ungulate species to drives). The differential mortality of species in the man-induced rinderpest epizootiċ of the late 1890s (Brooks and Macdonald 1983; Pienaar 1983; Macdonald 1988) introduces a large measure of uncertainty into the designation of any community as 'pristine' after this date. Finally, the widespread introduction of game-proof fences into southern Africa during the present century terminated the natural movement patterns of mobile grazers in the region, in some cases leading to marked die-offs of the species involved (Bigalke 1961; Child 1972; Owen and Owen 1980; Berry 1981; Pienaar 1983). It seems unlikely that we shall ever know accurately the details of these wild ungulate communities before man's considerable influence on them began.

What we do know is that within a relatively short period they were almost completely replaced over virtually the whole of South Africa by herds of domestic cattle and flocks of sheep and goats. With the advent of the borehole and windmill pump it very soon became possible to graze these domesticated herds virtually throughout South Africa on a year-round basis. The reduction in the diversity of large mammalian herbivores, and their altered patterns of dispersion are thought to have resulted in substantially altered intensities and patterns of defoliation of native plants (Macdonald 1976; Walker 1979). Although poorly quantified, it appears that this in turn has led to large-scale changes in the vegetation.

GRAZING-INDUCED VEGETATION CHANGES

Some of the major landscape-level changes that have, it has been proposed, taken place in response to these changes in ungulate communities are as follows:

— encroachment of karoo bushes into the grassland over some 32 000 km² (Tidmarsh 1948; Acocks 1964, 1975, 1979; Downing 1978);
— reduction in the grass cover of the karoo (Acocks 1964; Roux and Vorster 1983);
— increased dominance of unpalatable karoo bushes throughout the karoo (Roux and Vorster 1983);
— virtual disappearance of grasses from the renosterveld of the fynbos biome as the unpalatable renosterbos *Elytropappus rhinocerotis* assumed dominance (Cowling et al 1986);
— invasion of montane grasslands by fynbos bushes in the eastern Cape (Trollope 1973; Trollope in Downing 1978);
— the encroachment of *Acacia karroo* and other shrubs into the eastern Cape grasslands (Acocks 1975, 1979; Downing 1978; Roux and Van der Vyver 1988);
— replacement of palatable grass species by unpalatable species: by ngongoni grass *Aristida junciformis* in the Natal midlands (Venter 1968), by *Eragrostis* species in the eastern Orange Free State (Foran 1976), and by *Merxmuellera disticha* in the Cape midlands mountains (Roux and Van der Vyver 1988); and
— bush encroachment of vast areas within the region's savanna vegetation types (Van der Schijff 1964; Barnes 1979). Downing (1978) quotes an estimate of 405 000 km² so affected in South Africa and Namibia.

In all cases the changes are poorly characterized as to precise cause, extent and time of occurrence. In almost all cases the relative contribution of changes in herbivory and in fire regimes to the observed shifts in vegetation composition are unknown. It is likely that the effects of changes in these two factors are to a large extent compounded (Macdonald 1976).

WHAT DO THESE VEGETATION CHANGES INDICATE FOR ECOSYSTEM FUNCTIONING?

In most cases these vegetation changes increase the proportion of bare ground and the dominance of woody plants over grasses. Both these trends tend to reduce fire frequency and intensity. These reductions favour the woody plant component of the vegetation over grasses. The process thus becomes self-reinforcing. Similarly, these changes tend to increase rainfall runoff (Table 3.5) and the associated soil erosion (Haylett 1960; Du Plessis and Mostert 1965; Barnes and Franklin 1970). This is particularly so where stocking densities are high, and in South Africa the rangelands have consistently been overstocked since at least the 1930s (Downing 1978; Scotney et al 1988). For example, the entire karoo and adjacent eastern Cape agricultural regions are still considered to be stocked at 130% of their safe carrying capacities (Roux and Opperman 1986; Roux and Van der Vyver 1988). This overstocking has been associated not only with a massive loss of topsoil — both stock densities (Figure 3.5) and erosion (Rooseboom 1978; Roux 1986) peaking in the 1930s and 1940s — but has also resulted in a general lowering of the carrying capacity of the vegetation as indicated by changes in stock numbers (Downing 1978; Roux and Opperman 1986), and by lowered estimates of this capacity (Roux and Van der Vyver 1988; Scotney et al 1988). As the area removed from the natural grazing lands by transformation has increased throughout the historic period, and their carrying capacity has decreased, the extent of the overstocking of remaining rangelands has probably increased. However, improvements in planted pastures and supplementary feeding of livestock over the historic period might have offset these reductions in the total area available for rough grazing.

TABLE 3.5   Changes in runoff from natural vegetation in response to grazing by domestic livestock and baring of the soil in three trials in southern Africa

Site	Pretoria[1]		Glen[2]	Matopos[3]
Soil	Red sandy loam		Red sandy loam	Granite sand
% clay	31	31	15	5
% silt	16	16	2	8
% slope	4	7	5	4
Annual rainfall (mm)	720	740	500	680
Grazing treatment	% of rainfall running off			
Zero	2	2	3	4
Moderate intensity	3	5	4	6
High intensity	n/a	n/a	6	8
Soil bared	49	47	32	30

Sources: 1   Haylett (1960)
          2   Du Plessis and Mostert (1965)
          3   Barnes and Franklin (1970)

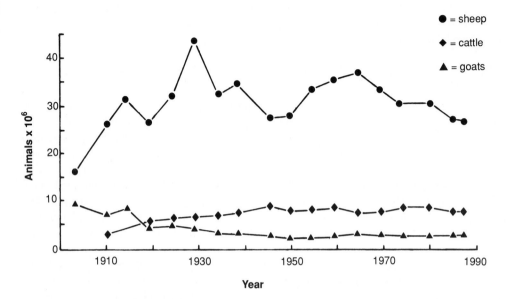

FIGURE 3.5   The numbers of domestic ungulates in the white-controlled portion of South Africa (Downing 1978; J L Cloete pers comm)

The situation described is not peculiar to the 'white' farming area of South Africa. Virtually identical scenarios have been recorded for the Transkei (McKenzie 1984) and for the subsistence agriculture areas of Zimbabwe (West 1969).

The effects of these changes on the conservation of biotic diversity in the subcontinent have never been fully assessed. One of the points that emerges from numerous independent analyses is that wetland habitats such as vleis are threatened on a subcontinental basis as a result of their destruction through over-utilization and the resultant erosion. For example, Begg (1987) has found that 58% of the original area of wetlands in the Mfolozi catchment has been destroyed over the last few centuries. Brooke (1984) found that 48% of South African bird species in the Red Data Book utilized moist grasslands or wetlands, and concluded that these habitats were particularly threatened by over-utilization and other forms of mismanagement.

Modification of native predator communities

Southern Africa has a remarkable diversity of both mammalian (Stuart 1981; Smithers 1983; Macdonald and Frame 1988) and avian (Macdonald and Gargett 1984; Maclean 1985) predators. Most large mammalian carnivores have been exterminated within the developed portions of the subcontinent (Smithers and Wilson 1979; Stuart et al 1985; Norton 1986). Several of the larger eagles have shown similar range reductions (Boshoff et al 1983; Brooke 1984; Tarboton and Allan 1984), as have certain of the scavenging birds (Boshoff et al 1983; Macdonald and Macdonald 1983; Brooke 1984).

REDUCTION IN THE DENSITIES OF RAPTORIAL BIRDS

Comparisons between the larger savanna reserves and adjacent man-modified areas have shown that most birds of prey, particularly the larger species, now occur at much lower densities in the latter areas. This has been demonstrated for the Kalahari Gemsbok National Park in the northern

Cape (Cade 1965; Siegfried 1968; Dean 1975; Liversidge 1984), the Kruger National Park in the Transvaal (Kemp 1980; Tarboton and Allan 1984), and the Hluhluwe–Umfolozi Game Reserve in Natal (Macdonald 1984).

Active persecution and accidental poisoning have generally been suggested as the reasons for the lower densities in farming areas. However, the wide range of species apparently showing reduced densities suggests that more fundamental ecological changes may be involved. Some possibilities in this respect are scrub encroachment, prey-density reductions (as a consequence of lowered productivity through desertification and/or the competitive effects of domestic livestock), and nest-site reductions (through the widespread felling of large trees, particularly along riparian strips which have been heavily transformed by agriculture). Whatever the causal mechanisms, the end result is that avian predator populations are now much reduced in density over large portions of the semi-arid regions of the subcontinent.

ECOLOGICAL CONSEQUENCES OF ALTERED PREDATOR POPULATIONS

With the reduction in populations of certain mammalian carnivores it has been suggested that certain other species have increased in density. Thus cheetah *Acinonyx jubatus* are thought to have increased on the ranchlands of Namibia as a result of the elimination of the larger carnivores (Joubert and Mostert 1975), and caracal *Felis caracal* in the Bedford District of the eastern Cape, as a result of the elimination of black-backed jackal *Canis mesomelas* (Pringle and Pringle 1979). In both cases, the carnivore that was supposed to have been historically rare has now become a 'problem animal' allegedly killing large numbers of domestic livestock. For example, by 1981 the caracal was considered the most important 'problem animal' in 82% of the Divisional Council areas within the Cape Province (Stuart 1981). Such historical shifts in predator community composition are, however, hard to verify.

Possibly of greater significance are suggestions that predator-prey relationships have been significantly destabilized as a result of man-induced changes in predator communities. For example, it has been postulated that the observed decrease in the interval between outbreaks of the brown locust *Locustana pardalina* in South Africa could, in part, be due to reductions in locust predators (Hockey 1988). Similarly, the reduction in predators, both avian and mammalian, of the rock hyrax *Procavia capensis* is thought to have enabled it to extend its range within the karoo where it now occurs on tiny outcrops of rock. The foraging of this native herbivore has been implicated in range deterioration there (Roux and Opperman 1986) while, even in small nature reserves within this modified area, native plant species are being subjected to chronic over-utilization by this prolific herbivore (Rebelo 1987; Macdonald, pers obs). It has been indicated that this type of change has massive implications for the long-term conservation of native species in the subcontinent (Diamond et al 1982).

CONCLUSION

Man has played an important role in modifying or transforming virtually every ecosystem in southern Africa. The quantitative extent of these changes is generally difficult to ascertain. However, it appears likely that they are significant in terms of the ability of the affected ecosystems to support populations of native species. Historic reductions in the ranges of many of these species indicate that it is essential that adequate areas be protected from man's influence if the diversity of southern Africa's biota is to be conserved.It is equally important that limits be established for the allowable extents of certain transformations (eg, the impoundment of rivers), and that certain landuse practices (eg, cultivation and domestic livestock grazing) be improved if essential ecological processes are to be maintained. A guiding principle in all such decisions should be that the practically 'non-renewable' soil resource must be protected.

One of the factors giving cause for concern is that many of the man-induced changes in southern African ecosystems are very recent. In some cases introductions of new technologies have occurred within the life span of some of the longer-lived species in native communities. An example of this is the recent mortality of large trees alongside tarmacadam roads in the savannas of the region (Milton and Macdonald 1988). A further example may be provided by the widespread mortality of large trees in the Kruger National Park during the last three decades (Viljoen 1988). Although this has been attributed largely to the effects of elephant populations (eg, Coetzee et al 1979; Viljoen 1988) there are certain species, for example, *Combretum imberbe*, in which this is apparently not the case, and here drought-induced changes in the water table have been suggested as the causal mechanism (Viljoen 1988). It has not yet been proposed that this abnormal lowering of the water table might have resulted, at least in part, from the extraction of underground water for the provision of artificial watering points. It is, however, known that pumping of underground water took place for the first time in the three decades preceding the observed mortalities (Smuts 1976). The implication of the fact that many of these man-induced changes are so recent is that their untoward effects may manifest themselves only in the future.

Currently many of the adverse effects of changes being wrought by man's activities are manifesting themselves most strongly in the wetlands, rivers and estuaries of the subcontinent. As southern Africa is predominantly semi-arid, these mesic habitat types are intrinsically a rare resource in the subcontinent. The scarcity of water means that already large demands are being placed on the surface and underground water resources of the subcontinent for human use. As the human population of the subcontinent continues its exponential rate of increase, so this demand can be expected to increase. The inevitable conflict between the demand for human use and the requirements of natural ecosystems serves to illustrate the fundamental need to limit the growth of the human population if there is to be any long-term chance of conserving southern Africa's rich natural heritage.

ACKNOWLEDGEMENTS

The author's research is funded by the National Programme for Ecosystem Research of the Foundation for Research Development. The following are thanked for making data available to the author: A Batchelor (Department of Development Aid), Dr O Bourquin (Natal Parks Board), Dr J Cloete (Department of Agriculture and Water Supply), B L Dawson (PU-NTC Research Institute for Reclamation Ecology), W R J Dean (University of Cape Town), B Gasson (University of Cape Town), K H Cooper (Wildlife Society of Southern Africa), G D La Cock (Chief Directorate of Nature and Environmental Conservation in the Cape), L Lotter (Transvaal Directorate of Nature Conservation) Professor R Lubke (Rhodes University), C McDowell (University of Cape Town), L G Pienaar (Department of Agricultural Economics and Marketing), P J Pretorius (Cape Town City Council) and N van der Spuy (Western Cape Regional Services Council). G J Breytenbach, Dr P W Roux and two anonymous referees provided useful criticisms of the initial draft. The manuscript was typed on wordprocessor by Pauline Solomon and my wife, Susan.

REFERENCES

ACOCKS J P H (1964). Karoo vegetation in relation to the development of deserts. In *Ecological Studies in Southern Africa*. (ed Davis D H S) Junk, The Hague. pp 100 –112.

ACOCKS J P H (1975). Veld Types of South Africa. (Second edn). *Memoirs of the Botanical Survey of South Africa* **40**, 1 – 128.

ACOCKS J P H (1979). The flora that matched the fauna. *Bothalia* **12**, 673 – 709.

ALLAN D G, GAMBLE K, JOHNSON D N, PARKER V, TARBOTON W R and WARD D M (1987). Report on the Blue Swallow in South Africa and Swaziland. Report of the Blue Swallow Working Group, Photostat report. Endangered Wildlife Trust, Johannesburg. 41 pp.

ALLAN D G and TARBOTON W R (1985). Sparrowhawks and plantations. In *Proceedings of the Birds and Man Symposium*. (ed Bunning L J) Witwatersrand Bird Club, Johannesburg. pp 167 – 177.

ALLANSON B R and GIESKES J T (1961). Investigations into the ecology of polluted waters in the Transvaal. Part 2. An introduction to the limnology of Hartebeespoort Dam with special reference to the effect of industrial and domestic pollution. *Hydrobiologia* **18**, 77 – 94.

ANONYMOUS (1985). *Landbou-ontwikkelingsprogram noord-westesubstreek.* Department of Agriculture, Winter Rainfall Region, Elsenburg.

ATTWELL R I G (1970). Some effects of Lake Kariba on the ecology of a floodplain of the mid-Zambezi valley of Rhodesia. *Biological Conservation* **2**, 189 – 196.

AVERY D M (1987). Micromammalian evidence for natural vegetation and the introduction of farming during the Holocene in the Magaliesberg, Transvaal. *South African Journal of Science* **83**, 221 – 225.

BALON E K (1978). Kariba: the dubious benefits of large dams. *Ambio* **7**, 40 – 48.

BARNES D L (1979). Cattle ranching in the semi-arid savannas of east and southern Africa. In *Management of semi-arid ecosystems.* (ed Walker B H) Elsevier, New York, NY. pp 9 – 54.

BARNES D L and FRANKLIN M J (1970). Runoff and soil loss on a sandveld in Rhodesia. *Proceedings of the Grassland Society of South Africa* **5**, 140 – 144.

BATCHELOR G R (1985). Die bewaringstatus van Acocks se veldtipes in Transvaal. Internal Report, Transvaal Nature Conservation Division, Pretoria. 7 pp.

BEGG G (1979). The estuaries of Natal. *Natal Town and Regional Planning Reports* **41**, Town and Regional Planning Commission, Pietermaritzburg.

BEGG G W (1987). The distribution extent and status of wetlands in the Mfolozi catchment. In Ecology and conservation of wetlands in South Africa. (Compilers Walmsley R D and Botten M L). *Ecosystem Programmes Occasional Report* **28**, 51 – 62. CSIR, Pretoria.

BERRY H H (1981). Abnormal levels of disease and predation as limiting factors for wildebeest in the Etosha National Park. *Madoqua* **12**, 242 – 253.

BIGALKE R C (1961). Some observations on the ecology of the Etosha Game Park, South West Africa. *Annals of the Cape Provincial Museum* **1**, 49 – 67.

BOSHOFF A F, VERNON C J and BROOKE R K (1983). Historical atlas of the diurnal raptors of the Cape Province. *Annals of the Cape Provincial Museums* **14**, 173 – 297.

BOUCHER C (1981). Western Cape Province lowland alien vegetation. In *Proceedings of a Symposium on coastal lowlands of the Western Cape.* (ed Moll E) University of Western Cape, Cape Town. 67 pp.

BOURQUIN O (1987). Professional Services Division Yearbook 1986 – 87. Internal Report, Natal Parks Board, Pietermaritzburg. 85 pp.

BOWMAKER A (1970). A prospect of Lake Kariba. *Optima* **20**, 68 – 74.

BREDENKAMP G J and VAN VUUREN D R J (1987). Note on the occurrence and distribution of *Aloe marlothii* Berger on the Pietersburg Plateau. *South African Journal of Science* **83**, 498 – 500.

BROEKHUYSEN C J (1965). An analysis of bird casualties on the roads in the south-western Cape Province. *L'oiseau et la Revue Francaise d'Ornithologie* **35**, 35 – 51.

BROOKE R K (1984). South African Red Data Book — birds. *South African National Scientific Programmes Report* **97**, CSIR, Pretoria. 1 – 213.

BROOKS P M (1978). Ungulate population estimates for Hluhluwe Game Reserve and northern corridor for July 1978, based on foot and helicopter counts. Natal Parks Board, roneoed report. 14 pp.

BROOKS P M and MACDONALD I A W (1983). The Hluhluwe-Umfolozi Reserve: an ecological case history. In *Management of large mammals in African conservation areas.* (ed Owen-Smith R N) Haum, Pretoria. pp 51 – 77.

BROOKS P M, MACDONALD I A W and WHATELEY A (1980). An assessment of the state of the gravel roads in Hluhluwe Game Reserve and northern corridor in April 1980, with suggestions for an improved approach to roads planning throughout Natal's reserves. Natal Parks Board, roneoed report. 15 pp.

BROWN C J and GUBB A A (1986). Invasive alien organisms in the Namib Desert, Upper Karoo and the arid and semi-arid savannas of western southern Africa. In *The ecology and management of biological invasions in southern Africa.* (eds Macdonald I A W, Kruger F J and Ferrar A A). Oxford University Press, Cape Town. pp 93 – 108.

BRUTON M N and VAN AS J (1986). Faunal invasions of aquatic ecosystems in southern Africa, with suggestions for their management. In *The ecology and management of biological invasions in southern Africa.* (eds Macdonald I A W; Kruger F J and Ferrar A A). Oxford University Press, Cape Town. pp 47 – 62.

CADE T J (1965). The status of the peregrine and other falconiforms in Africa. In *Population biology of the Peregrine falcon.* University of Wisconsin Press. pp 289 – 321.

CAMBRAY J A, DAVIES B R and ASHTON P J (1986). The Orange-Vaal River system. In *The ecology of river systems.* (eds Davies B R and Walker K F) Junk, Dordrecht. pp 89 – 122.

CHILD G (1972). Observations on a wildebeest die-off in Botswana. *Arnoldia* **5**, 1 – 13.

CHILD G and LE RICHE J D (1969). Recent springbok treks (mass movements) in southwestern Botswana. *Mammalia* **33**, 449 – 504.

CLINNING C F (1986). Transvaal grasslands. *Fauna and Flora* **44**, 3 – 5.

COETZEE B J, ENGELBRECHT A H, JOUBERT S C J and RETIEF P F (1979). Elephant impact on *Sclerocarya caffra* trees in *Acacia nigrescens* tropical plains thornveld of the Kruger National Park. *Koedoe* **22**, 39 – 60.

COETZEE J A, SCHOLTZ A and DEACON H J (1983). Palynological studies and the vegetation history of the fynbos. In Fynbos palaeoecology: a preliminary synthesis. (eds Deacon H J, Hendey Q B and Lambrechts J J N). *South African National Scientific Programmes Report* **75**, CSIR Pretoria. 156 – 173.

COLVIN I S (1984). Survey of conservation-worthy areas in Natal — procedure and methodology. Natal Parks Board, roneoed report. 18 pp.

COWAN G I (1987). S A Plan for Nature Conservation Annual Report (April 1986 – March 1987). Report of the Department of Environmental Affairs, Pretoria. 14 pp.

COWLING R M, PIERCE S M and MOLL E J (1986). Conservation and utilization of South Coast Renosterveld, an endangered South African vegetation type. *Biological Conservation* **37**, 363 – 377.

DAVIES B R and DAY J A (1986). *The biology and conservation of South Africa's vanishing waters.* Centre for Extramural Studies, University of Cape Town, Cape Town. 186 pp.

DAWSON B L (1986). Roads as a disturbing factor in Fynbos. Poster paper, 8th Annual Research Meeting of the Fynbos Biome Project, Cape Town, June 1986.

DEACON H J (1983). The peopling of the fynbos region. In Fynbos palaeoecology: a preliminary synthesis. (eds Deacon H J, Hendey Q B and Lambrechts J J N). *South African National Scientific Programmes Report* **75**, CSIR, Pretoria. 183 – 204.

DEACON J (1986). Human settlement in South Africa and archaeological evidence for alien plants and animals. In *The ecology and management of biological invasions in southern Africa*. (eds Macdonald I A W, Kruger F J and Ferrar A A). Oxford University Press, Cape Town. pp 3 – 19.

DEAN W R (1969). A checklist of the birds of Benoni. *South African Avifauna Series* **64**, 1 – 17.

DEAN W R J (1975). Dry season roadside raptor counts in the northern Cape, South West Africa and Angola. *Journal of the South African Wildlife Management Association* **5**, 99 – 102.

DEPARTMENT OF WATER AFFAIRS (1986). *Management of the water resources of the Republic of South Africa*. Government Printer, Pretoria. 483 pp.

DIAMOND J M et al (1982). Implications of island biogeography for ecosystem conservation. In Conservation of ecosystems: theory and practice. (eds Siegfried W R and Davies B R). *South African National Scientific Programmes Report* **61** CSIR, Pretoria. 46 – 60.

DOWNING B H (1978). Environmental consequences of agricultural expansion in South Africa since 1850. *South African Journal of Science* **74**, 420 – 422.

DU PLESSIS M C F and MOSTERT J W C (1965). Afloop en grondverliese by die landbounavorsingsinstituut Glen. *South African Journal of Agricultural Science* **8**, 1051 – 1060.

DU PLESSIS S F (1969). The past and present geographical distribution of the perissodactyla and artiodactyla in southern Africa. MSc thesis, University of Pretoria, Pretoria. 333 pp.

EBEDES H (1977). Anthrax epizootics in Etosha National Park. *Madoqua* **10**, 99 – 118.

EDWARDS D (1974). Survey to determine the adequacy of existing conserved areas in relation to vegetation types. A preliminary report. *Koedoe* **17**, 2 – 37.

ELWELL H A (1972). The influence of agricultural systems on rainfall runoff. *Rhodesian Agricultural Journal, Technical Bulletin* **15**, 13 – 26.

ELWELL H A (1975). Conservation implications of recently determined soil formation rates in Rhodesia. *Rhodesia Science News* **9**, 312 – 313.

FEELY J M (1979). Did iron-age man have a role in the history of Zululand's wilderness landscapes? *South African Journal of Science* **76**, 150 – 152.

FEELY J M (1987). The early farmers of Transkei, southern Africa, before AD 1870. *Cambridge Monographs in African Archaeology* **24**, 1 – 142.

FERRAR T and PITMAN D (1980). To drown or not to drown an invaluable asset. *The Farmer* **15**, 33 – 35.

FORAN D B (1976). The development and testing of methods for assessing the condition of three grassveld types in Natal. MSc thesis, University of Natal, Pietermaritzburg.

FORESTRY DEPARTMENT (1980). *Annual Report of the Department of Forestry 1980*. Department of Environment Affairs, Directorate of Forestry, Pretoria.

FRASER W (1987). The urban birds of Johannesburg. *Bokmakierie* **39**, 67 – 70.

FUGGLE R F and RABIE M A (eds) (1983). *Environmental concerns in South Africa*. Juta, Cape Town. 587 pp.

GASSON B (in press). *Metropolitan Cape Town Environmental Inventory*. Urban Problems Research Unit, University of Cape Town, Cape Town.

GIFFORD G F and HAWKINS R H (1978). Hydrologic impact of grazing on infiltration: a critical review. *Water Resources Research* **14**, 305 – 313.

GUY P R (1981). River bank erosion in the mid-Zambezi Valley, downstream of Lake Kariba. *Biological Conservation* **19**, 199 – 212.

HALL A V and ASHTON E R (1983). *Threatened plants of the Cape Peninsula*. Threatened Plants Research Group, University of Cape Town, Cape Town. 26 pp.

HALL A V and VELDHUIS H (1985). South African Red Data Book: Plants — Fynbos and Karoo Biomes. *South African National Scientific Programmes Report* **117**, CSIR, Pretoria. 1 – 160.

HALL M (1979). The Umfolozi, Hluhluwe and corridor reserves during the iron-age. *Lammergeyer* **27**, 28 – 40.

HAYLETT D G (1960). Runoff and soil erosion studies at Pretoria. *South African Journal of Agricultural Science* **3**, 379 – 394.

HEEG J and BREEN C M (1982). Man and the Pongolo floodplain. *South African National Scientific Programmes Report* **56**, CSIR, Pretoria. 1 – 117.

HENDERSON L and WELLS M J (1986). Alien plant invasions in the grassland and savanna biomes. In *The ecology and management of biological invasions in southern Africa*. (eds Macdonald I A W, Kruger F J and Ferrar A A). Oxford University Press, Cape Town. pp 109 – 117.

HEYDORN A E F and TINLEY K L (1980). Estuaries of the Cape. Part 1. *CSIR Research Report* **380**. 97 pp.

HEYNS R (1986). *South Africa 1986*. Bureau for Information, Pretoria. 1049 pp.

HOCKEY P A R (1988). The brown locust — war or peace? *South African Journal of Science* **84**, 8 – 10.

HUNTLEY B J (1978). Ecosystem conservation in southern Africa. In *Biogeography and ecology of southern Africa*. (ed Werger M J A) Junk, The Hague. pp 1333 – 1384.

HUTCHISON I P G (1976). Lake St Lucia — mathematical modelling and evaluation of ameliorative measures. *Hydrological Research Unit Report* **1176**, University of the Witwatersrand, Johannesburg.

JARMAN P J (1968). The effect of the creation of Lake Kariba upon the terrestrial ecology of the middle Zambezi Valley, with particular reference to larger mammals. PhD thesis, University of Manchester, Manchester. 318 pp.

JARMAN P J (1972). Seasonal distribution of large mammal populations in the unflooded middle Zambezi Valley. *Journal of Applied Ecology* **9**, 283 – 299.

JOHNSTON J H (1979). Identification of priority areas for nature conservation in the Transvaal. Internal report, Directorate of Nature Conservation, Transvaal Provincial Administration.

JOUBERT E and MOSTERT P K N (1975). Distribution patterns and status of some mammals in South West Africa. *Madoqua* **9**, 1 – 23.

KEMP A C (1980). The importance of the Kruger National Park for bird conservation in the Republic of South Africa. *Koedoe* **23**, 99 – 122.

KLEIN R G (1974).On the taxonomic status, distribution and ecology of the blue antelope, *Hippotragus leucophaeus* (Pallas 1766). *Annals of the South African Museum* **65**, 99 – 143.

KRUGER F J and BIGALKE R C (1984). Fire in fynbos. In *Ecological effects of fire in South African ecosystems.* (eds Booysen P D V and Tainton N M.) Springer, Berlin. pp 67 – 114.

LIGGITT B (1983). The invasive alien plant *Chromolaena odorata* with regard to its status and control in Natal. *Monograph no 2,* Institute of Natural Resources, University of Natal, Pietermaritzburg.

LIVERSIDGE R (1978). It was all exaggeration. *African Wildlife* **32**, 26 – 27.

LIVERSIDGE R (1984). The importance of national parks for raptor survival. *Proceedings of the 5th Pan-African Ornithological Congress.* pp 589 – 600.

MACDONALD I A W (1976). A suggested format for teaching management and conservation of veld in semi-arid southern Africa. *Proceedings of the Symposium on Conservation Education, Skukuza, September 1976.* Wildlife Society of Southern Africa. pp 75 – 100.

MACDONALD I A W (1979). The effects of gulley erosion on the vegetation of Hluhluwe Game Reserve. Paper presented at the Symposium/Workshop on the Vegetation Dynamics of the Central Complex in Zululand, Hluhluwe Game Reserve, 10 – 12 August. 5 pp.

MACDONALD I A W (1983). Alien trees, shrubs and creepers invading indigenous vegetation in the Hluhluwe-Umfolozi Game Reserves complex in Natal. *Bothalia* **14**, 949 – 959.

MACDONALD I A W (1984). An analysis of the role of the Hluhluwe-Umfolozi Game Reserves Complex in the conservation of the avifauna of Natal. *Proceedings of the 5th Pan-African Ornithological Congress.* pp 601 – 637.

MACDONALD I A W (1986). Do redbreasted sparrowhawks belong in the karoo? *Bokmakierie* **38**, 3 – 4.

MACDONALD I A W (1988). The history, impacts and control of introduced species in the Kruger National Park, South Africa. *Transactions of the Royal Society of South Africa* **46**, 251 – 276.

MACDONALD I A W , CLARK D L and TAYLOR H C (1989). Alien plant control in the Cape of Good Hope Nature Reserve. *South African Journal of Botany* **55**, 56 –75.

MACDONALD I A W and FRAME G W (1988). The invasion of introduced species into nature reserves in tropical savannas and dry woodlands. *Biological Conservation* **44**, 67 – 93.

MACDONALD I A W and GARGETT V (1984). Raptor density and diversity in the Matopos, Zimbabwe. *Proceedings of the 5th Pan-African Ornithological Congress.* pp 287 – 308

MACDONALD I A W and JARMAN M L (eds) (1985). Invasive alien plants in the terrestrial ecosystems of Natal, South Africa. *South African National Scientific Programmes Report* **118**, CSIR, Pretoria. 1 – 88.

MACDONALD I A W, JARMAN M L and BEESTON P (eds) (1985). Management of invasive alien plants in the Fynbos Biome. *South African National Scientific Programmes Report* **111**, CSIR, Pretoria. 1 – 140.

MACDONALD I A W and MACDONALD S A (1983). The demise of the solitary scavengers in Southern Africa — the early rising crow hypothesis. In *Proceedings of the Birds and Man Symposium.* (ed Bunning L J) Johannesburg. pp 321 – 335.

MACDONALD I A W and RICHARDSON D M (1986). Alien species in terrestrial ecosystems of the fynbos biome. In *The ecology and management of biological invasions in southern Africa.* (eds Macdonald I A W, Kruger F J and Ferrar A A). Oxford University Press, Cape Town. pp 77 – 91.

MACDONALD I A W , RICHARDSON D M and POWRIE F J (1986). Range expansion of the hadeda ibis *Bostrychia hagedash* in southern Africa. *South African Journal of Zoology* **21**, 331 – 342.

MACLEAN G L (1985). *Roberts' Birds of Southern Africa.* John Voelcker Bird Book Fund, Cape Town. 848 pp.

MARKER M E and EVERS T M (1976). Iron-age settlement and soil erosion in the eastern Transvaal, South Africa. *South African Archaeological Bulletin* **31**, 153 – 165.

MARTIN A K (1987). Comparison of sedimentation rates in the Natal Valley, south-west Indian Ocean, with modern sediment yields in east coast rivers of southern Africa. *South African Journal of Science* **83**, 716 – 724.

MCDOWELL C R (1988). Factors affecting the conservation of Renosterveld by private landowners. PhD thesis, University of Cape Town, Cape Town.

MCKENZIE B (1984) Utilization of the Transkeian landscape — an ecological interpretation. *Annals of the Natal Museum* **26**, 165 – 172.

MENTIS M T (1970). Estimates of natural biomasses of large herbivores in the Umfolozi Game Reserve area. *Mammalia* **34**, 363 – 393.

MILTON S J and MACDONALD I A W (1988). Tree deaths near tar roads in the northern Transvaal. *South African Journal of Science* **84**, 164 – 165.

MOLL E J and BOSSI L (1984). A current assessment of the extent of the natural vegetation of the fynbos biome. *South African Journal of Science* **80**, 355 – 358.

MORRIS J W (1976) Automatic classification of the highveld grassland of Lichtenburg, south-western Transvaal. *Bothalia* **12**, 267 – 292.

NORTON P M (1986) Historical changes in the distribution of leopards in the Cape Province. *Bontebok* **5**, 1 – 9.

O'KEEFFE J H (1986a). Ecological research on South African rivers — a preliminary synthesis. *South African National Scientific Programmes Report* **121**, CSIR, Pretoria. 1 – 121.

O'KEEFFE J H (ed)(1986b). The conservation of South African rivers. *South African National Scientific Programmes Report* **131**, CSIR, Pretoria. 1 – 115.

OWEN M and OWEN D (1980). The fences of death. *African Wildlife* **34**, 25 – 27.

PETERSEN W, NEWBERRY R and JACOBSEN N (1985). *Cordylus giganteus* is alive and well and living at Rietpoort. *Fauna and Flora* **42**, 26 – 29.

PIENAAR U DE V (1968). The ecological significance of roads in a national park. *Koedoe* **11**, 169 – 174.

PIENAAR U DE V (1983). Management by intervention: the pragmatic/economic option. In *Management of large mammals in African conservation areas.* (ed Owen-Smith R N) Haum, Pretoria. pp 23 – 36.

POORE D (1978) Ecosystem conservation. In Sourcebook for a world conservation strategy (second draft). *General assembly paper GA 78/10 add 4* IUCN, Morges, Switzerland. 25 pp.

PORTER R N (1981). *A preliminary impact assessment of the environmental effects of proposed dams in the Mfolozi catchment, Natal.* Natal Parks Board. 443 pp.

PRESTON-WHYTE R A and TYSON P D (1988) *The atmosphere and weather of southern Africa.* Oxford University Press, Cape Town. 375 pp.

PRINGLE J A and PRINGLE V L (1979). Observations on the Lynx *Felis caracal* in the Bedford district. *South African Journal of Zoology* **14**, 1 – 4.

RAAL P A (1986). The Transvaal threatened plants programme. *Fauna and Flora* **44**, 17 – 21.

REBELO A G (1987). Management implications. In A preliminary synthesis of pollination biology in the Cape flora. (ed Rebelo A G). *South African National Scientific Programmes Report* **141**, CSIR, Pretoria. 193 – 211.

ROSEBOOM A (1978). Sediment afvoer in Suider-Afrikaanse riviere. *Water S A* **4**, 14 – 17.

ROURKE J P, FAIRALL P and SNIJMAN D A (1981). William Frederick Purcell and the flora of Bergvliet. *Journal of South African Botany* **47**, 547 – 566.

ROUX P W (1986). Grondbewaring. *Bylaag E van die Direkteursvergadering, Pretoria 3 – 5 November, 1986.* Departement Landbou en Water Voorsiening, Pretoria.

ROUX P W and OPPERMAN D P J (1986). Soil erosion. In The karoo biome: a preliminary synthesis. Part 1. (eds Cowling R M, Roux P W and Pieterse A J H). *South African National Scientific Programmes Report* **124**, CSIR, Pretoria. 92 – 111.

ROUX P W and VAN DER VYVER J (1988). The agricultural potential of the eastern Cape and Cape midlands. In *Towards an environmental plan for the eastern Cape.* (eds Bruton M N and Gess F W). Rhodes University, Grahamstown. pp 100 – 125.

ROUX P W and VORSTER M (1983). Vegetation change in the Karoo. *Proceedings of the Grassland Society of South Africa* **18**, 25 – 29.

SCHARFETTER H (1987). Timber resources and needs in southern Africa. *South African Journal of Science* **83**, 256 – 259

SCHOEMAN J L and SCOTNEY D M (1987). Agricultural potential as determined by soil, terrain and climate. *South African Journal of Science* **83**, 260 – 268.

SCOTNEY D M et al (1988). The agricultural areas of southern Africa. In Long-term data series relating to southern Africa's renewable natural resources.(eds Macdonald I A W and Crawford R J M). *South African National Scientific Programmes Report* **157**, CSIR, Pretoria. 316 – 336.

SEELY M K and WARD J (1988). The Namib Desert. In Long-term data series relating to southern Africa's renewable natural resources (eds Macdonald I A W and Crawford R J M). *South African National Scientific Programmes Report* **157**. CSIR, Pretoria. 268 – 279.

SERFONTEIN J (1976). Beperkings op landbou in berggebiede. *Proceedings of the conference on Mountain Environments,* Habitat Council, Pretoria.

SHANNON L V and ANDERSON F P (1982). Applications of satellite ocean colour imagery in the study of the Benguela Current system. *South African Journal of Photogrammetry, Remote Sensing and Cartography* **13**, 153 – 169.

SIDNEY J (1965). The past and present distribution of some African ungulates. *Transactions of the Zoological Society of London* **30**, 1 – 397.

SIEGFRIED W R (1965). A survey of wildlife mortality on roads in the Cape Province. *Cape Department of Nature Conservation, Investigational Report* **6**, 1 – 20.

SIEGFRIED W R (1966). Casualties among birds along a selected road in Stellenbosch. *Ostrich* **37**, 146 – 148.

SIEGFRIED W R (1968). Relative abundance of birds of prey in the Cape Province. *Ostrich* **39**, 253 – 258.

SKEAD C J (1980). *Historical mammal incidence in the Cape Province Vol 1. The western and northern Cape.* Cape Department of Nature and Environmental Conservation, Cape Provincial Administration, Cape Town. 903 pp.

SKEAD C J (1987). *Historical mammal incidence in the Cape Province Vol 2. The eastern half of the Cape Province, including the Ciskei, Transkei and East Griqualand.* Department of Nature and Environmental Conservation, Cape Provincial Administration, Cape Town.

SMITHERS R H N (1983). *The mammals of the southern African subregion.* University of Pretoria, Pretoria. 736 pp.

SMITHERS R H N and WILSON V J (1979). Check list and atlas of the mammals of Zimbabwe Rhodesia. *Museum Memoirs* **9**, Trustees of the National Museums and Monuments, Salisbury, Zimbabwe Rhodesia. 193 pp.

SMUTS G L (1976). Population characteristics and recent history of lions in two parts of the Kruger National Park. *Koedoe* **19**, 153 – 164.

SNELL M L (1979). The vulnerable blue swallow. *Bokmakierie* **31**, 74 – 78.

SNELL M L (1988). Local extinction of the blue swallow at Nyanga. *Honeyguide* **34**, 30 – 31.

STOCKING M (1984). Rates of erosion and sediment yield in the African environment. *International Association of Hydrological Science Publications* **144**, 285 – 293.

STUART C T (1981). Notes on the mammalian carnivores of the Cape Province, South Africa. *Bontebok* **1**, 1 – 58.

STUART C T, MACDONALD I A W and MILLS M G L (1985). History, current status and conservation of large mammalian predators in Cape Province, Republic of South Africa. *Biological Conservation* **31**, 9 – 17.

SWYNNERTON C F M (1918). Some factors in the replacement of the ancient East African forest by wooded pasture land. *South African Journal of Science* **15**, 493 – 518.

TALKENBERG W F M (1982).An investigation of the environmental impact of surface diamond mining along the arid west coast of South Africa. Unpublished MSc thesis, School of Environmental Studies, University of Cape Town, Cape Town.

TARBOTON W R (1987). The Nyl floodplain: its significance, phenology and conservation status. In Ecology and conservation of wetlands in South Africa. (compilers Walmsley R D and Botten M L) *Ecosystem Programmes Occasional Report* **28**, CSIR, Pretoria. 101 – 114.

TARBOTON W R and ALLAN D G (1984). The status and conservation of birds of prey in the Transvaal. *Transvaal Museum Monograph* **3**, 115 pp.

THORNTON J A (ed) (1982). Lake McIlwaine: the eutrophication and recovery of a tropical man-made lake. *Monographiae Biologicae Volume* **49**, Junk, The Hague. 251 pp.

TIDMARSH C E M (1948). Conservation problems in the Karoo. *Farming in South Africa* **23**, 519 – 530

TINLEY K L (1985). Coastal dunes of South Africa. *South African National Scientific Programmes Report* **109**, CSIR, Pretoria. 1 – 297.

TOMS R (1985). The "Parkmore Prawn". *African Wildlife* **39**, 200 – 202.

TROLLOPE W S W (1973). Fire as a method of controlling macchia (Fynbos) vegetation on the Amatole mountains of the eastern Cape. *Proceedings of the Grassland Society of South Africa* **8**, 35 – 41.

TYSON P D, KRUGER F J and LOUW C W (1988). Atmospheric pollution and its implications in the eastern Transvaal Highveld. *South African National Scientific Programmes Report* **150**, CSIR, Pretoria. 111 pp.

VAN DER SCHIJFF H P (1964). 'n Hervaluasie van die probleem van bosindringing in Suid-Afrika. *Tydskrif Natuurwetenskap* **4**, 67 – 80.

VENTER A D (1968). The problems of *Aristida junciformis* encroachment into the veld of Natal. *Proceedings of the Grassland Society of South Africa* **3**, 163 – 165.

VERSFELD D B and VAN WILGEN B W (1986). Impact of woody aliens on ecosystem properties. In *The ecology and management of biological invasions in southern Africa.* (eds Macdonald I A W, Kruger F J and Ferrar A A). Oxford University Press, Cape Town. pp 239 – 246.

VILJOEN A J (1988). Long-term changes in the tree component of the vegetation in the Kruger National Park. In Long-term data series relating to southern Africa's renewable natural resources. (eds Macdonald I A W and Crawford R J M). *South African National Scientific Programmes Report* **157**, CSIR, Pretoria. 310 – 315.

WALKER B H (1979). Game ranching in Africa. In *Management of semi-arid ecosystems.*(ed Walker B H) Elsevier, New York, NY. pp 55 – 81.

WALKER B H, EMSLIE R H, OWEN-SMITH R N and SCHOLES R J (1987). To cull or nor to cull: lessons from a southern African drought. *Journal of Applied Ecology* **24**, 381 – 401.

WEISSER P J (1982). Vegetation and conservation priorities in the dune area between Richards Bay and Mlalazi Estuary. Unpublished report, Botanical Research Institute, Pretoria. 111 pp.

WELLS M J, DUGGAN K and HENDERSON L (1980). Woody plant invaders of the central Transvaal. In *Proceedings of the Third National Weeds Conference of South Africa, Cape Town.*(eds Neser S and Cairns A L P), Balkema, Cape Town. 11 – 23.

WELLS M J, ENGELBRECHT V M, BALSINHAS A A and STIRTON C H (1983). Weed flora of South Africa, 3: more power shifts in the veld. *Bothalia* **14**, 967 – 970.

WEST O (1969). Fire: its effect on the ecology of vegetation in Rhodesia and its application in grazing management. *Proceedings of the Veld Management Conference, May 1969, Bulawayo.* Government Printer, Salisbury. pp 59 – 64.

WEST O (1971). Fire, man and wildlife as interacting factors limiting the development of climax vegetation in Rhodesia. *Proceedings of the Tall Timbers Fire Ecology Conference* **11**, 121 – 145.

WHITLOW J R (1980a). Agricultural potential in Zimbabwe. *Zimbabwe Rhodesia Agricultural Journal* **77**, 97 – 106.

WHITLOW J R (1980b). Land use, population pressure and rock outcrops in the tribal areas of Zimbabwe Rhodesia. *Zimbabwe Rhodesia Agricultural Journal* **77**, 3 – 11.

WILMAN M (1946). *Preliminary check list of the flowering plants and ferns of Griqualand West.* Deighton Bell, Cambridge.

WORLD RESOURCES INSTITUTE (1986). *World Resources 1986: an assessment of the resource base that supports the global economy.* World Resource Institute and the International Institute for Environment and Development, Basic Books, New York, NY. 353 pp.

ZIMMERMAN H G, MORAN V C and HOFFMAN J H (1986). Insect herbivores as determinants of the present distribution and abundance of invasive cacti in South Africa. In *Ecology and management of biological invasions in southern Africa* (eds Macdonald I A W, Kruger F J and Ferrar A A). Oxford University Press, Cape Town. pp 269 – 274.

PART 2

Human dependence on biotic diversity

CHAPTER 4

Conservation: a controlled- versus free-market dialogue

M T Mentis

INTRODUCTION

The aim of this chapter is to analyse the concept of conservation. The focus is on conservation rather than biodiversity, and the context is essentially South African. The presentation is in the form of a dialogue among the three Galilean interlocutors, Sagredo, Salviati and Simplicio (Galilei 1630). Sagredo represents the educated layman for whose favourable opinion the other two are striving. Simplicio and Salviati are the adversaries, being in this case controlled marketeer and free marketeer, respectively.

A brief explanation of the title is warranted. 'Controlled market' implies a socialistic system — the subordination of individual freedom in the interests of the community. 'Free market' is intended to convey a *laissez-faire* system arising from free enterprise under an open society and common law. The dialogue has been written against the background of topical South African futuristic literature (eg Louw and Kendall 1986; Sunter 1987).

In composing this dialogue I have tried to maintain a logical flow. The headings are for the reader's convenience and lie outside the dialogue itself. The referencing is not exhaustive and is intended only to lead the unfamiliar reader into the literature.

WHAT IS CONSERVATION?

SAGR. It has been said that formerly conservation had an unfortunate aesthetic-ethical label attached to it, but it is has now been developed into an unambiguous philosophy (Hanks 1983).

SIMP. Over the past century a strong philosophy has developed (Nicholson 1971). We might regard conservation as having been approached from two directions. The first has been an appeal for stability and zero population growth (ZPG). This is exemplified by Leopold's (1966) land ethic in which conservation is a state of harmony between men and land, and the various propositions for a steady-state economy (Daly 1973). The second approach has been an appeal for altruism, that the individual must constrain his own actions, to his own immediate detriment, for his later benefit, or for the benefit of other individuals of present and future generations. Such is implicit in the quasi-economic view of an efficient resource policy which will achieve the greatest good for the greatest number. Also, there is the recent definition of conservation by the World Conservation Strategy (IUCN 1980) implying a use of resources that will yield greatest sustainable present benefit yet not prejudice the interests and aspirations of future generations.

SALV. But none of this is to gainsay that conservation is an ethic, a code prescribing but one among many possible types of moral conduct. There is no need to be apologetic about aesthetics and ethics, as if some other label — for example, scientific — were superior. Everything man does is value-driven, based on some or other ethical or moral standpoint (Sperry 1983). Above all else, surely we should cherish our sets of values? They are what we live by.

WILL CONSERVATION ARISE SPONTANEOUSLY?

Getting back to nature

SAGR. Perhaps we should then consider the different points of view about conservation by addressing a range of questions. For example, is conservation going to arise spontaneously? If it will, we presumably have no reason to consider the matter further. If it will not, is it knowingly achievable? How is achievement to be attempted? Is conservation justifiable, rational and scientific? Might I ask then why a stable-state economy and ZPG should not arise spontaneously?

SIMP. Indeed, why not? Recognition by the populace at large — perhaps with some education — must bring people to their senses. The ever-rising population and GNP lead only to rising resource consumption and accelerated depletion. We cannot live in disregard of the laws of Mother Nature. We must inevitably learn to live within the constraints. We must become less consumptive of energy and materials, and use these more efficiently. We must develop a sustainable way of life. Simply, we must restore the balance of nature and the state of harmony between men and land.

SALV. I venture, Simplicio, that you take an overly simplistic view of Mother Nature. The world is changing. That much was convincingly argued by the ancient Greek philosopher, Heraclitus. To arrest the change — from a supposedly former utopian state — was argued by Plato. But, in a socio-political context, Popper (1966) has opposed Plato as a suppressor of free thought and the truth, and as a defender of lying, tabooist superstition and brutal violence. To quote Popper, 'There is no return to a harmonious state of nature. If we turn back, then we must go the whole way — we must return to the beasts.'

SAGR. Yet, as you say Salviati, Popper argues from the socio-political viewpoint. Do his arguments have a biophysical parallel?

SALV. The conservation argument **is** in the socio-political arena. One way — not necessarily the only and correct way — of viewing the biophysical situation is as follows. Living systems are open systems that are far-from-equilibrium, ever changing and ever evolving in at least a partially irreversible way. They are dissipative structures that require an energy throughput for their maintenance as the far-from-equilibrium systems that they are (Prigogine and Stengers 1984). The appearance of order that has been part and parcel of the emergence of the complexity we know as life has involved periodic reorderings under far-from-equilibrium circumstances. The localized appearance of low entropy systems has been at the expense of increased entropy elsewhere. For example, the prosperity of the First World has been built on an exploitation of the Third World. If the present conservation problem is to be resolved, I submit that it will require more order — a state still further from equilibrium — than we have now. This is to be achieved not by arresting and reversing change, and going backwards — that in fact conflicts with time's arrow and Prigogine's thermodynamic irreversibility. What is necessary — at least as a plausible alternative — is still more energy-sapping and disharmonious dissipative structures for a 'conserving' organization to society.

SIMP. But the conservation movement is built on the idea of a different meaning to harmony. People are realizing that man's impact on the biosphere threatens to displace these far-from-equilibrium states. I am saying that the maintenance of these dissipative structures is Aldo Leopold's harmony between men and land.

SALV. Much of the biotic diversity we see about us — the spatial and temporal patterns of organisms — arises from disturbance which may be viewed, at least in the classical sense, as disharmony. If the disturbance is withdrawn, communities simplify, possibly by virtue of species interactions which proceed to conclusion, and the competitively superior species coming to dominate. Alternatively, if disturbance is extreme, few species will tolerate and survive the harsh circumstances. Disturbance or disharmony is thus a two-edged sword, the control of which can maintain or decrease diversity.

SIMP. But that does not address the issue of imposing limits on our demand for energy and materials, and of curtailing their use. For example, there must surely come a time when our input of energy and matter into the biosphere will disturb the global system of stocks and flows. If we release large amounts of heat, and if we change the atmospheric composition — increase the carbon dioxide level — the heat budget might be so altered as to make life uncomfortable or even impossible.

SALV. I concede your point, Simplicio, at least in the extreme. Yet it is incongruous for biologists, for example, to impose a regime of absolute stasis on us. This is quite unnatural. The evolution of life has been marked by the emergence of ever more dissipative structures. Thus to argue for a stable-state economy as part of getting back to nature conflicts with what is our perception of nature.

The limits of altruism

SAGR. Yet this still does not convince me that man will not spontaneously achieve the 'conserving' organization and the con-commitant ZPG, the sustainable use of resources, and harmonious existence. We have but to appeal to our fellow men to cooperate.

SALV. You are now resting the case on altruism. Individual sacrifice for the benefit of others or for posterity is a laudable action, at least in some cultures. But it is not one universally blessed by Mother Nature. Recall the legend of the Spartan mother advising her son going to war (Hardin 1978): 'Return with your shield, or on it.' Where are the heroes of yester-year? Where is Sparta now? No, altruism is not universally practised, and you can see why. It has been selected for particular circumstances only, in which the interests of the individual and the group coincide. Hence we have parental care and social insects. Spontaneous conservation demands from the individual an altruism against which his Darwinian heritage rebels. Conservation will not arise spontaneously, not beyond the degree that is commonly perceived to be of individual interest in the foreseeable future.

The right of organisms to life

SIMP. But Salviati, you ignore the fact that organisms other than man have a right to life on this planet. Not only is it in our interests to preserve them; it is our duty.

SALV. Whence come these rights and duties? They are no more than man-made conventions. They are not immutable laws of nature. This was the substance of the debate among the ancient Greek philosophers. They established this distinction. It was they who distinguished between the immutable laws by which the cosmos functions, and the system of rules or conventions by which man's society operates. It was they who recognized that the conventions by which the antecedent closed society lived were not unquestionable laws by which people had to abide. It was they who first subjected these conventions to analysis in open debate, and recognized that the conventions could be changed. That is what prompted the open society. And Simplicio, by just such analysis, I suggest you might not live to tell of the right of a mamba to your kitchen, or the AIDS-virus to your bloodstream.

IS CONSERVATION KNOWINGLY ACHIEVABLE?

Stasis, complexity and stability

SAGR. Yet it still seems that to conserve some of the biotic diversity is indeed in our interests. We therefore need to examine the question of whether we can know when conservation has or has not been achieved.

SIMP. We could say that conservation is knowingly achieved when ZPG and stability of resource use are attained.

SALV. But Simplicio, we have already argued that the world is changing. Do you want to go back the whole way — to return to the beasts? If you agree to this, your conservation is obstructionist to mainstream society and conservation will never be achieved, let alone be knowingly achieved. On the other hand if you accept change, you must be aware of the implications. The present functioning of our civilization is not sustainable. For example, in South Africa our energy supply is based on coal which is projected to run out over the next century. We might use our coal resources and mineral wealth to build a new society that can function on fission and later fusion power, and that can live by its wits rather than off nature's bounteous windfall (Sunter 1987). All this coincides with Prigogine's theory of dissipative structures. We might be prepared not only to accept change but actively to engineer it — to develop fresh dissipative societal structures in keeping with the altering circumstances — so as to make possible a viable South Africa, if that is what we want. The kinds of change that may be necessary may not be achievable by maintaining the so called conservation status — by retaining the present biotic diversity and pattern of biospheric processes. Constraining the possibly necessary changes by conventions of ZPG (as opposed to other plausible population policies) and obstinately maintaining (as opposed to changing) biotic diversity and biospheric processes could be an inversion of priorities. It is sheer madness, if not national suicide. The priority is to develop a 'winning' nation that can subsist comfortably off its wits (Sunter 1987). Stability and maintenance (of biotic diversity, biospheric process and the status quo) are therefore not criteria of a 'winning' status. Indeed, they risk being the very antithesis of it.

SIMP. But Salviati, you are assuming that the status of the winning nation is the right thing. What makes this view any better than any other? You are implying that 'sheer madness' is any view that does not coincide with Sunter's and yours.

SALV. It is just one view. But the claim it has to being 'better' is that South Africans like it. Sunter has presented his case before more than 300 audiences, and he continues to be inundated with invitations from across the wide socio-political spectrum of South Africa. By comparison, the preachings of the Ehrlichian prophets of doom invite limited audience. Apparently South Africans want to hear the optimistic view of what the future might bring. They want to know what will benefit social welfare. They aspire to the 'good life'. This implies accumulation of material wealth and comforts, which is going to require economic growth.

SIMP. Salviati, you are seemingly uncaring and callous about the fate of the environment and of species. The remaining wild species are a storehouse of genetic material, a resource for future crops and plant extracts. We cannot let any species become extinct. And there are the ecological processes which we must preserve for our own survival.

SALV. Nuclear holocaust or not, a massive extinction of species over the next few centuries seems inevitable. As the species dwindle, man may become alarmed. If the alarm is sufficient (ie species are of use and value in that they are wanted and the supply does not overwhelm the demand) remedial measures will be a paying proposition. Otherwise, who cares? What does it matter? Those who claim to care advance debatable points of view. Oh, we need to preserve a storehouse of raw genes for posterity. Yet what difference is it going to make to the future if n or 2n species are exterminated now? Might the advances in genetic engineering and tissue culture make good (or better!) any immediate losses? Evidently you venture not. You plead that the art of intelligent tinkering is to keep all the parts. But systems are not universally so tightly interconnected. I have thrown away parts of my car and it goes better than before. I do not know that I lead a lesser life for want of a dodo. Nevertheless, I do admit that a certain recklessness may be hazardous. In terms of the current theory of connective stability (O'Neill et al 1986) it is evidently not the number of species, or the diversity *per se*, that confers stability. Rather, it is the manner of interconnection. 'Overconnected' systems are too fragile. The displacement of one component leads to system collapse. At the other extreme, 'underconnected' systems have insufficient feedback for perturbations to be damped. Therefore, to view conservation status as maintaining biotic diversity over the medium term is to attach a certain meaning to conservation. This meaning implies that conservation deteriorates if diversity is lost. Yet conceivably such loss — say the extermination of the AIDS-virus — may be to the benefit of society. So, perhaps, despite the fact that conservation is not achieved, the quality of our lives may improve. In other words, conservation (the maintenance of biotic diversity) and development (the creation of Sunter's winning nation) could conflict.

SIMP. I do not like your example of the AIDS-virus. Herein lies a potential solution to our population problem. Unless a vaccine or a cure or both are found soon, many of the insolubles will be solved. AIDS promises to stave off the day of reckoning for the things wild and free that remain. Conserving all strains of organisms — predatory, parasitic and pathological — **is** maintaining the balance of nature. The persistence of tsetse fly has been hailed as the greatest conservator in Africa. Long live the tsetse fly, and may the virologists suffer long and hard to eradicate AIDS! Stop the clock!

SALV. Simplicio, you are returning to the beasts. You are on treacherous ground. You are using the very arguments, the very deceitful, brutal and non-humanitarian strategies, that prompted Popper to annihilate Plato's thesis. You are also missing the point about diversity and stability. The two are not directly related.

SIMP. Your theory of connective stability is but one interpretation of the facts. Is it any more valid than the one that contradicts it?

SALV. There are indeed other viewpoints, and it is useful to have competitive theories. The explanatory or predictive power of a single theory in isolation is difficult to evaluate. Comparison of various theories enables us to pick the most appropriate conceptions, whatever the criteria of appropriate are. Furthermore, I am proposing the theory of connective stability, not as the correct view but as one which plausibly is more general than the traditional one of 'complexity equals stability'. The latter theory certainly does fit some observation sets, but it fails in respect of others (O'Neill et al 1986). The former theory demands that we focus on the way ecological systems are put together, and on what is the nature and extent of the interactions, rather than on the components (eg the species) themselves. Perhaps, then, the present emphasis on species in conserving biodiversity is a red herring. We need not now assert which of the approaches is the right one. I suggest, however, that we might bear both in mind.

The greatest good

SAGR. Salviati, you knock the very cornerstone of the World Conservation Strategy, the thesis that conservation as it is traditionally understood in terms of stability, maintenance and harmony is compatible with the development that will be needed to make South Africa a winning nation.

SALV. Yes, in a way I am doing that. If South Africa persists along the current course — Sunter's low road — we shall continue to experience growing competition for, and depletion of, resources. But a move towards a more libertarian society — the high road — may prompt economic growth, industrialization, urbanization and a slowing in the rate of population growth. For one thing, large areas of land marginally suitable for agricultural production may be vacated and become available for preserving things wild and free. Improved methods of food production, fueled by economic growth and the development of biotechnology, could mean that we would be able to meet our food needs on the relatively small area of land of high agricultural capability. In this scenario economic growth and development are the catalysts for improving conservation status. So the wish to maintain the life-support functioning of the biosphere might be made more attainable by pursuing economic growth, in apparent contradiction to the traditional conservation approach of stasis.

SAGR. In view of the way you argue, can the criterion of conservation not be 'the greatest good for the greatest number'?

SALV. How do we achieve this greatest good (Tisdell 1983)? What does it mean? Many happy people now, few later? Or few happy people now and many contented ones later? Or some happy people now and later? Or the maximum number of souls that go to heaven?

Maximizing present benefits and catering for posterity

SIMP. Clearly the principle of the greatest good is too vague. But the World Conservation Strategy has provided us with a guide that meets the objections. We must maximize present sustainable benefits, yet cater for the interests and aspirations of future generations.

SALV. That guide is in fact no more helpful. It asks us to maximize more than one thing simultaneously. Now, whereas we can compromise, it is rarely possible to maximize more than one thing at a time.

SAGR. No, but the Strategy does not say 'maximize' for posterity.

SALV. My attack is in this respect extreme. But it makes no difference to the conclusion of my argument, whether you use 'maximize' or 'cater for'. Supposing we try to cater for posterity and yet try to maximize present sustainable benefits. This will require trading off current against future resources and opportunities. What is the current value of resources of the future? Indeed, what are the resources of the future? The future is unknown. For example, suppose that at some future time a synthetic steak is invented — tastier, cheaper and more nutritious than the real thing. What will be the raw materials for this synthetic steak? How do we now value these unknown raw materials in relation to the present resources of soil and veld? How do we conduct our trade-offs to guide us in our action to cater for posterity? How do we decide what future options we are foreclosing?

How do we cater for posterity? What, in any case, are the aspirations of generations to come? It is not as if ecosystems simply have thresholds this side of which things hang together, and the other side of which everything falls apart. By and large there are graded sequences of response to impact. Without knowing the relative values of present and future resources we cannot calculate the optimal position along these gradations — the position that will maximize present sustainable benefits yet cater for posterity. In any case, what is sustainable? Our present soil resource and the way we are using it? Many of our cultivated fields have half-lives currently estimated in decades or less. Can we make the use of these fields sustainable? Not without aggravating human suffering. Many will simply starve. And is our use of our energy resources sustainable? Coal provides currently the cheapest option. Its export also brings in badly needed foreign exchange. Shut the coal mines now and not only will it be raw bacon and eggs for breakfast and a shivering winter on the highveld, but we foreclose the opportunity to build a society that will be able to run on some other fuel resource.

Controlled-market and free-market options

SAGR. Salviati, you argue the World Conservation Strategy not as a logical analysis and an unambiguous philosophy, but as an exposition of subjective, heartfelt convention. To be sure, some objectives are stated, but we mortals left to implement its laudable policies face an impasse with regard to developing attainable goals, the achievement of which can be measured. I foresee grave practical difficulties. For example, if public funds are voted for conservation, how are we ever to assess whether the funds are indeed being expended efficiently on conservation? How might the politician convince his electorate that a good job is being done? It seems that we can only resort to a system of conventions that the officers of conservation must observe. But how do we develop and implement these conventions?

SIMP. We appoint, as conservation officers, qualified people who know the right things to do. And we let them manage conservation for us.

SALV. Who is qualified? Those that have been through the school? There are several schools. Which one is right?

SIMP. The schools of science are right. Science aims to describe the cosmos in terms of value-neutral facts independent of preconceptions. Thus it is the men of science that we must employ for conservation.

SALV. Simplicio, you live in the mistaken tradition of the Baconian-Cartesian-Newtonian paradigm. Two centuries ago Immanuel Kant asserted that things cannot be known in themselves; that our conception of reality is no more than a mental construct. In about 1840 William Whewell, the English philosopher-biologist, wrote 'There is a mask of theory over the whole face of Nature'. Human activities are value-driven and facts theory-laden. 'Write down what you observe.' How is this possible without preconception of what is important; without defining a procedure of observation? And each procedure of description itself needs a description, so that a complete description is hindered by an infinite regress. We therefore always start with preconceptions. Indeed, we are born with preconceptions — our genetic predispositions. To be sure, these preconceptions or values can and do change (Sperry 1983). We view the world through tinted glasses. The coloured facts impinge upon our values which might then change, and tint the glasses in a different way. It is therefore not possible to 'to tell it like it is', only to tell it as we see it. And different individuals see it differently.

Simplicio, you advocate a hard-core scientism that sees genuine knowledge residing in science. But this evidently stands on feet of clay (Stent 1978). If science claims value-neutrality, how can it pretend to prescribe on values? But even if it claims no such neutrality, and accepts the theory-laden nature of its wisdom, whose values are right, or even superior? The values of science and of ecology are not necessarily the ones society cherishes. Thus the men of science are not necessarily equipped to give society what it wants.

So Simplicio, you are advocating a brand of authoritarianism whereby the appointed officers of conservation would impose the particular doctrines of this elitist schooling on society. We must submit to their heartfelt conventions that, for example, 10% of the country should be proclaimed as nature reserve, that no species are to become extinct, that ecological processes are not to be meddled with, and so on. On what rational grounds do you decide that for 10% of the country to be a reserve is of optimal benefit to present and future generations? What assurance can you give that in all cases the loss of a species will not be outweighed by other societal benefits? With what will maintaining ecological processes reward us, now and in the future? Who defines what are in fact benefits and rewards? The conventions of the conservation creed do not provide us with a logical basis for trading off what we might gain or lose in the future. They may very well not coincide with making South Africa a winning nation. Yet, maybe you, Simplicio, as a socialistic conservationist, cherish your conservation ideal of keeping things wild and free as more important than the citizen's wish to belong to a winning nation. If you stick to your view you must accept that your brand of conservation departs from the societal mainstream and is bound to fail.

SIMP. You keep arguing as if Joe Public wants to live in a winning country. That is not necessarily so.

SALV. Convince me otherwise, Simplicio. Sunter's book is a best seller. His 'road show' is a box-office sell-out.

SIMP. But Salviati, you have previously argued that conservation will not arise spontaneously. Now you are rejecting any imposition of discipline regarding the way we might use the environment. Yet expert opinion is that we are headed for environmental disaster. Are you proposing that we should let such disaster happen?

SALV. No. In my view there are several options in environmental regulation. At the one extreme is your 'hard' option, controlled under statute law. At the other extreme is a 'weak' option, where everything is up for grabs, and let the strongest man win. I assume that we do not wish to debate this option. In between is a 'soft' option that operates under common law. This is the one I advocate. In terms of this option the state creates and manages the legal machinery for the operation of common law. Preserved are the rights of individuals to (1) do whatever they like provided this does not interfere with the wishes of others to do as they like; (2) secure personal benefits and accumulate property by their own efforts; and (3) defend in a court of law rights (1) and (2). Specifically, in relation to the environment: any development or act that threatens environmental quality or amenity must be preceded by a public statement of intent; failure to make such public statement can be condoned, but if a jury subsequently finds the entrepreneur negligent he shall be liable to pay compensation to the injured party; objection to a public statement of intent will be followed by an environmental impact assessment (EIA), the assessors to be appointed by the community to be affected and funded by the entrepreneur; the EIA will enjoy a public hearing; representatives appointed by the community to be affected will be the final arbitrators; all conditions of development specified by the arbitrators on moderating the impact will be binding on the entrepreneur; and failure to meet these conditions will make the entrepreneur liable for compensation to the injured parties.

SIMP. But who says this view is any better than my 'hard' socialistic one? To what ultimate source do you appeal to judge that my view is inferior to yours?

SALV. You are quite right. Neither of the alternatives is non-ideological, and they are on an equal footing. I can appeal to no ultimate criterion in order to subordinate your preference. We can only debate the differences. And one difference is that you wish to put your hand in my pocket to have your view imposed on me. I am not asking you to fund the imposition of my view on society. I am prepared to compete in an open market.

SIMP. But there are countless examples of resource depletion and environmental degradation under your free-market libertarianism. The whole problem is too big and too important to be left to free enterprise. We shall never resolve these problems without a strong central body to coordinate action.

SALV.Yes, there has been resource depletion and the like under free enterprise. But has controlled-market environmentalism fared any better? Take agriculture in South Africa as an example. By virtue of the countless billions of rands invested in it in terms of state support of research, extension, drought subsidies, conservation schemes, control boards and so on it should be a showpiece. But look at it! The farmer debt is R14 billion. Farmers face bankruptcy and are leaving the land. Whole farming regions are collapsing. The karoo is being desertified. The false karoo is advancing eastwards. Annually 400 million tonnes of soil are being washed into the sea.

Bearing in mind that the conservation ethic is a system of conventions, it seems that the only way in which it is logically possible to assess its degree of achievement is by the support society does or does not accord it. This does of course lay me open to being branded a 'philosopher king' of a type, but I am not trying to impose my view on others, and have it implemented with their money. It would be quite wrong of me to assert that my proposal is value-neutral. It is not. But I suggest that it recognizes that there may be several value-sets, and it allows them to compete in an open market. In contrast, your socialism, Simplicio, demands that there be authorities who know what is best for us. Such authorities might be well-intentioned, but can they be omniscient?

SAGR. Salviati, you imply a satisfying distinction. We appear to have two means of setting and measuring conservation goals. Firstly we have what Simplicio favours, the socialistic option. This requires the setting of goals, which is problematic in itself. But the authority imposes his creed upon us. This has the merit of perhaps saving species and establishing nature reserves. But with what general agreement by society, and to what benefit in the attainment of the status of a winning nation, seem dubious and haphazard. Secondly, the alternative of Salviati is the free-market option. Here we free-wheel, our course decided by the ballot box and free enterprise, and swayed to keeping or losing biotic diversity by open debate on the consequences of individual development projects. It is public perception of these consequences and the way they are seen to affect the greatest sector of society that will determine the course. This might not maintain the biotic diversity that the socialistic option could, but it might better serve the purpose of becoming a winning nation. Specific conservation goals here are not ends in themselves. Indeed, society does not have to try the evidently impossible— to articulate them. Conservation 'goals' are subordinated by the goal of society's setting its own course, whatever that course might be.

HOW IS CONSERVATION ACHIEVED?

SAGR. If you agree with my view of these two options, can we proceed with the next question which is how to achieve conservation?

Correcting ecological myopia

SIMP. We can achieve conservation by study of the problems, by well-aimed information campaigns, by education starting at kindergarten, and by appealing to resource users to act with constraint in the interest of others. If we cannot thereby correct the ecological myopia, we can use the law and force people to comply with environmental legislation.

SALV. Yet I suggest, Simplicio, that man is ecologically myopic for good reason. The average planning horizon is for the immediate, foreseeable future. The distant future is unknown. Studying the problems is very necessary. It may help us refine our theory of connective stability — or some such substitute — and thereby guide us to avoid popularly undesirable ecosystem configurations. But beware of the shortcomings. Studies usually take years. It then takes a generation or more to change society's values. By that time, in this rapidly changing world of ours (Toffler 1970), there will have come into being a new system configuration, a new suite of problems, requiring a new mind-set for solution. What you are suggesting, Simplicio, implies that we must apply brakes to the rate of change. In other words, you are suggesting that we constrain development, the attainment of the status of winning nation. If you really want to stop development because it threatens conservation, it could equally well be argued that people should not be educated. Education risks equipping exploiters with increased understanding of the resources, thereby improving their ability to exploit or extract and hastening the day of resource-depletion. As for any impassioned appeal, you must know the limits of altruism. In a conflict, when the individual resource user has his back to the wall, he will act in his own rather than in society's interests.

SIMP. Nevertheless, there are things we know well enough to enforce compliance among individuals if they do not acquiesce to education, information and appeal. We can compel the grazier to stock within the grazing capacity; we can enforce some form of birth control; we can enforce the retention or creation of stable ecosystem configurations — kinds of system interconnections which even you, Salviati, admit are worth establishing.

SALV. How? There is no single criterion of overstocking (Mentis 1984). The economically optimal rate to stock depends on the grazier's objectives, the starting state of the system, the type of management applied, and the prevailing economic environment. No grazier's situation is like any other's. But even supposing there were a standard, it is physically impossible for the state to check on every grazier. Just as it is hardly possible to post a policeman under every bed. And as for the desirable ecosystem configurations, supposing they are revealed, how do you propose that they be achieved? Imposing severe constraints — by totalitarian decree — on the resource user, the entrepreneur, risks stifling the individualism, initiative and innovation we need.

SIMP. Salviati, you seem to place great faith in the omniscience of the public at large. Most of them are but poorly informed regarding the threats to the environment. How then can you expect them to make wise judgements on the future of their environment?

SALV. Certainly, we cannot expect every person to be an expert on the environment, and a public education exercise is warranted. Indeed, as a academic I believe it to be my duty to make known to society the implications and consequences of alternative decisions and actions, in as dispassionate a manner as possible. But I know I cannot be entirely dispassionate — completely value-neutral. Therefore I believe that others of my ilk, who feel duty bound to advertise the implications of society's actions, must also be permitted access to the public. As for the public's being uneducated or ill-informed on matters environmental, we can but try. If people remain unreceptive or 'ignorant' it may be because they do not regard the issues as being as relevant as we do, or because the experts are so varied in their opinions that, in any event, the best course is not clear.

Socio-economic fiddling

SIMP. Salviati, you imply that the implementation of environmental legislation will encounter difficulties. There are subtle or indirect ways of enforcing compliance. By means of differential subsidy, specially designed taxing systems and similar methods we may be able to control the amount of stock the grazier runs, the way resources are managed and exploited, and the number of children couples have.

SALV. For decades we have had a succession of drought relief schemes, conservation subsidies and stock withdrawal programmes. None of them seems to have amounted to anything but subsidized overgrazing. All these tactics have done has been to tide the resource user over the lean years, to discourage savings in the good years, to negate incentive for managing the resource for sustainable use, and to keep the incompetent farmer on the land. The government's 'aid' has simply reduced the cost of exploitation by the exploiter, and we know from experience in the fishing industry that when this happens the chances of 'overfishing' are increased (Clark 1985).

SIMP. But all that may be true simply because the details rather than the principles of the cure were wrong. Socio-economic fiddling has the great merit of having fairly immediate effect, especially in comparison with research and education.

SALV. But we rarely seem to get the details right. Perhaps this is the inevitable result of the fact that none of us is omniscient, so that often market controls are object lessons in the law of unintended consequence.

SIMP. It seems, Salviati, that you are not primarily interested in conservation goals.

Conservation in an open society

SALV. Correct. It is not the success or achievement of conservation *per se* that concerns me. What I am arguing for is a deal that is fair to all. This is possible in an open society and under a free-enterprise system. Under these conditions it is not a matter of the rulers imposing their values upon the ruled, rather that the majority view of society determines what rules shall exist and how they shall be administered. For example, if the public at large does not want bushbuck females to be hunted, or thinks it cruel to hunt with a bow and arrow, or does not mind having a nuclear power station, it is not for the ecologists to say that there are no good biological reasons for these wishes, and thereby overrule popular feeling. To be sure, there can and should be open debate on the relative merits of this or that law or action. It may well sway public opinion in favour of the long or the short view. The role of the economist, ecologist, or other expert in this case is to try to explain and make known, as dispassionately as possible, the societal consequences of the alternative decisions and actions. Under these circumstances the amount of conservation achieved will be close to that which average opinion seeks. The criterion of correctness of the ultimate decision has only incidentally something to do with the way conservation is served. Rather it has to do with the degree to which society decides its own fate. Whether this actually maximizes social benefit is not critical — in fact, such a goal is poorly testable. I say it is not critical because we have no system able to guarantee maximum social benefit. The free-market option is not perfect either. Indeed, it requires some degree of 'malfunction' for error to be perceived and corrected. This may be a small sacrifice in comparison with the controlled-market option, which is a burden to the taxpayer and is often instigated to serve the interests of privileged groups (Louw and Kendall 1986).

IS CONSERVATION JUSTIFIABLE, RATIONAL AND SCIENTIFIC?

SAGR. At the outset of this debate I was inclined to think that conservation was at last coming of age, and that fine-tuned technology — although perhaps not as refined as that which landed a man on the moon — was in the wings. I see now, thanks to your argument, Salviati, that such a technology is remote. If we are to go by Simplicio's socialistic option, we shall be at pains to define how we are to design, implement and knowingly achieve a conservation strategy. If we adopt Salviati's free-market option, we, as a society, do not have to bother about defining conservation goals. We simply let society have its way, whatever that is. Yet that leaves me wondering about another question. It seems, at least on the face of it, that conservation is justifiable and quite rational. Can it be scientific as well?

SALV. Conservation is justifiable, at least as far as anything can be justified. Man's impact threatens his survival. It seems that all alertable minds have been alerted to this, and there is no need to flog a dead horse. Yet although many things – such as perpetual motion machines – are justifiable, that does not mean they are attainable.

With regard to rationality, conservation is, as we have said, an ethic. I do not know that moral codes and values have to be ultimately rational. For example, 'I like blondes'. I surely do not have to rationalize this predilection to the ultimate degree for it to be personally gratifying and socially acceptable. Likewise, valuing fresh air, clean water, and peace and quiet loses no marks because one cannot provide a complete rationale for such preferences.

As to whether conservation is scientific, this does depend on one's definition of science. In my view things have to be falsifiable to be scientific. They have to be amenable to testing, which can conceivably demonstrate inconsistency between expectation or prediction on the one hand, and observation on the other. Additionally, we could consider a field scientific if it involves establishing falsifiable statements from metaphysical ones. If this approach is accepted, conservation is poorly falsifiable. Its goals are ambiguous, non-operable or debatable — as you like. This does not necessarily imply that conservation will remain at this level, but at the moment conservation is a religion that invokes clairvoyance or crystal-gazing.

SIMP. I have two objections to this view. First, if I want to perpetuate, say, a particular species, I can use science to do so. There is after all a budding discipline called conservation biology. Second, Salviati, you are arrogantly asserting that what is not science is intellectually barren.

SALV. How to preserve a particular species is one issue. That can be the business of science. Whether to preserve it is another issue which requires very much of a value judgement which might be made by your authoritarian decree, Simplicio, or in my market-place. Neither method is scientific, unless you want to change my definition of science. Of course, my critics are quick to condemn my non-scientific label for conservation as purely an artefact of my definition of science. But none of them can, will or does propose any other definition which might render conservation 'scientific'. I say, therefore, that conservation biology is a slapdash juxtaposition of words that imply very different procedures. In terms of my world-view the 'conservationists' ethos is quite legitimate. I in no way wish to imply that it is intellectually barren. Nor do I wish to play down the primacy of emotions and value systems (Sperry 1983). I have great respect for a policy of 'anything goes' (Feyerabend 1975), provided society supports it. Yet the conservation biologists, at least as educators, might beware of teaching only this world-view. Rather, as educators and as critics of society, we are duty bound to offer as broad a perspective as we can muster.

CONCLUSION

SAGR. Simplicio and Salviati, I thank you for your spirited debate. The opposing views you propose are of the sort which cannot be ruled right or wrong. I consider that individual world-views are different by virtue of individual differences in genetic constitution and environment. The conceptual framework that enables one person to get along in the world may well not enable another person to do the same. Yet, of course, in society the different world-views are in conflict.

Simplicio, it can be said of your view that it is the conventional one, and one which has made some progress. Naturally, we cannot say that more or less conservation would have been achieved had the state not intervened to protect the environment. Some might assert that without this intervention we would have no Kruger National Park and no large wild mammals left. Others would contend that the healthy status of much wildlife on some 85% of South African land — that which is private property — shows that free enterprise is a potent conservation 'force'. We may accept that the one 'force' acting in the absence of the other would have behaved differently, but how differently is debatable. However, even if we were to accept unreservedly that the controlled-market approach has excelled in the past, it is the future that we must consider. In doing this I must turn to you, Salviati.

Free-market environmentalism might not be the common view. Some may admit it is an idea ahead of its time, but it is currently relevant to the South African situation. If the government were to change tomorrow, the new regime might not be as sympathetic to conservation as the present one. In any case we seem to be in an era of growing financial constraint on nature conservation budgets. What setbacks might the past achievements then suffer, and what remedies might the conservationists now consider? I believe, Salviati, that your view has one great merit. It sees conservation in the broad socio-political arena, a pragmatic view regarding which conventional conservation is weak. Whether the government conservation agencies take on a more free-market image, or the provision of wildlife services becomes predominantly private, matters little perhaps. What strikes me, however, is that the present command economy in relation to natural resources is a microcosm of the social ills that afflict South Africa. A totalitarian and paternalistic government, and its instruments, have set themselves up as authorities on what is good for us, and on the way our natural resources are to be used. By way of proliferating statute law and bureaucratic red tape, a majority has been excluded from competing for a fair share of the cake (Louw and Kendall 1986). It seems to me quite incongruous that conservationists, mostly democratically minded people, persist in supporting autocratic conservation agencies. More cogently, the conventional conservationist believes in keeping things 'natural', yet he opposes free-market environmentalism, when Adam Smith's hidden hand is the self-same one that holds 'the balance of nature' (Fiske 1988).

REFERENCES

CLARK C W (1985). *Bioeconomic modelling and fisheries management.* Wiley and Sons, New York, NY.

DALY H (ed)(1973). *Towards a steady-state economy.* Freeman, San Francisco, Ca.

FEYERABEND P (1975). *Against method.* Humanities Press, London.

FISKE S (1988). A free market in natural resources. In A controlled versus free market environmentalism. (ed Mentis M T). *South African Institute of Ecologists Bulletin,* (in press).

GALILEI G (1630). *Dialogue concerning the two chief world systems — Ptolemaic and Copernican.* Translated by S Drake 1953. University of California Press, Berkley, Ca.

HANKS J (1983). *Conservation for development.* Endangered Wildlife Trust, Bedfordview.

HARDIN G (1978). *Stalking the wild taboo.* Kaufman, Los Altos, Ca.

IUCN (1980). *The world conservation strategy: living resource conservation for sustainable development.* IUCN — UNEP — WWF, Gland. 44 pp.

LEOPOLD A (1966) *A sand county almanac.* Óxford University Press, New York, NY.

LOUW L and KENDALL F (1986). *South Africa: the solution.* Amagi, Bisho, Ciskei.

MENTIS M T (1984). Optimising stocking rate under commercial and subsistence pastoralism. *Journal of the Grassland Society of Southern Africa* **1(1)**, 20 – 24.

NICHOLSON M (1971). *The environmental revolution: a guide for the new masters of the world.* Hodder and Stoughton, London.

O'NEILL R V, DE ANGELIS D L, WAIDE J B and ALLEN T F H (1986). A hierarchical concept of ecosystems. *Monographs in Population Biology* **23**. Princeton University Press, Princeton, NJ.

POPPER K R (1966). *The open society and its enemies.* (Fifth ed) Routledge and Kegan Paul, London.

PRIGOGINE I and STENGERS I (1984). *Order out of chaos.* Heinemann, London.

SPERRY R (1983). *Science and moral priority.* Blackwell, Oxford.

STENT G S (1978). *Paradoxes of progress.* Freeman, San Francisco, Ca.

SUNTER C (1987). *The world and South Africa in the 1990s.* Human and Rousseau, Cape Town.

TISDELL C A (1983) An economist's critique of the World Conservation Strategy, with examples from Australian experience. *Environmental Conservation* **10**, 43 – 52.

TOFFLER A (1970). *Future shock.* Bantam Books, Toronto.

CHAPTER 5

Indigenous plant use: balancing human needs and resources

A B Cunningham

CONSERVATION, HUMAN NEEDS AND BIOTIC DIVERSITY

Modern conservation theory has broadened from the past emphasis on strictly policed protected areas, primarily for large mammals, to an emphasis on sustainable resource use, maintenance of ecological processes, and genetic diversity (IUCN 1980). This recent approach to conservation coincides with the realization that in the foreseeable future, greater human needs and numbers will be the dominant feature of human ecology in Africa. It is also widely acknowledged that the conservation of wildlife in proclaimed reserves cannot be sustained in the long term without the acceptance and support of the surrounding human population (Fraser-Darling 1960; Huxley 1961; Tinley 1979; Infield 1986). The planning of 'resource areas' as buffer zones around core conservation areas (Tinley and van Riet 1981), and the establishment of farm 'conservancies' on commercial farmland in Natal reflect the necessity of maintaining biotic diversity and reducing landuse conflict.

The use of plant resources to meet human needs is an important aspect of the multiple-use of land in many parts of southern Africa. Indigenous plants, for example, are a source of fuel, medicine, building material, craftwork materials, dye, income and food supplements. Black South Africans represent the largest (72%) and fastest growing sector of the total South African population, their numbers increasing from 3,5 million in 1904 to 20,6 million in 1980 with a growth rate of 2,35% per annum (Grobbelaar 1985). The rural sector of this population is particularly dependent on indigenous plant resources, although urbanization does not preclude the use of wild plant resources. Herbal medicines, fuel wood, wild spinaches and mat-rush (*Juncus kraussii*) sleeping mats are all items of trade from rural source areas supplying an urban demand.

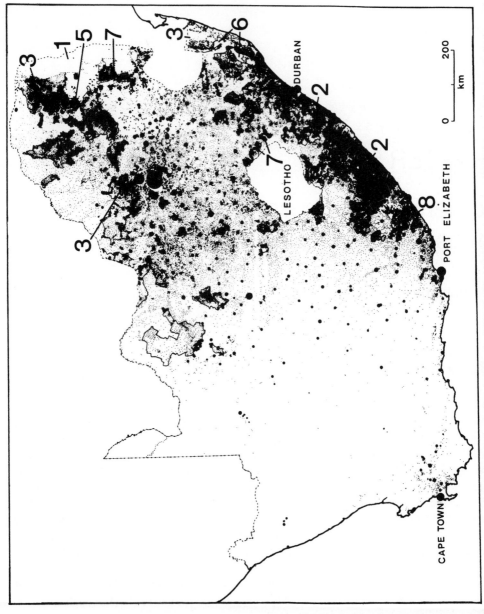

FIGURE 5.1   Distribution of the rural and urban black population in South Africa in 1970 (Smith 1976) in relation to selected conservation areas in the eastern half of the region, recorded in Greyling and Huntley (1984), noting the number of national parks and reserves in each cluster

Despite the close proximity of high population-density rural areas to some important national parks and reserves (Figure 5.1), it is only recently that studies in South Africa have focused on improving local support for conservation areas (Infield 1986). Viewing conservation areas from the perspective of a subsistence farmer or pastoralist is a useful exercise for park managers, for it also provides important insight into pro-active management requirements necessary to reduce land use conflict. African customary knowledge of resources is also the key to plants with potential value as new crops, medicines and other plant products to a wider spectrum of society. This knowledge, accumulated over thousands of years, is disappearing at an ever accelerating rate with the passing of skilled herbalists, cultivators and pastorialists of the 'ecosystem generation'. The implications of this loss of knowledge regarding useful plant products, and indicator plants used by traditional cultivators (Allan 1965) and applied in mapping surveys (Trapnell 1953) are inestimable.

BIOGEOGRAPHIC DIVERSITY AND PLANT USE

In terms of species-area ratios, the southern African flora has a much higher value (0,0081) than humid, tropical floras such as those of Brazil (0,0044) and Asia (0,0041) (Gibbs Russell 1985, 1987). Even if the extremely species-rich Cape flora is excluded from this assessment, the species-area ratio for the remainder of southern Africa is still higher (0,0061) than that of the humid tropics and nearly twice that of Australia (0,0032) (Gibbs Russell 1987). It is not surprising therefore that major works on medicinal (Watt and Breyer-Brandwijk 1962) and edible plants (Fox and Norwood-Young 1982) need to be updated and expanded.

The number of indigenous plant species used by rural people is a reflection of the botanical knowledge of the resource users and the species diversity within the southern African region. Only 37 edible species are reportedly used by OvaHimba and OvaTjimba nomadic pastoralists in the arid savanna of Namibia which has a comparatively low species diversity (Malan and Owen-Smith 1974). In comparison, over 100 indigenous plant species provide a source of fruits, spinaches and beer to both rural Pedi and Tembe-Thonga people (Quin 1959; Cunningham 1985), over 60 species to Tsonga people (Liengme 1981), and over 220 species to Swazi people (Ogle and Grivetti 1985) in the vegetation of the moister eastern half of southern Africa, which has a comparatively high species diversity.

The link between regional species diversity and customary botanical knowledge is not restricted to edible species alone. Traditional practitioners from the Maputaland and Pondoland regions of the moist east coast are nationally renowned among black South Africans for their knowledge and ability as herbalists and diviners. Among conservation biologists, these areas are renowned for their high species richness and heterogeneity of habitats.

REALIZING THE POTENTIAL OF INDIGENOUS PLANT RESOURCES.

Nineteenth- and early twentieth-century colonial botanists were instrumental in using the customary knowledge of African people as a key to tracking down potentially important plants (eg, Medley-Wood and Evans 1898). Potential sources of rubber for example, such as *Landolphia kirkii* and *Euphorbia tirucalli*, were tested, commercially exploited, and considered for plantations (Sim 1903, 1920). The selection of marula (*Sclerocarya birrea* subsp *caffra*) cultivars (Von Teichmann 1983; Goosen 1985) and cultivation of plants for essential oils (Piprek et al 1982) are encouraging exceptions to the lack of recent applied work on genetic enhancement or the cultivation of indigenous plant resources with potential. This apparent lack of interest contrasts markedly with the knowledge of and value placed on them by 'unskilled' rural people. The lack of applied work is astonishing, since extensive data have now been accumulated by the National Food Research

Institute on the nutritional values of wild foods (Wehmeyer 1966; van der Merwe et al 1967; Lee 1973; Cunningham 1985) and on the active ingredients of medicinal plants (Watt and Breyer-Brandwijk 1962).

Paradoxically, most applied work on potentially important indigenous plants has been done elsewhere. Cultivation of the Kalahari marama bean *Tylosema esculenta,* which has been internationally recognized as an important crop plant (NAS 1979), has been implemented in Texas (Bousquet 1982) but not in South Africa. Plantations of *Euphorbia tirucalli* have also been developed recently — on the island of Okinawa by a Japanese plastics company (Calvin 1979).

DIRECT USE OF INDIGENOUS PLANTS

Indigenous plant species and, more recently, exotic species such as the black wattle *Acacia mearnsii* supply many of the basic needs of people in southern Africa (Liengme 1981, 1983a; Gandar 1983, 1984; Cunningham 1985, 1988a). Fuel wood, fencing and building materials account for the highest volume of plant material used annually (Liengme 1983b; Gandar 1984; Eberhard 1986). Although harvesting wood for this purpose is selective for hardwood species, it is less species specific than for the use of edible and medicinal plants or for more minor uses such as craftwork materials (fibres and dyes). Fuel wood gathered from forest, woodland and exotic plantations accounts for 51% of the domestic energy use in South Africa (Basson 1987). Fuel wood consumption also represents the biggest consumption of plant biomass, with rates varying from 0,27 to 1,12 t capita^{-1} yr^{-1} (Gandar 1984), and an estimated total consumption of 12,9 million t yr^{-1} (Eberhard 1986).

Using indigenous species as building material (Figure 5.2) for rapidly growing housing requirements has several advantages. Firstly, they are a renewable source of building material. Secondly, harvesting, sale and construction generates income and preserves traditional skills (Johnson 1982; Cunningham and Gwala 1986). Thirdly, thatched housing is insulated, which is well suited to African conditions (Knuffel 1976). Wood consumption varies with building style and the availability of materials. Wood consumption for Tsonga hut construction in mopane savanna was 0,23 t family^{-1} yr^{-1} (Liengme 1983b), whereas 0,16 to 0,18 t family^{-1} yr^{-1} was used in Tembe-Thonga hut construction, owing to the use of *Phragmites* reeds rather than laths for wall construction (Cunningham 1985). The complex stockades surrounding Ovambo homesteads require even larger volumes of wood (mainly *Colophospermum mopane*). Although the effects of this form of wood use in changing savanna woodland to grassland have not been quantified, they can be seen clearly from aerial photographs.

Wild plant foods are an important supplement to the predominantly starchy staple diet common to many communities in southern Africa. Wild spinaches classified as weeds by First World agriculturalists, are an important source of nicotinic acid (Hennessy and Lewis 1971; Ogle and Grivetti 1985) to subsistence agriculturalists (Figure 5.3), and even to residents of peri-urban areas (Whitbread 1986). Also important are: the protein from *Sclerocarya birrea* subsp *caffra* kernels (Figure 5.4) (Quin 1959; Burger et al 1987); vitamin C from fruits (Quin 1959; Wehmeyer 1966; van der Merwe et al 1967; Cunningham 1985; Ogle and Grivetti 1985); and palm wine (Cunningham and Wehmeyer 1988).

Traditional medicinal plants are important to both rural and urban communities for physiological, psychological or religious purposes (Figure 5.5). Medical care is limited in the tribal 'homeland' areas of South Africa; the overall ratio of medical doctors to the total population was 1:17 400 in 1976, but there have been ratios as low as 1:119 000 in Qwaqwa and 1:30 000 in Lebowa in 1982 (Savage 1985). In contrast, the ratios of traditional practitioners to the total population are high. In Venda, Arnold and Gulumian (1984) estimated a traditional

Figure 5.2   Hardwood laths and *Strelitzia* fibre used in hut construction
Figure 5.3   Daily food in the Sand forest region — *Pyrenacantha scandens* spinach (imifino)
Figure 5.4   Cracking *Sclerocarya birrea* (marula) kernels, an important winter food
Figure 5.5   Zulu diviner collecting *Plectranthus grallatus* tubers

practitioner:total population ratio of between 1:720 and 1:1 200. In Zimbabwe, Gelfand et al (1985) estimated a traditional practitioner:total population ratio of 1:575, a level considered comparable to that of Natal/KwaZulu despite cultural differences (Cunningham 1988a). Well-trained traditional healers fill an important gap in the health-care system which has previously been neglected (Holdstock 1979; Buhrmann 1984). Holdstock (1978) has estimated that 80 to 85% of Soweto residents consult traditional practitioners. More than 400 indigenous and 20 exotic plant species used as traditional medicines by Zulu people are sold in urban areas of Natal/KwaZulu (Cunningham 1988). In Umlazi, one of the largest 'townships' in the Durban area, 30% of a random sample of residents had used the highly toxic medicinal plant *Callilepis laureola* (impila) (Wainwright et al 1977), and this is not the most popular species sold in Umlazi (Cunningham 1988a). Use of medicinal plants is even higher in rural areas in Natal — all respondents interviewed in a random sample in the Estcourt area used medicinal plants (Ellis 1986).

Although difficult to assess, the replacement value and importance of indigenous plants in providing income in impoverished areas should not be underestimated. Milton and Bond (1986), for example, have estimated that thornveld savanna is worth at least R425 family^{-1} yr^{-1} or R48 ha^{-1} yr^{-1} to inhabitants of the Msinga area in Natal/KwaZulu. Areas of productive wetlands (*Phragmites* reeds and *Juncus kraussii* mat-rush saltmarsh), *Hyphaene* palmveld and thatching grass all have utilitarian and direct economic value to rural people (Cunningham 1985, 1987, 1988b). In Maputaland, northern Natal/KwaZulu, the cutting and sale of *Phragmites* reeds generated over R38 000 within a 12-month period, providing building materials and self-employment for about 60 people. An estimated 500 people were self-employed in the tapping, sale, transport and resale of palm wine in the same area, generating about R160 000 within a 12-month period (Cunningham 1985). Despite low individual profits from the sale of palm wine, craftwork or reeds, with rising unemployment (Maasdorp and Whiteside 1987), these plant resources provide an important opportunity for self-employment.

RESOURCE VALUES, CUSTOMARY CONSERVATION AND DIVERSITY

It is generally accepted by white South Africans that conservation in South Africa began, about 100 years ago, with the proclamation of the first reserve in the Pongola area in 1894. This view is increasingly being called into question as a result of evidence from widespread indigenous societies (Johannes 1978; Scudder and Conelly 1985). Customary conservation practices have important implications for the maintenance or creation of habitat diversity in southern Africa. In two cases from the early nineteenth century, for example, Sarili, the paramount chief of the Gcaleka people and the Zulu king, Shaka, established exclusive hunting rights over the Dwesa-Manubi Forest in the Transkei (Cawe 1986), and in the Mfolozi area in Natal/KwaZulu (Hall 1977), respectively. Both of these areas are proclaimed reserves today.

Although a 'conservation behaviour programme' was formulated by Ferrar (1983), and the management of resources in communally owned areas is a widely known problem, little attention has been paid to customary conservation practices in South Africa. Chapman (1987) suggests that for a natural resource to be intentionally managed in any society, the following conditions are necessary:

— the resource must be of value to the society;
— the resource must be perceived to be in short supply and vulnerable to over-exploitation by man; and
— the socio-political nature of the society must have the necessary structure for resource management.

Chapman's beliefs are certainly borne out by what Klee (1980) terms intentional and inadvertent customary conservation practices. In southern Africa these range from seasonal restrictions on gathering medicinal plants (Cunningham 1988a) to the widespread social conventions that prevent the felling of fruit-bearing trees (eg *Vangueria infausta* (wild medlar), *Sclerocarya birrea* subsp *caffra* (marula), *Trichilia emetica* (Natal mahogany)). These practices play an important role in the maintenance of biotic diversity in communal lands where access to fuel wood, thatch grass, grazing, water and game animals is considered to be a tenurial right. Tenurial regulations which control access to resources and allow 'privatization' of palm wine-tapping areas or fruit trees are particularly noteworthy in this respect (Cunningham 1985). The maintenance of favoured fruit, fodder or shade-providing trees significantly buffers the effects of agricultural clearing on woody vegetation and is widespread in southern Africa (Cunningham 1985; Campbell 1986). In Zimbabwe, for example, Campbell (1986) found that the three most frequently used fruit trees (*Diospyros mespiliformis* (jakkalsbessie), *Azanza garkeana* (snot apple) and *Strychnos cocculoides* (monkey orange)) remained relatively constant in cover between climax woodland and cleared agricultural lands. As a result, their proportional contribution to total woody cover increased from 0,5% in uncleared woodland to 5% in cultivated fields.

Customary African practices resulting in maintaining biotic diversity for other reasons is also recorded. Protection of vegetation at grave sites, for religious and possibly aesthetic reasons, is well known in southern Africa (Cunningham 1985; Webb and Wright 1986). In the nineteenth century, for example, specific Zulu regiments were called up year after year to burn fire-breaks around the grave sites of Zulu kings, overgrown sites being considered sanctuary for game animals (Webb and Wright 1986). All these factors play a role in maintaining diversity in a landscape greatly changed by man.

ANTHROPOGENIC LANDSCAPES AND SPECIES DIVERSITY

It is now widely acknowledged, primarily through the studies of archaeologists, that man has had a major influence over the past 110 000 years (Beaumont et al 1978) on the vegetation along the moist eastern seaboard of southern Africa, particularly during the Iron Age (Feely 1980, 1985; Hall 1984). Under low human population densities, the diversity created through the use of fire, agriculture or pastoralism, results in a range of vegetation types in various stages of recovery after clearing and burning.

From a conservation viewpoint, the creation of a patchy distribution of vegetation stands of different ages is a positive attribute. Each patch at a different stage of recovery represents a different habitat, thereby increasing the diversity of plant and animal species within a landscape. From the perspective of the human user, greater vegetation diversity is also a positive attribute, attracting game to recently burnt patches and providing gatherers with a wider range of plant products in the more diverse vegetation. Plant residues from archaeological remains in the fynbos biome, for example, indicate that Holocene hunter-gatherers were aware of the high below-ground production of geophytes and used fire as a tool to increase the production of Iridaceae (Deacon et al 1983). This 'fire-stick' farming is still practised by herd-boys on the Tongaland coastal plain today, with the deliberate burning of palm veld savanna and coastal grasslands to stimulate the production of *Salacia kraussii, Parinari curatellifolia*, and *Eugenia albanensis* which are favoured for their fruits (Cunningham 1985, 1988c).

Removal of disturbance can also create problems in an anthropogenic landscape. In the Hluhluwe–Umfolozi Game Reserve, for example, the absence of man and elephant, and a low fire frequency combined with overstocking of game resulted in a 33% increase in forest and a 227% increase in scrub area between 1937 and 1982, with a marked change in faunal composition (King 1987).

DECREASING DIVERSITY, CULTURAL CHANGE AND RESOURCE MANAGEMENT

Changes in landuse and in human and livestock population numbers have resulted in rapid changes in vegetation cover (Acocks 1975). The direct use of plant resources for fuel, building, dyes, fruits or medicinal plants is interlinked with this vegetation change. Frequent fires, overgrazing, large-scale monocultural projects, and a higher density of subsistence agriculturalists than the land can support, result in the breakdown of the mosaic of habitat types, simplifying vegetation and reducing species diversity. Decreased habitat diversity not only reduces the scientific and aesthetic values of a landscape, it reduces the range of plant resources which provide a buffer against rural poverty during climatic extremes.

Gathering in agro-pastoral societies has traditionally been a response to drought and is common practice in areas of marginal agricultural potential (Colson 1979; Grivetti 1979), leading Felgate (1982) to comment that 'the striking feature of Thonga subsistence activity is that food collecting and agriculture play an equally important role in the Thonga subsistence economy'. Grivetti (1979), for example, considered that the principal factor contributing to Tswana nutritional success at the peak of the drought in the Kalahari was a diversified food base with an emphasis on wild food plants. Removal of remaining trees and baring of the landscape under high human population densities, and decreased biotic diversity reduces landuse options, paving the way to a 1980s scenario of starvation in Ethiopia and the Sudan (Owen 1979).

FIGURE 5.6   Radial depletion of building materials with subsequent re-coppice of *Colophospermum mopane* (mopane) trees in semi-arid savanna

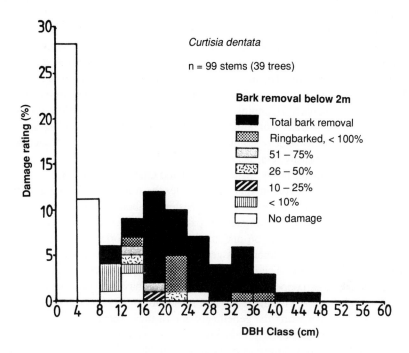

FIGURE 5.7   Damage to *Curtisia dentata* (assagaai) trees in eMalowe State Forest, Transkei, as a result of bark removal for medicinal purposes. DBH = trunk diameter at breast height (Cunningham 1988a).

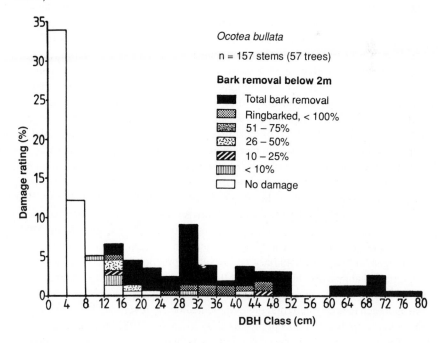

FIGURE 5.8   Damage to *Ocotea bullata* (stinkwood) trees in eMalowe State Forest, Transkei. DBH = trunk diameter at breast height (Cunningham 1988a).

In the same way that edible wild foods provide a buffer against starvation during drought, economically important species provide a buffer against unemployment during cyclical economic depressions. Basket weavers, palm wine tappers, reed cutters, and commercial gatherers and hawkers of medicinal plants all have something in common — a low level of formal education and limited formal employment opportunity. Simkins (1985) estimates that by 2000, 45% of the 34,7 million black South Africans will be urbanized. The exploitation of economically important plant resources in response to demand from urban settlements (for example, medicinal plants) can be expected to increase (Cunningham 1988a). Reduction of woody cover is already being experienced in rural areas where the demand for fuel wood exceeds supply. Cutting live wood reduces the availability of browse and worsens erosion (Gandar 1983). Vegetation recovery is also affected, since the brushwood 'islands' protecting perennial grasses and seedlings of palatable woody plants are finally burnt for fuel (Malan and Owen-Smith 1974).

Radial depletion of woody plants characterizes the surrounds of villages in the western section of southern Africa (Figure 5.6). Along the eastern seaboard, where settlement is more scattered and human carrying capacity is higher, the pattern of resource depletion is not as clear, but it occurs nonetheless. Rapid urbanization and the continued importance of traditional medicines to a growing urban black population, has generated a country-wide trade from rural source areas to urban markets and shops. The rare medicinal plants *Siphonochilus natalensis* (wild ginger) and *Warburgia salutaris* (pepper-bark tree), for example, have been completely overexploited in the Natal/KwaZulu region with the result that most supplies for the urban demand now come from the Transvaal, exacerbating the genetic erosion of these species. Forest trees providing popular medicinal barks are particularly vulnerable to overexploitation owing to slow growth rates and restricted distribution of forests. For example, ringbarking of most large *Ocotea bullata* (stinkwood) and *Curtisia dentata* (assegaai) trees in Afromontane forest in Natal/KwaZulu has resulted in a shift to exploiting these species in Transkei (Figures 5.7 and 5.8) (Cunningham 1988a). The subsequent tree die-off has a marked effect on forest structure, resulting in canopy gap formation far exceeding that caused by natural disturbance.

Unlike the harvesting of low-diversity, high-biomass vegetation types (*Phragmites australis* reeds, *Juncus kraussii* mat-rush) or encroaching species (eg *Dichrostachys cinerea* (sickle-bush)), the exploitation of scarce or slow-growing species requires intensive management effort if sustainable use is to be achieved (Cunningham 1985). With current economic constraints, intensive management of indigenous forests in tribal 'homelands', similar to that practised in the southern Cape, seems unlikely. Legislation and patrolling by forest guards may slow the rate of exploitation, but do not provide a solution. If other sources of energy, building materials and medicinal plants are not available, managers will face increasing conflict with people using these resources. Providing other sources of supply outside the boundaries of these areas can do much to decrease tensions which may threaten the long-term future of protected areas.

The social consequences of this downward trend are felt first by the poorest sector of the community, since they walk further or pay more for fuel wood, craftwork materials and medicinal plants (Gandar 1983, 1984; Cunningham and Milton 1987; Cunningham 1988a). Patrolling, providing woodlots, fencing, and funding extension workers require money and in any event are only stopgap measures. Customary conservation practices, which represent a form of limited access and a change from the 'tragedy of the commons', break down under commercial exploitation and scarcity.

An approach to integrated rural development aimed at reversing these trends is the 'community land company approach' concept proposed by Reynolds (1981) for managed multiple-use areas. This approach — which deserves support — is being implemented in Zimbabwe through the participation of local shareholders in managed resource areas. In the long term, successful management depends on whether or not the social and demographic issues that threaten any African conservation proposals are addressed. If they are not addressed, conservation of biotic diversity faces a bleak future. Kaplan (1976) found that an effect of formal education was a highly significant reduction in the use of traditional medicines. Lack of formal education, on the other hand, reduces the chance of formal-sector employment. Maasdorp and Whiteside (1987) have suggested a massive public works programme aimed at reducing unemployment. Planting wood-lots or medicinal plants would certainly be more constructive than depleting these resources. It would also be a less risky and more lucrative occupation than poaching stinkwood bark.

CONSERVATION: WHO PAYS THE COSTS?

Whether the costs to society are measured in monetary terms, or in the quantity of habitat or species lost, there is little doubt that these costs are increasing exponentially. The only way out of this downward spiral lies in lowering population densities and simultaneously enhancing habitat quality and carrying capacity. Merely shifting the focus in forestry policy from commercial afforestation to resources currently exploited from the wild would provide employment, increase resource availability, and protect indigenous forest remnants.

The development of plantations as an alternative to the exploitation of wild stocks has been proposed for fuel wood, building material and medicinal plants. These proposals have met with little support as yet. Gandar (1983) estimated that if half of KwaZulu's fuel wood needs of two million t yr^{-1} were to be satisfied by woodlots, 125 000 ha would be needed. However, 94% of the 27 000 ha under afforestation in KwaZulu is under commercial timber production, and only 0,1% of the annual KwaZulu budget was committed to woodlot development. The cultivation of medicinal plants was proposed over 40 years ago by Gerstner (1946), who specified *Warburgia salutaris* as one of the tree species most in need of cultivation. Nothing was done, and today the pepper-bark tree is considered endangered in South Africa (Cunningham 1988a), and is heavily exploited elsewhere in Africa (FAO 1986).

Little can be achieved without the support of policy-makers and planners. If the conservation goals set by the IUCN (1980) are to be reached, neither the 'TBVC countries' and tribal 'homelands', nor the social and political processes shaping them, can be seen in isolation. Whether support for conservation efforts in these areas is increased or not has implications for the entire southern African region. Reduction in vegetation cover, for example, affects conservation areas adjacent to or downstream from degraded lands. High population concentrations in 'homelands' and Trust Lands are the result of population growth, influx control, and the relocation of black South Africans totalling 42% of the population but who have been directed onto 13% of the land (Grobbelaar 1985; Christopher 1986). The forced removal of over three million people since 1960 for ideological reasons (Anonymous 1983) has had disastrous ecological and social consequences. It has also caused suspicion regarding the relocation of people for ecological reasons, for example, from mountain catchments such as the Upper Tugela region. What is sorely needed is, first, management action guided by a clear and far-sighted policy on man-land relationships for the entire region. Secondly, we need a repeatable method of evaluating response to management action so that management effort can be modified when necessary.

CONCLUSIONS

In hindsight, ethnobotany in southern Africa has come full circle. In the late nineteenth century, the use of customary African knowledge was the key to identifying species with commercial potential to colonial governments. Little emphasis was placed on either the social or the environ-mental consequences of exploitation. Ethnobotanical studies now form an important part of the justification of conservation as a form of landuse — by all sectors of society. Maintenance of biotic diversity is also maintenance of choice in lifestyle. The direct costs of resource depletion not only affect the rural poor, they are borne by people who might have used the resources in the future. Ultimately the costs are imposed on the whole of society, both now and in the future.

ACKNOWLEDGEMENTS

This chapter has benefited from the comments of Dr K L Tinley and Dr M Plotkin, while my own perceptions expressed here have been immeasurably sharpened by discussions and field-work with herbalists, craft-workers and resource users in southern Africa. I greatly appreciate the financial support for my research from the Human Needs, Resources and Environment Section, HSRC; Ecosystems Programmes, CSIR; Natal Parks Board; KwaZulu Bureau of Natural Resources; Conservation section, Department of Development Aid; National Nyanga's Association; Herb Traders' Association; and Wildlife Society of Southern Africa.

REFERENCES

ACOCKS J P H (1975). Veld Types of South Africa. (Second edn). *Memoirs of the Botanical Survey South Africa* **40**. Government Printer, Pretoria.

ALLAN W (1965). *The African Husbandman*. Oliver and Boyd, London.

ANONYMOUS (1983). Forced removals in South Africa. *Surplus people project report* **1**. Multicopy Centre, University of Natal, Pietermaritzburg. 120 pp.

ARNOLD H-J and GULUMIAN M (1984). Pharmacopoeia of traditional medicine in Venda. *Journal of Ethnopharmacology* **12**, 35 – 74.

BASSON J A (1987). Energy implications of accelerated urbanisation. *South African Journal of Science* **83**, 284 – 290.

BEAUMONT P B, DE VILLIERS H and VOGEL J C (1978). Modern man in sub-Saharan Africa prior to 49 000 years BP: a review and evaluation with particular reference to Border Cave. *South African Journal of Science* **74**, 409 – 419.

BOUSQUET J (1982). The morama bean of the Kalahari desert as a potential food crop, with a summary of current research in Texas. *Desert Plants* **3**, 213 – 215.

BUHRMANN V (1984). *Living in Two Worlds*. Human and Rousseau, Cape Town.

BURGER A E C, DE VILLIERS J B M, DU PLESSIS L M (1987). Composition of the kernel oil and protein of the marula seed. *South African Journal of Science* **83**, 733 – 735.

CALVIN M (1979). Petroleum plantations and synthetic chloroplasts. *Energy* **4**, 851 – 869.

CAMPBELL B M (1986). The importance of wild fruits for peasant households in Zimbabwe. *Food and Nutrition* **12**, 38 – 44.

CAWE S G (1986). A quantitative and qualitative survey of the inland forests of Transkei. MSc thesis, University of Transkei, Transkei.

CHAPMAN M D (1987). Traditional political structure and conservation in Oceania. *Ambio* **16**, 201 – 205.

CHRISTOPHER A J (1986). The inheritance of apartheid planning in South Africa. *Land Use Policy* **3**, 330 – 335.

COLSON E (1979). In good years and in bad: food strategies of self-reliant societies. *Journal of Anthropological Research* **35**, 18 – 29.

CUNNINGHAM A B (1985). The resource value of indigenous plants to rural people in a low agricultural potential area. PhD thesis, University of Cape Town, Cape Town.

CUNNINGHAM A B (1987). Commercial craftwork: balancing out human needs and resources. *South African Journal of Botany* **53**, 259 – 266.

CUNNINGHAM A B (1988a). Development of a conservation policy on the herbal medicine trade in southern Africa: Zulu medicinal plants. *Investigational Report* **29**, Institute of Natural Resources, University of Natal, Pietermaritzburg. 133 pp.

CUNNINGHAM A B (1988b). Leaf production and utilization in *Hyphaene coriacea*: management guidelines for commercial harvesting. *South African Journal of Botany* **54**, 189 – 195.

CUNNINGHAM A B (1988c). Collection of wild plant foods in Tembe Thonga society: a guide to Iron Age gathering activities? *Annals of the Natal Museum* **29**, 433 – 466.

CUNNINGHAM A B and GWALA B R (1986). Plant species and building methods used in Tembe Thonga hut construction. *Annals of the Natal Museum* **27**, 491 – 511.

CUNNINGHAM A B and MILTON S J (1987). Effects of basket weaving industry on Mokola palm and dye plants in North-western Botswana. *Economic Botany* **41**, 386 – 402.

CUNNINGHAM A B and WEHMEYER A S (1988). Nutritional value of palm wine from *Hyphaene coriacea* and *Phoenix reclinata* (Arecaceae). *Economic Botany* **42**, 301 – 306.

DEACON H J, SCHLOTZ A and DAITZ L D (1983). Fossil charcoals as a source of palaeoecological information in the fynbos region. In Fynbos palaeoecology: a preliminary synthesis. (eds Deacon H J, Hendey Q B and Lambrechts J J N) *South African National Scientific Programmes Report* **75**, CSIR, Pretoria. pp 174 – 182.

EBERHARD A A (1986). Energy consumption patterns in underdeveloped areas in South Africa. *Report* **94**, Energy Research Institute, University of Cape Town, Cape Town. 120 pp.

ELLIS C G (1986). Medicinal plant use — a survey. *Veld and Flora* **72**, 72 – 73.

FAO (1986). *Some medicinal forest plants of Africa and Latin America*. FAO Forest Resources Development Branch, Forest Resources Division FAO, Rome.

FEELY J M (1980). Did Iron Age man have a role in the history of Zululand's wilderness landscape? *South African Journal of Science* **76**, 150 – 152.

FEELY J M (1985). Smelting in the Iron Age of Transkei. *South African Journal of Science* **81**, 10 – 11.

FELGATE W S (1982). The Tembe Thonga of Natal and Mozambique: an ecological approach. *Occasional Report* **1**, Department of African Studies, University of Natal, Pietermaritzburg. 181 pp.

FERRAR A A (1983). A South African perspective on conservation behaviour — a programme description. *South African National Scientific Programmes report* **76**, CSIR, Pretoria. 36 pp.

FOX F W and NORWOOD-YOUNG M E (1982). *Food from the Veld.* Delta books, Johannesburg.

FRASER-DARLING F (1960). *Wildlife in an African Territory.* Oxford University Press, London.

GANDAR M V (1983). Wood as a source of fuel in South Africa. *Monograph* **4**, Institute of Natural Resources, University of Natal, Pietermaritzburg.

GANDAR M V (1984). Firewood in KwaZulu: quantities and consequences. *Proceedings of the Conference on Energy for Underdeveloped areas.* Energy Research Unit, University of Cape Town, Cape Town pp 4,1 – 4,15.

GELFAND M, MAVIS S, DRUMMOND R B and NDEMERA B (1985). *The traditional medical practitioner in Zimbabwe.* Mambo Press, Harare.

GERSTNER J (1946). Some factors affecting the perpetuation of our indigenous flora. *Journal of the South African Forestry Association* **13**, 4 – 11.

GIBBS RUSSELL G E (1985). Analysis of the size and composition of the southern African flora. *Bothalia* **15**, 613 – 630.

GIBBS RUSSELL G E (1987). Preliminary floristic analysis of the major biomes in southern Africa. *Bothalia* **17**, 213 – 227.

GOOSEN H (1985). Die maroela word getem. *South African Panorama* **30**, 21 – 25.

GREYLING T and HUNTLEY B J (1984). Directory of southern African conservation areas. *South African National Scientific Programmes Report* **98**, CSIR, Pretoria. 311 pp.

GRIVETTI L E (1979). Kalahari agro-pastoral hunter-gatherers: the Tswana example. *Ecology of Food and Nutrition* **7**, 235 – 256.

GROBBELAAR J A (1985). The population of Natal/KwaZulu 1904 - 2010. *Natal Town and Regional Planning Report* **65**, Natal Town and Regional Planning Commission, Pietermaritzburg. 80 pp.

HALL M (1977). Shakan pitfall traps: hunting technique in the Zulu kingdom. *Annals of the Natal Museum* **23**, 1–12.

HALL M (1984). Prehistoric farming in the Mfolozi and Hluhluwe valleys of south-east Africa: an archaeo-botanical survey. *Journal of archaeological Science* **11**, 223 – 235.

HENNESSY E F and LEWIS O A M (1971). Anti-pellagragenic properties of wild plants used as dietary supplements in Natal (South Africa). *Plant Foods in Human Nutrition* **2**, 75 – 78.

HOLDSTOCK T L (1978). Panel discussion. Proceedings of a conference on traditional medicine. *The Leech* 48, 22.

HOLDSTOCK T L (1979). Indigenous healing in South Africa: a neglected potential. *South African Journal of Psychology* **9**, 118 – 124.

HUXLEY J (1961). The conservation of wildlife and natural habitats in Central and East Africa. UNESCO, Paris.

INFIELD M M (1986). Wildlife resources, utilization and attitudes towards conservation: a case study of the Hluhluwe and Umfolozi Game Reserves in Natal/KwaZulu. MSc thesis, University of Natal, Pietermaritzburg.

IUCN (1980). *World Conservation Strategy: Living resource conservation for sustainable development.* IUCN/UNEP/WWF, Gland, Switzerland.

JOHANNES R E (1978). Traditional marine conservation methods in Oceania and their demise. *Annual Review of Ecology and Systematics* **9**, 349 – 364.

JOHNSON C T (1982). The living art of hut building in Transkei. *Veld and Flora* **68**, 109 – 110

KAPLAN A L (1976). The marketing of branded medicine to the Zulu consumer. PhD thesis, University of Natal, Durban.

KING N D (1987). Vegetation dynamics in Hluhluwe Game reserve, Natal, with emphasis on the bush encroachment problem. MSc thesis, University of Natal, Pietermaritzburg.

KLEE G A (1980). *World systems of traditional resource management.* John Wiley and Sons, New York, NY.

KNUFFEL W E (1976). *The construction of the Bantu grass hut.* Akademische Druk, Graz.

LEE R B (1973). Mongongo: the ethnography of a major wild food resource. *Ecology of Food and Nutrition* **2**, 307 – 321.

LIENGME C A (1981). Plants used by the Tsonga people of Gazankulu. *Bothalia* **13**, 501 – 518

LIENGME C A (1983a). A survey of ethnobotanical research in southern Africa. *Bothalia* **14**, 621 – 629.

LIENGME C A (1983b). A study of wood use for fuel and building in an area of Gazankulu. *Bothalia* **14**, 245 – 257.

MAASDORP C G and WHITESIDE A W (1987). Absorbing labour in South Africa — the employment challenge. *South African Journal of Science* **83**, 302 – 305.

MALAN J S and OWEN-SMITH G L (1974). The ethnobotany of Kaokoland. *Cimbebasia Ser* **B 2**, 131 – 178.

MEDLEY-WOOD J and EVANS M S (1898). *Natal Plants.* Vol 1. Bennett and Davis, Durban.

MILTON S J and BOND C (1986). Thorn trees and the quality of life in Msinga. *Social Dynamics* **12**, 64 – 76.

NAS (1979). *Tropical legumes: resources for the future.* Academy of Sciences, Washington DC. 331 pp.

OGLE B R and GRIVETTI L E (1985). Legacy of the chameleon: edible wild plants in the Kingdom of Swaziland, southern Africa. A cultural, ecological, nutritional study. Part 4 — nutritional analysis and conclusions. *Ecology of Food and Nutrition* **17**, 41 – 64.

OWEN D F (1979). Drought and desertification in Africa: lessons from the Nairobi conference. *Oikos* **33**, 139 – 151.

PIPREK S R K, GRAVEN E H and WHITFIELD P (1982). *Some potentially important indigenous aromatic plants for the eastern seaboard areas of southern Africa. In Aromatic plants: Basic and applied aspects.* Martinus Nijhoff, The Hague, pp 255 – 263.

QUIN P J (1959). *Food and feeding habits of the Pedi.* Witwatersrand University Press, Johannesburg.

REYNOLDS N (1981). The design of rural development: proposals for the evolution of a social contract suited to conditions in southern Africa. Pts 1 and 2. *SALDRU papers* **41** and **42**, University of Cape Town, Cape Town.

SAVAGE M T D (1985). Health. In Basic needs in rural areas. *South African National Scientific Programmes Report* **116**, CSIR, Pretoria. pp 49 – 66.

SCUDDER T and CONELLY T (1985). *Management systems for riverine fisheries.* IDA report, FAO, Rome. 89 pp.

SIM T R (1903). India rubber in Tongaland. *The Agricultural Journal and Mining Record* **6**, 25 – 30.

SIM T R (1920). South African Rubber. *The South African Journal of Industries Bulletin* **53**, Government Printer, Pretoria 1 – 28.

SIMKINS C (1985). Projecting African population distribution and migration to the year 2000. *RSA 2000* **7**, 41 – 46.

SMITH D M (1976). Separation in South Africa: 1. People and policies. *Occasional Paper* **6**, Department of Geography, Queen Mary College, London.

TINLEY K L (1979). The maintenance of wilderness diversity in Africa. Voices of the Wilderness. *Proceedings of the First World Wilderness Congress, Johannesburg* pp 29 – 42.

TINLEY K L and VAN RIET W R (1981). Tongaland: Zonal ecology rural land-use proposals. *Report to the Department of Co-operation and Development.* Pretoria. 186 pp.

TRAPNELL C G (1953). *The soils, vegetation and agricultural systems of north-eastern Rhodesia.* Government Printer, Lusaka.

VAN DER MERWE A LE R, BURGER I M and WEHMEYER A S (1967). *Suid-Afrikaanse veld-kosse: 1. Makhatini vlakte, Noord Natal.* National Food Research Institute, CSIR, Pretoria.

VON TEICHMANN I (1983). Notes on the distribution, morphology, importance and uses of the Anarcardiaceae. 2. The importance and uses of *Sclerocarya birrea* (the marula). *Trees in South Africa* **35**, 1 – 7.

WAINWRIGHT J, SCHONLAND M M and CANDY H J (1977). The toxicity of *Callilepis laureola. South African Medical Journal* **52**, 313 – 315.

WATT J M and BREYER-BRANDWIJK M G (1962). *Medicinal and poisonous plants of southern and eastern Africa.* (Second edn). E and S Livingstone, Edinburgh.

WEGG C de B and WRIGHT J B (1986). *The James Stuart Archive of recorded oral evidence relating to the history of the Zulu and neighbouring peoples.* Vol 4. University of Natal Press, Pietermaritzburg.

WEHMEYER A S (1966). The nutrient composition of some edible wild fruit found in the Transvaal. *South African Medical Journal* **40**, 1102 – 1104.

WHITBREAD M W (1986). Preliminary studies on the utilization and adaptation of indigenous and introduced grain amaranths (*Amaranthus* spp.) in the Natal and KwaZulu midlands. MSc thesis, University of Natal, Pietermartizburg.

CHAPTER 6

The contribution of veld diversity to the agricultural economy

N M Tainton, P J K Zacharias, M B Hardy

INTRODUCTION

Approximately 80% of the 86 million ha of land owned by commercial farmers and a large proportion of the remaining 27 million ha in South Africa, much of it under subsistence agriculture, is currently uncultivated, natural vegetation (veld). The primary use of this land is for livestock production and indeed veld remains the backbone of this most important agricultural industry, generating as it does almost 50% of the income derived from agriculture in South Africa (De Klerk 1986). The mohair, karakul and wool industries are reliant almost entirely on veld, while it also makes a substantial contribution to the feeding of beef cattle and mutton sheep, and in some regions also to the feeding of dairy cattle. To its role in the agricultural sector must be added the important role veld plays in the social and economic welfare of Third World (subsistence) agriculturalists. Here livestock production is dependent primarily on veld. Its role in the feeding of domestic and subsistence livestock is therefore crucial and it demands consideration in any programme designed to improve the conservation status of the vegetation of South Africa.

The vegetation of southern Africa has the highest species richness of any floral region in the world (Gibbs Russell 1987). The complex physiography together with the wide range of climatic conditions (Tyson 1986) is probably responsible for the high species diversity reflected in Acocks's map of Veld Types (Acocks 1975), both within and between habitat types. From an agronomic standpoint the pastoral industry is affected, at the farm scale, by diversity both within and between communities (eg forage flow, toxicity and selection by animals). It is also affected by differences between habitats (eg aspectal differences, landscape position, and season of use) and at the regional level by large-scale vegetation differences which determine the production system and choice of suitable animal species for the area.

The aim of this paper is to discuss the interrelationships and responses between the pastoral industry and these scales of plant diversity.

FACTORS AFFECTING ANIMAL PERFORMANCE ON VELD

The performance of animals on natural grazing depends on both the inherent ability of the animal to make effective use of the nutrients it harvests during grazing or browsing, and on the capacity of veld to provide the nutrients required by the animal.

The ability to make effective use of the nutrients harvested during grazing and browsing is known to vary between animal classes (eg between mono-gastric animals and ruminants), species and breeds, and between individuals of the same breed, according to their genetic constitution. This aspect will not, however, be considered further in this paper. Rather, emphasis will be placed on the inherent capacity of veld to provide domestic animals with an adequate intake of nutrients during different seasons.

Nutrient intake

The intake of feed by herbivores will set the absolute limit to nutrient absorption by the animal (Bransby 1981). This will in turn depend on the amount of feed available, and on the acceptability, nutritive value and digestibility of that feed (Minson 1982).

AVAILABILITY

Feed availability has frequently been shown to have a major impact on animal performance, apparently through its influence on intake. This relationship is expressed in the Jones and Sandland (1974) model which relates animal performance (commonly average daily liveweight gain) to stocking rate (Hart 1978) and needs no further elaboration in this paper.

On veld, animal performance is governed by the relationship between animal number and the amount of grazable material available. This may often comprise only a fraction of the total herbage (Mentis 1982; Hardy 1986) owing to the wide range in palatability exhibited by veld species (Theron and Booysen 1966; Barnes et al 1984).

ACCEPTABILITY

Acceptability is defined here as the sum of those factors which operate to determine whether and to what degree animals will consume a particular feed. It depends partly on the palatability of the feed, as influenced by the specific attributes of the feed itself (eg physical and chemical deterrents), but is also influenced by a number of other factors, many of which pertain to the type and condition of the animal itself. In any event it is not possible to quantify either acceptability or palatability. Both will vary according to the specific circumstances prevailing in any situation. Nonetheless, it is common knowledge that forage species and plant communities differ widely in both these characteristics, giving rise to wide ranges in intake by foraging animals.

NUTRITIVE VALUE

Whereas it is possible to determine accurately the nutrient content of forages, it is not possible to predict accurately the ability of forages to meet the nutrient needs of animals. This is so particularly for ruminants, in which the availability of nutrients for absorption from the intestinal tract is dependent on both physical and chemical action by micro-flora within the rumen. Laboratory techniques are, nonetheless, available which permit a ranking of forages for their potential to provide the major nutrients known to be required by animals (Tilley and Terry 1963; Minson 1982;

Van Soest 1983; Zacharias 1986). These again show a wide range in the apparent capacity of different species and varieties (Theron and Booysen 1966; Schoeman 1968) to meet adequately the nutrient needs of animals during different seasons (Kirkman 1988; Zacharias 1988), based both on their nutrient content and on differences in the availability of these nutrients as determined by their digestibility.

DIGESTIBILITY

Digestibility in tropical and subtropical forages is, in general, considerably lower than in temperate forages (Minson 1982). Because of this, nutrient availability rather than nutrient content is often a major limitation to animal performance off veld in southern Africa (Du Toit et al 1940). Hence, while chemical analyses may show equivalent nutrient contents in forage produced by different species, such forages may differ appreciably in their capacity to supply animals with nutrients (Van Soest 1965). Also, even within the same species, digestibility may change appreciably from one season to the next (eg, from spring to mid-summer) (Kirkman 1988; Shackleton pers comm). Such seasonal change in digestibility does, however, vary across the major vegetation types used for livestock production in South Africa.

Seasonality of nutrient availability

In South Africa vegetation is grouped into three classes based on the period of the year during which it can carry livestock in a productive state (Figure 2.3 in Tainton 1981a). These groups are the so-called 'sweetveld', 'sourveld' and 'mixedveld' (Pole-Evans 1920; Scott 1947). The sourveld comprises mainly humid highlands and sandy mesic areas and is defined as being able to maintain animals in a productive state for six to eight months. Sweetveld includes most of the more arid vegetation types and maintains animals in a productive state for 12 months of the year. Of course there is an intergrade between these which we refer to as mixedveld (Scott 1947). The differences between these is not based solely on the presence of different species but on the nutritive characteristics of the individual species which may occur in both sweetveld and sourveld (Zacharias 1988). A further complication is that the 'sweet' character of the veld may be owing to animals' using different components of the vegetation at different times. The definitions are clearly agronomic and as yet have no ecological base, although this question is being addressed (Zacharias 1988).

HUMID GRASSLANDS

Typically these grasslands are either sour or mixed, except where climatic conditions are such that growth normally takes place throughout the year (Tainton 1981a). Where the rainfall is restricted to the summer season, the veld is normally capable of providing nutrients in an available form to ruminants only during the active growing season (sourveld) (Hardy 1986) or at best for a period extending also into the early dormant season (mixedveld) (Jones and Arnott 1977; Bredon and Stewart 1978). However, where rain falls in both summer and winter, the tropical/subtropical species typically provide summer grazing and the temperate species winter grazing, so that the veld is by definition sweet (Tainton 1981a). Here, as in areas with a seasonally restricted rainfall, there are large acceptability differences between species and between communities at any time, so selection occurs widely.

ARID TO SEMI-ARID GRASSLAND

Such veld is typically classified as sweet to sweet-mixed, since even mature herbage is able to meet the nutrient needs of livestock. Arid to semi-arid grassland therefore has the potential to meet the year-round nutrient needs of animals (Roberts 1981).

SAVANNA

The herbaceous component of South African savannas may be sour, mixed or sweet. The same pattern of seasonal nutrient supply as applies to grassveld therefore applies here. Among the shrubs and trees, also, seasonal patterns of nutrient supply are evident, such as between deciduous woodland, with a restricted season of available browse, and evergreen woodland, which provides a year-round supply of browse. There are, however, also indications that the acceptability of trees and shrubs may change over the season (Owen-Smith 1988).

KAROO

With the exception of the karoo mountains, which provide useful forage mainly during the summer season, the karoo provides useful grazing year-round. As a general rule the grasses (where they occur) provide much of the summer forage, whereas karoo bushes normally provide the preferred grazing in winter (Roux 1966). However, karoo bushes have the potential to provide year-round grazing.

FYNBOS

This veld has little potential for livestock production. What forage is produced is available during the late autumn, winter and spring, so the summer season is normally one of severe shortages of forage.

VARIATIONS WITHIN AND AMONG PLANT SPECIES IN THEIR ABILITY TO SUPPLY NUTRIENTS TO LIVESTOCK

Intra-species variation across veld and land types

Although this field has not been well researched, common experience and the little research that has been done (Theron and Booysen 1966; Kirkman 1988; Zacharias 1988) suggest that there may be large differences in the acceptability, nutritive value and digestibility of plants of the same species growing under different conditions. It is not yet clear whether these differences are linked to genetic variation or whether they result more directly from environmental effects (Gibbs Russell and Spies 1988; Spies and Gibbs Russell 1988). In any event, environmental diversity will normally give rise to quite substantial differences in the forage value of a particular species.

Inter-species variation within veld and land types

Species of the South African veld vary widely in their potential to provide useful forage to livestock (Du Toit et al 1940;Theron and Booysen 1966; Walker 1980a). This has given rise to the widely used classification of veld species into desirable, moderately desirable and undesirable species, or equivalent groupings (Vorster 1982; Barnes et al 1984). Each category is then scored according to its relative forage value. So, for example, such species as *Aristida junciformis, Elionurus muticus, Diheteropogon filifolius* and *Rendlia altera* within a highland sourveld community typically provide little or no grazing to livestock, which must then rely on such associated species as *Themeda triandra, Heteropogon contortus* and *Tristachya leucothrix* for their forage. So too among the trees of the savanna and the shrublets of the karoo. Indeed, such differences in the forage value of different species within single communities are used in the assessment of the value of veld for livestock production (Vorster 1982; Barnes et al 1984; Stuart-Hill 1987).

FORAGE SELECTION AMONG DOMESTIC LIVESTOCK SPECIES

Small animals overcome their greater relative demand for energy, compared with large animals, primarily by being more selective. This they are able to do because their generally small mouths allow them to be more discerning when foraging (Hoffman 1973), and because they experience a more favourable relationship between time available for foraging and their absolute intake than do large animals. However, plants favoured by one animal species are invariably favoured by other animal species as well (Mentis 1981), and selection is likely to be between plant species as well as between different parts of the same plant.

It is not surprising, therefore, that the intake of sheep has been found to have a higher digestible organic matter, nitrogen (Langlands and Sanson 1976; Zeeman et al 1983, 1984; Lyle et al undated), water-soluble carbohydrate, and a lower fibre content (Dudzinski and Arnold 1973) than that of cattle. Also, sheep grazing mixed karoo have been shown to select less strongly for grasses than do cattle (Botha et al 1983), those on Utah range more strongly for forbs than grass, compared with cattle (Cook et al 1967), and those on Californian range more strongly for legumes and leafy material than cattle (Van Dyne and Heady 1965). All evidence suggests, therefore, a major difference in the selection patterns of cattle and sheep. Even different breeds of sheep have been shown to exhibit different grazing habits (Roux and De Kock 1971).

Goats, like sheep, are known to be highly selective, but according to Zeeman et al (1984) Boer goats are less selective for different species than merino sheep. Goats do, however, spend much more of their foraging time browsing than do sheep (Wilson 1957; Campbell et al 1962, Du Toit 1972). Horses and donkeys, on the other hand, are essentially grazers, and are generally believed to be considerably less selective than cattle, although no quantitative data have been found to substantiate this.

INFLUENCE OF PLANT SPECIES DIVERSITY ON LIVESTOCK PRODUCTION OFF VELD

Variations within and among plant species in their nutritional value to livestock and the seasonality of nutrient availability have already been discussed. It follows that, together with forage selection patterns and the preference exhibited by various species of livestock for different types of forage, the great diversity of plant species and communities encountered in southern Africa affects the distribution and performance of livestock in the region.

The influence of species richness and community diversity on livestock production

High species richness and community diversity have a positive influence on livestock production only if different components of the sward lead to an improved seasonal availability of forage. So, for example, summer rains in the karoo stimulate grass growth and a supply of summer forage, whereas autumn rains provide little growth stimulus to the grasses, but promote the growth of karoo shrublets which are the preferred forage for sheep in winter (Roux 1968). This is also the case in the arid grassveld and savannas, where there are indications that at least some of the grasses respond differently to early and late season rains (McNaughton 1985; Walker et al 1986). For the vast areas of humid grassveld and savanna, the seasonal distribution of forage production depends largely on the growth and tillering patterns of grasses. Swards dominated by *Eragrostis curvula,* *E plana, Sporobolus africanus* and *S pyrimidalis* (the 'Mtshiki' species) produce high-quality forage in the early grazing season, but as these species mature rapidly (and lose their nutritional value as they mature) they provide a relatively short grazing season. Grassland areas dominated by species such as *Themeda triandra* and *Tristachya leucothrix* maintain a relatively high quality of forage for a longer period in the grazing season by continuing to produce new tillers after flowering (Tainton 1981b; Mentis 1982; Hardy 1986). The forage-flow potential for livestock production of swards dominated by *T triandra* is therefore higher than those dominated by 'Mtshiki' species.

The influence of community diversity on livestock production

Community diversity commonly arises at the farm scale owing to variations in topography (slope and aspect), soil and microclimate (moisture and temperature). In practice, provided these major community types can be isolated from one another (by fencing into camps), and the different community types can assist in lengthening the grazing season, such diversity can have a positive influence on livestock production. Indeed, where such diversity exists it is commonly taken into account in the planning of veld grazing systems, as will be shown later in this paper.

The influence of regional and biome-level diversity on livestock production

Cattle are the preferred livestock type in the grassveld. Sheep generally perform well on such grassveld, but they prefer short grass and have the ability to graze very close to the ground, which often leads to veld degradation through their tendency to graze small patches intensively (Edwards 1981).

In the savanna areas cattle are again to be preferred as grazers, but goats should usually be introduced here to provide the browsing component of the livestock system (Kelly et al 1978; Aucamp et al 1984). The introduction of browsers appears to be important here in order to optimize red meat production (Stuart-Hill 1987). Browsers may also be introduced to manipulate the grass:bush ratio (Walker 1980a, b).

The karoo, on the other hand, is well suited to small stock (sheep, goats, karakul), except in the mountains which are inherently more grassy and where cattle are the preferred livestock type. Indeed, on the vast karoo plains, woolled sheep have been the most important grazing animals since 1790 (Roberts 1981). More recently, changed economic circumstances, and perhaps to an extent also the change in the composition of the vegetation, have led to an increase in the number of Angora goats and dual-purpose sheep in the area.

Where sour, mixed and sweet grassveld or savanna are in close proximity, a system of trek farming is often practised (eg Natal Midlands — incorporating the highland sourveld (sourveld), tall grassveld (mixedveld) and valley bushveld (sweetveld). Here sheep and cattle graze sourveld in spring and early summer, mixedveld in late summer and autumn, and sweetveld in the winter.

THE INFLUENCE OF FIRE AND GRAZING BY DOMESTIC LIVESTOCK ON PLANT SPECIES DIVERSITY

Fire has long been an important factor in structuring southern African vegetation. The grassland, savanna and fynbos biomes are particularly well adapted to fire, whereas fire is a rare occurrence in the forest or karoo biomes (Booysen and Tainton 1984). Several authors have suggested that both grazing (and/or grazing and browsing) and fire were important factors in the development and maintenance of grassland and savanna vegetation prior to the colonial era (Trollope 1974; Stuart-Hill and Mentis 1982). Also, of all the environmental factors affecting plant species diversity, fire and grazing are the two most easily manipulated by man. Management systems have therefore been devised to ensure that the response of vegetation to grazing and fire inputs will maximize long-term animal production. These fire and grazing management options are discussed later. Here we concentrate on the detrimental effects of grazing and fire on southern African vegetation.

The most dramatic impression of the effects of management on plant communities is obtained where a fence line separates two areas which have a long history of different grazing management. Despite similar soil, aspect, slope and soil-available moisture the plant communities are often so distinct that the fence line is evident from some distance away. Where the fence separates grazed and ungrazed areas the contrast is most obvious. The grazed area is often dominated by grass species, whereas the ungrazed area is dominated by scrub or forest species. Where the fence separates heavily grazed from lightly grazed, or cattle grazed from sheep grazed areas, the effects are more subtle but nonetheless often evident (Crawley 1983).

Livestock production systems may result in a decrease in plant vigour (in the short term), a change in species composition to less palatable species (in the long term), or loss of plant cover. The utilization of vegetation by livestock *per se*, however, is not the only factor to consider. A reduction or increase in the frequency and intensity of fire, and variations in the distribution and amount of rainfall, have a significant impact. It is combinations of these factors that have influenced plant species diversity in southern Africa.

Humid grasslands

As discussed previously, humid grasslands are typically sour and mixed, and are best suited to cattle grazing. Overgrazing results in a change in species composition from swards dominated by *Themeda triandra* and *Tristachya leucothrix* to swards dominated by pioneer grass species such as 'Mtshiki'. This species change is often accompanied by declining species richness. In most situations plant cover remains relatively high. A decline in livestock production potential rather than soil loss is the main consequence (Tainton 1981a).

In certain areas where grazing results in insufficient fuel to carry a fire, scrub and forest precursor species become dominant. Encroachment by fynbos species such as *Elytropappus rhinocerotis* and *Euryops* sp, and the aggregation of *Cliffortia linearifolia* and *Erica brownleeae* in various Veld Types of the eastern Cape (Acocks 1975), have been ascribed to overgrazing, the destruction of plant cover, and the elimination of fire (Story 1952; Trollope 1970; Acocks 1975; Danckwerts 1979).

Vegetation exhibiting high species richness and community diversity is susceptible to selective utilization, particularly by sheep. Selective grazing in humid grasslands leads to an increase in abundance of 'wire' grasses. These include *Elionurus muticus, Diheteropogon filifolius, Rendlia altera,* and *Aristida junciformis* which are all extremely unpalatable to livestock (Opperman et al 1974; Tainton 1981a).

Under-utilization by livestock and, more importantly, the elimination of fire from the system leads to a dominance of forest precursor species such as *Leucosidea sericea* and *Rhus dentata* (Trollope 1979 in Danckwerts 1979; Tainton 1981c; Everson 1985).

Arid and semi-arid grasslands

In these areas of low and erratic rainfall, overgrazing, selective grazing and a lack of fire have caused major changes in the composition and structure of the vegetation. Selective grazing and overgrazing leads to a sward dominated by unpalatable pioneer grasses such as *Aristida congesta* subsp *barbicollis, Tragus racemosus, Sporobolus nitens,* and shrubs. Overgrazing and the exclusion of fire have resulted in the dramatic encroachment of karoo and savanna species into arid grasslands.

Acocks (1975) described the process of invasion and subsequent aggregation of xerophytic dwarf shrubs in the grassland areas of the south-western Orange Free State and north-eastern Cape as the most striking and alarming change that has taken place in the vegetation of the country in modern times. An area of some 66 000 km^2 of former grassland has been encroached on by species such as *Chrysocoma tenuifolia, Pentzia incana, Felicia muricata* and *F filifolia* (Huntley 1984). The replacement of grassveld by karoo is undesirable because it is accompanied by reduced cover, increased runoff, and soil erosion (Snyman and Van Rensburg 1986). In some cases areas invaded by karoo are so severely eroded that regeneration to productive grassland may not be possible (Tidmarsh 1957). However, some attempts at reclamation have been successful (Howell D 1976; Howell L N 1976; Danckwerts 1979).

Overgrazing and a reduction in the incidence and intensity of fires has promoted the encroach-ment into and aggregation of savanna species in grassland (Trollope 1974, 1983). Overgrazing reduces the competitive ability of grasses, producing a more favourable environment for the germination and establishment of shrubs and trees. *Acacia karroo* is the most widespread of the encroaching species, having encroached into the arid and semi-arid grasslands of the western Transvaal, central Orange Free State, eastern Cape and Natal (Acocks 1975), but a number of other species also take part in this process (eg *Dichrostachys* spp, *Euclea* spp, *Maytenus* spp).

Savanna

Trollope (1984) has detailed the ecological effects of fire in savanna ecosystems. Essentially, a reduction in the intensity and frequency of fires results in an increase in the density of trees. Grazing has a twofold effect. First, it reduces the competitive ability of the grass layer, thus producing a more favourable environment for the germination and establishment of tree seedlings. Secondly, it reduces the amount of fuel available for fires of sufficient intensity to reduce or prevent further aggregation of trees (Trollope 1984). Trollope (1981) points out that savanna areas are used mainly for cattle ranching. Commercial goat farming has been confined mainly to the valley bushveld and false thornveld of the eastern Cape, but it is gaining popularity elsewhere. Therefore, the present diversity of savanna vegetation is often a consequence of the utilization of only the grass component.

In arid savannas productive and highly acceptable genera such as *Panicum, Setaria, Digitaria, Cenchrus* and *Themeda* have been widely replaced by less productive pioneer species of *Aristida, Eragrostis, Chloris, Urochloa* and *Tragus*. In the semi-arid Kalahari thornveld (Acocks 1975) more than one million hectares had been encroached, primarily by *Acacia* spp, by the late 1960s (Donaldson 1969).

Grazing and the infrequent use of fire have resulted in a change in density rather than in species richness of the woody components of moist savannas (Trollope 1981). Large-scale changes in the woody component have also occurred in succulent variations of the valley bushveld. Here a significant result of grazing and browsing animals has been a reduction in the populations of the highly acceptable and productive *Portulacaria afra,* and an increase in unacceptable species such as *Euphorbia bothae* and *E coerulescens* (Trollope 1981).

Karoo

Fire has not been an important factor in the development of karoo vegetation. Compared with the fynbos and other biomes in southern Africa, the potential fuel loads in the karoo are low and are concentrated in usually widely dispersed, often semi-succulent, dwarf shrubs with low aerial cover (Edwards 1984).

Karoo veld is characterized by high seasonal and annual variations in forage quality and quantity (Vorster and Roux 1983). Cowling (1986) suggests that before large numbers of domestic livestock were introduced, indigenous ungulates would have responded in number and migratory patterns to the variability of forage production. After a severe drought it would seem likely that some time would have elapsed between the first good rains and the build up of indigenous herbivore numbers to pre-drought levels. The period of relatively low herbivore pressure and rapid plant growth could have been crucial for seed production and seedling establishment. The influx of domestic livestock and the fencing of properties, thus confining livestock to a particular area, has undoubtedly caused over-utilization of the natural resources. Roux and Vorster (1983) have reported dramatic vegetation changes and they stress that these changes can be ascribed primarily to anthropogenic factors, of which grazing by domestic livestock is by far the foremost.

BIOTIC DIVERSITY IN MANAGEMENT PLANNING

Because of the different seasonal patterns of acceptability of the forage produced by different species, high diversity usually implies a wide range of acceptability of plants to grazing animals. This, in turn, leads to selective grazing and it is the adverse effects of such selection that have dominated much of the research on grazing management in South Africa. In essence, there are two possible situations. In either of these situations there is likely to be a wide range of acceptability, either of different species or of different communities within each paddock. The first is where veld has either a high species richness and/or community diversity. Secondly, it may be possible to separate communities that differ in acceptability into different paddocks so that the vegetation is essentially uniform within each paddock (eg the group-camp system (Roux and Skinner undated)).

Vegetation with high species richness or community diversity

Research into the management of such vegetation has taken two approaches. The first has been to reduce, as far as is practicable, the degree to which animals graze selectively, whereas the second has aimed at reducing the detrimental effects of the selection that does take place on the production of forage and on the composition of the sward.

MANAGEMENT DESIGNED TO REDUCE THE EXTENT TO WHICH ANIMALS SELECT

Attempts to reduce the extent to which animals graze selectively within paddocks have culminated in the grazing procedure commonly referred to as non-selective grazing (NSG) (Acocks 1966). Here the veld is intensively camped (12 paddocks per group of animals) and large numbers of animals are grazed in relatively small paddocks in order to force uniform utilization of all the vegetation within a paddock. Even such unpalatable species as *Elionurus muticus* and *Diheteropogon filifolius* may be relatively well utilized under certain conditions (Bailey and Mappledoram 1983), but even with high grazing pressures it is not usually possible to achieve utilization of such species as *Aristida junciformis*.

Intra-species selection may, however, also occur, especially where grazing pressures are light. Here animals may concentrate their grazing on the young regrowth of previously grazed plants, and selectively avoid plants of the same species which have grown out to maturity.

In many vegetation types, programmes designed to force uniform utilization of a sward require occasional heavy, non-selective defoliation in order to remove low-quality and typically unacceptable foliage which accumulates not only during extended rest periods, but during the grazing phase as well. Fire is widely used to achieve such non-selective defoliation, and is normally followed by a period during which animals graze relatively non-selectively. After a fire, therefore, relatively uniform grazing can usually be achieved in swards of high species richness or community diversity, provided the stocking pressure is high. In time, however, selection patterns emerge, so it is necessary to use fire at regular intervals to recondition the sward.

MANAGEMENT DESIGNED TO REDUCE THE UNDESIRABLE CONSEQUENCES OF SELECTIVE GRAZING

In recognition of the tendency animals have of repeatedly regrazing the same plants or the same patches of vegetation, rotational grazing systems are designed to ensure that plants, having been grazed once, are permitted to recover a reasonable leaf area before they are again exposed to grazing. This period, commonly termed the period of absence, varies according to the expected rate of recovery of the grazed plants in any situation. However, where grazing periods (periods of occupation) within a paddock are relatively long, animals may repeatedly regraze plants within such grazing periods. To counteract this, short duration grazing (SDG) is often advocated (Savory 1967). The objective of SDG is to ensure that animals are removed from a paddock before recently grazed plants have regrown sufficiently to be regrazed.

The influence of selective grazing on the species composition of a sward is determined not only by the frequency with which favoured plants are grazed, but also by the intensity at which they are grazed. It is therefore now widely recommended that animals be removed from any paddock before the most favoured (key) species have been utilized intensively. This situation has given rise to a set of recommendations embodied in the system of controlled selective grazing (CSG) (Pienaar 1968).

In spite of all attempts to discourage selective grazing, it is normally inevitable. The consequence of this is, in time, a range in vigour both spatially within paddocks, and between different species. This further encourages selective grazing, since animals often tend to graze the less vigorous and therefore less mature plants from within a sward. A common feature of most grazing systems, therefore, is an occasional extended rest period designed to promote a recovery in the vigour of plants weakened by repeated grazing (Booysen 1966; Edwards 1981).

MANAGEMENT DESIGNED TO ENCOURAGE SELECTIVE GRAZING IN ORDER TO MODIFY SPECIES COMPOSITION

Even unpalatable perennial species in southern African vegetation require occasional defoliation if they are to remain vigorous and competitive (this is apparently an evolutionary adaptation to the long history of fire in this vegetation (Edwards 1984)). If defoliation does not take place, most plants become moribund and eventually die. To make use of this requirement for occasional defoliation to modify the species composition of a sward, the CSG approach to management also specifies that animals should be removed from a paddock before they have commenced grazing the unpalatable plants in the sward. This, associated with only moderate use of the favoured plants, provides a competitive advantage to the key forage-producing species.

THE ROLE OF FIRE IN MANAGEMENT

In agricultural practice where the emphasis of management is on the production of forage for domestic livestock, fire is frequently used to favour key forage-producing species which are typically well adapted to such intense defoliation. Fire is confined to the late dormant period (August and September in the summer rainfall areas); as has frequently been shown, fire at other times during the season often leads to a replacement of these species by other species less useful to livestock (Tainton 1981c; Tainton and Mentis 1984; Zacharias and Tainton 1988).

IMPROVED SEASONAL AVAILABILITY OF FORAGE TO LIVESTOCK

Where different components of the sward exhibit distinctly different growth patterns, increased diversity may assist in improving the year-round grazing capacity of veld. Examples of where this occurs have been cited in earlier discussions.

When this does occur, management needs to take into account the potential advantages to livestock feeding that then become available. This is normally achieved by deliberate rationing of the fodder produced by different species so that each component is reserved for feeding at the most appropriate time. For example, semi-arid savanna dominated by late-season species (eg *Digitaria eriantha*) may be deliberately rested out in the spring and summer so as to provide the greatest possible amount of forage for the late season. In estimating the appropriate stocking rates, the excessive grazing pressure that may be experienced by the preferred species should be borne in mind. Indeed, the forage produced by these species alone should be taken into account in estimating appropriate stocking rates. Much of the material produced by the sward will then not contribute to livestock production.

Management of areas of high habitat or landscape diversity

Where such diversity gives rise to a mosaic of communities with different forage characteristics and management requirements, it becomes necessary for these communities to be separated into individual camps. This permits the design of management programmes that can both make the most effective use of the forage produced by each community, and meet the specific management needs of each community (eg Roux 1968). As a result of this, livestock production should benefit, both immediately and in the long term.

CONCLUSIONS

Species diversity plays a role in livestock production off veld by virtue of the variable forage-producing characteristics and management requirements of different species. It appears that the greater the diversity, the greater the likely variability in the above two characteristics among the plants represented in a community, and the more complex the management required to achieve optimum levels of utilization. However, diversity is but one of the factors that influence livestock production off veld. Nonetheless, it has a major influence on management decisions, which in turn have a major influence on diversity. In essence, though, management is seldom if ever designed with the specific objective of increasing diversity. What management aims to do is promote the maintenance or development of a sward composed primarily of good forage-producing species. Whether such swards are made up of only one species or of a number of good forage-producing species is of no concern to the pastoralist. Such an attitude has important consequences if, as McNaughton (1985) has argued, both stability and resilience are positively related to species diversity, whether this is reflected in species richness or equitability. It is important to note, however, that the advantages that high plant species diversity imparts to the mixed game system described by McNaughton do not necessarily apply to a livestock system based on only one or two types of animal. Because of the relative simplicity of the system, it is possible to identify those species that make a major contribution to livestock production. Not surprisingly, it is only these species that are of concern to the pastoralist. Given the choice, most pastoralists would no doubt prefer to mimic the system commonly adopted by those who develop pastoral systems based on cultivated pastures. Pastures are commonly sown in monospecific stands which are more easily managed than mixed stands. Diversity, with its attendant influence on seasonal forage supply, is achieved across, rather than within, pastures by the planting of a number of species, each in a pure stand.

In essence, then, species diversity is not seen by the pastoralist as an important characteristic in vegetation used for livestock production. In view of the large proportion of southern Africa's land surface devoted to pastoral production, this must cause some concern for the future stability and resilience of much of the southern African vegetation.

REFERENCES

ACOCKS J P H (1966). Nonselective grazing as a means of veld reclamation. *Proceedings of the Grassland Society of South Africa* **1**, 33 – 39.

ACOCKS J P H (1975). Veld Types of South Africa. *Memoirs of the Botanical Survey of South Africa* **40**.

AUCAMP A J, DANCKWERTS J E and VENTER J J (1984). The production potential of an *Acacia karroo* community utilised by cattle and goats. *Journal of the Grassland Society of Southern Africa* **1(1)**, 29 – 32.

BAILEY A W and MAPPELDORAM B D (1983). Effects of spring grazing on yield of three grasses in the Highland Sourveld of Natal. *Proceedings of the Grassland Society of Southern Africa* **18**, 95 – 100.

BARNES D L, RETHMAN N F G, BEUKES B H and KOTZE G D (1984).Veld condition in relation to grazing capacity. *Journal of the Grassland Society of Southern Africa* **1(1)**, 16 – 19.

BOOYSEN P de V (1966). A physiological approach to pasture utilization. *Proceedings of the Grassland Society of Southern Africa* **1**, 77 – 85.

BOOYSEN P de V and TAINTON N M (eds)(1984). *Ecological effects of fire in South African Ecosystems.* Springer Berlin.

BOTHA P, BLOM C D, SYKES E and BAARENHOORN A S J (1983). A comparison between the diets of small and large stock on mixed karoo veld. *Proceedings of the Grassland Society of Southern Africa* **18**, 101 – 105.

BRANSBY D I (1981). Forage quality. In *Veld and pasture management in South Africa.* (ed Tainton N M) Shuter and Shooter and Natal University Press, Pietermaritzburg.

BREDON R M and STEWART P G (1978). *Feeding and management of dairy cattle in Natal,* Part 1 (Revised ed) Natal Agricultural Research Institute, Department of Agricultural Technical Services, Pietermartizburg.

CAMPBELL Q P, EBERSOHN J P and VON BROEMBSEN H H (1962). Browsing by goats and its effects on the vegetation. *Herbage Abstracts* **32**, 273 – 275.

COOK C W, HARRIS C F and YOUNG M C (1967). Botanical and nutritive content of diets of cattle and sheep under single and common use on mountain range. *Journal of Animal Science* **26**, 1169.

COWLING R M (1986). A description of the karoo biome project. *South African National Scientific Programmes Report* **27**, CSIR, Pretoria.

CRAWLEY M J (1983). *Herbivory: the dynamics of animal-plant interactions*, Blackwell Scientific Publications, Oxford.

DANCKWERTS J E (1979). Recession of the grassland formation in South Africa. Unpublished report. Dohne Agricultural Research Station, P/bag X15, Stutterheim, South Africa.

DE KLERK C H (1986). 'n Ondersoek na faktore wat in die weg staan van die aanvaardiging van aanbevele veldbeheerpraktyke. Unpublished report. Department of Agriculture and Water Affairs, Pretoria.

DONALDSON C H (1969). *Bush encroachment with special reference to the blackthorn problem of the Molopo area.* Government Printer, Pretoria.

DUDZINSKI M L and ARNOLD C W (1973). Comparisons of diets of sheep and cattle grazing together on sown pastures on the Southern Tablelands of New South Wales by principal components analysis. *Australian Journal of Agricultural Research* **24**, 899 – 912.

DU TOIT P F (1972). The goat in the bush-grass community. *Proceedings of the Grassland Society of Southern Africa* **7**, 44 – 50.

DU TOIT P J, LOUW J G and MALAN A I (1940). A study of the mineral content and feeding value of natural pastures in the Union of South Africa (final report). *Onderstepoort Journal of Veterinary Science and Animal Industry* **14**, 123 – 327.

EVERSON C S (1985). Ecological effects of fire in the montane grasslands of Natal. PhD thesis. University of Natal, Pietermaritzburg.

EDWARDS D (1984). Fire regimes in the Biomes of South Africa. In *Ecological effects of fire in South African Ecosystems.* (eds Booysen P de V and Tainton N M) Springer, Berlin.

EDWARDS P J (1981). Application of grassland management principles — sour grassveld. In *Veld and pasture management in South Africa.* (ed Tainton N M) Shuter and Shooter and Natal University Press, Pietermaritzburg.

GIBBS RUSSELL G E (1987). Preliminary floristic analysis of the major biomes in southern Africa. *Bothalia* **17**, 213 – 227.

GIBBS RUSSELL G E and SPIES J J (1988). Variation in important South African pasture grass. I. Morphological and geographical variations. *Journal of the Grassland Society of Southern Africa* **5(1)**, 15 – 21.

HARDY M B (1986). Grazing dynamics in Highland Sourveld. MSc Agriculture thesis. University of Natal, Pietermaritzburg.

HART R H (1978). Stocking rate theory and its application to grazing on Rangelands. *Proceedings of the International Rangelands Congress* **1**, 547 – 554.

HOFFMAN R R (1973). The ruminant stomach. East African Monographs *Biology* **2(1)**, 1 – 354.

HOWELL D (1976). Observations of the role of grazing in revegetating problem patches of veld. *Proceedings of the Grassland Society of Southern Africa* **11**, 59 – 64.

HOWELL L N (1976). The development of a multi-camp system for a farm in the southern Orange Free State. *Proceedings of the Grassland Society of Southern Africa* **11**, 53 – 59.

HUNTLEY B J (1984). Characteristics of South African biomes. In *Ecological Effects of Fire in South African Ecosystems.* (eds Booysen P de V and Tainton N M) Springer, Berlin. pp 2 – 17.

JONES R I and ARNOTT J K (1977). *Natal farming guide: Section F. Fodder program planning.* Natal Agricultural Research Institute, Department of Agricultural Technical Services, Pietermaritzburg.

JONES R J and SANDLAND R L (1974). The relation between animal gain and stocking rate: derivation of the relation from results of grazing trials. *Journal of Agricultural Science, Cambridge* **83**, 335 – 342.

KELLY R D, SCHWIN W F and BARNES D L (1978). *An exploratory study of the production potential of goats in the lowveld gneiss woodland.* Annual report 1977/78, Division of livestock and pasture management. Government Printer, Salisbury, Rhodesia.

KIRKMAN K P (1988). Factors affecting the seasonal variation of forage quality in summer rainfall areas of South Africa. MSc Agriculture thesis. University of Natal, Pietermartizburg.

LANGLANDS J P and SANSON J (1976). Factors affecting nutritive value of the diet and the composition of rumen fluid of grazing sheep and cattle. *Australian Journal of Agricultural Research* **27**, 691 – 707.

LYLE A D, BREDON R M, SWARD C and MILLER P M (undated). *Final report on project no-Ko 5/2.* Department of Agricultural Technical Services, Pretoria.

MCNAUGHTON S J (1985). Ecology of a grazing ecosystem: the Serengeti. *Ecological Monographs* **53(3)**, 259 – 294.

MENTIS M T (1981). The animal as a factor in pasture and veld management. In *Veld and pasture management in South Africa.* (ed Tainton N M) Shuter and Shooter and Natal University Press, Pietermaritzburg.

MENTIS M T (1982). A simulation of the grazing of sour grassveld. PhD thesis. University of Natal, Pietermaritzburg.

MINSON D J (1982). Effects of chemical and physical composition of herbage upon intake. In *Nutritional limits to animal production from pastures.* (ed Hacker J B) Commonwealth Agricultural Bureau. pp 167 – 182.

OPPERMAN D P C, ROBERTS B R and NEL L O (1974). *Elyonurus argentius* Nees — a review. *Proceedings of the Grassland Society of Southern Africa* **9**, 123 – 131.

OWEN-SMITH N and COOPER S M (1988). Plant herbivore relations. *Journal of the Grassland Society of Southern Africa* **5**, 72 – 75.

PIENAAR A J (1968). Beheerde selektiewe beweiding. *Landbou-weekblad,* 40 – 41.

POLE-EVANS I B (1920). The veld: its resources and dangers. *South African Journal of Science* **17**, 1 – 34.

ROBERTS B R (1981). Application of grassland management principles — sweet and mixed grassveld. In *Veld and pasture management in South Africa.* (ed Tainton N M) Shuter and Shooter and Natal University Press, Pietermaritzburg.

ROUX P W (1966). Die uitwerking van seisoenreenval en beweiding op gemengde karooveld. *Proceedings of the Grassland Society of Southern Africa* **1**, 103 – 110.

ROUX P W (1968). Principles of veld management in the Karoo and the adjacent dry sweet grassveld. In *The small stock industry in South Africa.* Department of Agricultural Technical Services, Pretoria. 318 – 340.

ROUX P W and DE KOCK G C (1971). Benutting van *Kochia brevifolia* deur skape. *Proceedings of the Grassland Society of Southern Africa* **6**, 108 – 117.

ROUX P W and SKINNER T E (undated). The group-camp system. *Farming in South Africa.* Agricultural Research Institute, Karoo Region.

ROUX P W and VORSTER M (1983). Vegetation change in the Karoo. *Proceedings of the Grassland Society of Southern Africa* **18**, 25 – 29.

SAVORY C A R (1967). High intensity short duration grazing. *Rhodesian Farmer,* 6 – 8.

SCHOEMAN N S C (1968). Studies on the measurement of the digestibility of forage celluloses. MSc Agriculture thesis, University of Natal, Pietermaritzburg

SCOTT J D (1947). Veld management in South Africa. *Department of Agriculture Bulletin* **278**, Government Printer, Pretoria.

SNYMAN H A and VAN RENSBURG W L J (1986). Effect of slope and plant cover on run-off, soil loss and water use efficiency of natural veld. *Journal of the Grassland Society of Southern Africa* **3(4)**, 117 – 121.

SPIES J J and GIBBS RUSSELL G E (1988). Variation in important South African pasture grasses. II. Cytogenetic and reproductive variation. *Journal of the Grassland Society of Southern Africa* **5(1)**, 22 – 25.

STORY R (1952). A botanical survey of the Keiskamahoek district. *Memoirs of the Botanical Survey of Southern Africa* **27**.

STUART-HILL G C (1987). Refinement of a model describing forage production, animal production and profitability as a function of bush density in the Flase Thornveld of the Eastern Cape. *Journal of the Grassland Society of Southern Africa* **4(1)**, 18 – 24.

STUART-HILL G C and MENTIS M T (1982). Coevolution of African grasses and large herbivores. *Proceedings of the Grassland Society of Southern Africa* **17**, 122 – 128.

TAINTON N M (1981a). The ecology of the main grazing lands of South Africa. In *Veld and pasture management in South Africa.* (ed Tainton N M) Shuter and Shooter and Natal University Press, Pietermaritzburg. pp 25 – 56

TAINTON N M (1981b). The grass plant and its reaction to treatment. In *Veld and pasture management in South Africa.* Shuter and Shooter and Natal University Press, Pietermaritzburg. pp 215 – 238.

TAINTON N M (1981c). Veld burning. In *Veld and pasture management in South Africa.* (ed Tainton N M) Shuter and Shooter and Natal University Press, Pietermaritzburg. pp 363 – 381.

TAINTON N M, EDWARDS P J and MENTIS M T (1980). A revised method for assessing veld condition. *Proceedings of the Grassland Society of Southern Africa* **15**, 37 – 42.

TAINTON N M and MENTIS M T (1984). Fire in grassland In Ecological effects of fire in South African ecosystems (eds Booysen P de V and Tainton N M) *Ecological Studies* **48**. Springer, Berlin.

THERON E P and BOOYSEN P de V (1966). Palatability in grasses. *Proceedings of the Grassland Society of Southern Africa* **1**, 111 – 120.

TIDMARSH C E M (1957). Veld management in Karoo and adjacent grassveld regions. *Handbook for farmers in South Africa III.* Government Printer, Pretoria 624 – 635.

TILLEY J M A and TERRY R A (1963). A two-stage technique for the in vitro digestion of forage crops. *Journal of the British Grassland Society* **18**, 104 – 111.

TROLLOPE W S W (1970) A consideration of macchia (fynbos) encroachment in South Africa and an investigation into methods of eradication in the Amatola Mountains. MSc Agriculture thesis, University of Natal, Pietermartizburg.

TROLLOPE W S W (1974). Role of fire in preventing bush encroachment in the Eastern Cape. *Proceedings of the Grassland Society of Southern Africa* **9**, 67 – 72.

TROLLOPE W S W (1981). Application of grassland management principles — savanna. In *Veld and pasture management in South Africa.* (ed Tainton N M) Shuter and Shooter and Natal University Press, Pietermaritzburg.

TROLLOPE W S W (1983). Control of bush encroachment with fire in the arid savannas of south-eastern Africa. PhD thesis, University of Natal, Pietermaritzburg.

TROLLOPE W S W (1984) Fire in Savanna. In *Ecological effects of fire in South African Ecosystems.* (eds Booysen P de V and Tainton N M) Springer, Berlin.

TYSON P D (1986). *Climatic change and variability in Southern Africa.* Oxford University Press, Cape Town.

VAN DYNE G M and HEADY H H (1965). Dietary chemical composition of cattle and sheep grazing in common on a dry annual range. *Journal of Range Management* **18(2)**, 78 – 86.

VAN SOEST P J (1965). Voluntary intake in relation to chemical composition and digestibility. *Journal of Animal Science* **24**, 834 – 843.

VAN SOEST P J (1983). *Nutritional ecology of the ruminant.* O and B Books, Corvallis, Oreg.

VORSTER M (1982). The development of the ecological index method for assessing veld condition in the karoo. *Proceedings of the Grassland Society of Southern Africa* **17**, 84 – 89.

VORSTER M and ROUX P W (1983). Veld of the Karoo areas. *Proceedings of the Grassland Society of Southern Africa* **18**, 18 – 24.

WALKER B H (1980a). A review of browse and its role in livestock production in Southern Africa. In *Browse in Africa; the current state of knowledge.* (ed Le Houerou H N) ILCA, Addis Ababa, Ethiopia. pp 7 – 24.

WALKER B H (1980b). Stable production versus resilience: a grazing management conflict? *Proceedings of the Grassland Society of Southern Africa* **15**, 79 – 84.

WALKER B H, MATTHEWS D A and DYE P J (1986). Management of grazing systems - existing versus and event-orientated approach. *South African Journal of Science* **82**, 172

WILSON P H (1957). Studies on the browsing and reproductive behavious of the East African dwarf goat. *East African Agricultural Journal* **23**, 138 – 147.

ZACHARIAS P J K (1986). The use of the cellulase digestion procedure for indexing the dry matter digestibility of forages. *Journal of the Grassland Society of Southern Africa* **3**, 117 – 121.

ZACHARIAS P J K (1988). The factors affecting the seasonal variation of *Themeda triandra* in various ecological zones in Natal (In preparation). Department of Grassland Science, University of Natal, Pietermaritzburg.

ZACHARIAS P J K and TAINTON N M (1988). *A baseline survey of the grasslands of Royal Natal National Park.* Department of Grassland Science, University of Natal, Pietermaritzburg. 50 pp.

ZEEMAN P J L, MARIAIS P G and COETZEE M J (1983). Nutrient selection by cattle, goats and sheep on natural karoo pasture. 1. Digestibility of organic matter. *South African Journal of Animal Science* **13(4)**, 236.

ZEEMAN P J L, MARIAIS P G and COETZEE M J (1984). Nutrient selection by cattle goats and sheep on natural karoo pasture. 2. Nitrogen. *South African Journal of Animal Science* **14(4)**, 169 – 72.

CHAPTER 7

Molecular mechanisms of diversity and horizontal genetics, genetic engineering and biotechnology

D R Woods

INTRODUCTION

A conference on conserving biotic diversity at the end of the twentieth century would be incomplete without a paper on molecular mechanisms of diversity and gene-transfer systems. Developments in molecular genetics during the last few years have enhanced and modified our understanding of these basic genetic systems. For example, the genetic material, DNA (deoxyribonucleic acid), was considered to be a very stable molecule, and mutation provided the small changes on which recombination and selection could act. The discovery of transposition elements (jumping or selfish genes) has altered our ideas regarding the stability of DNA and ways in which relatively large changes can be brought about.

The concept of vertical genetics, and the transfer of genetic material from parents to offspring, is well known. However, the less well-known concept of horizontal genetics, which involves gene transfer between related (bacteria) or totally unrelated (bacteria and plants) organisms, is an important natural phenomenon. The extent of horizontal genetic systems among organisms is unknown at present, but it may be far more extensive than indicated by the few examples that have been discovered to date.

Gene cloning, coupled with gene-transfer systems, enables scientists to manipulate organisms genetically and carry out horizontal genetics. This obviously has implications for biotechnology which will affect agriculture, industry and medicine. However, recombinant-DNA technology and genetic engineering will also have an impact on assessing, preserving and enhancing biotic diversity. This paper gives an overview of the molecular mechanisms and impact of horizontal genetics, genetic engineering and biotechnology.

GENE STRUCTURE AND FUNCTION

Structure of DNA

Although the genetic material of viruses may be either DNA or RNA (ribonucleic acid), the genetic material of prokaryotes and eukaryotes is DNA (Figure 7.1). DNA is a polymer containing the

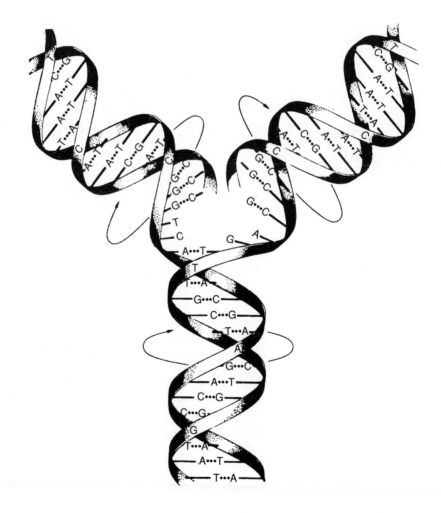

FIGURE 7.1 Structure and replication of the DNA double helix (2-DNA). The two chains are joined together by hydrogen bonds between pairs of bases. Adenine (A) is always paired with thymine (T), and guanine (G) with cytosine (C). Specific base pairing ensures that two identical 2-DNA molecules are formed from the parent 2-DNA molecule.

genetic code of four letters or bases (Freifelder 1987; Watson et al 1987). It is the sequence of these four bases arranged in triplets that forms the genetic code.

It is reasonable and convenient to consider the average size of a gene coding for a protein (structural gene) to be approximately 1 000 base pairs (bp), specifying a protein of approximately 330 amino acids. Using these approximations, bacteria, fungi, insects, frogs, birds and mammals have the maximum theoretical coding capacity for approximately 4×10^3, 6×10^4, 8×10^4, 2×10^7, 1×10^6, and $2,7 \times 10^6$ genes respectively (Herskowitz 1973). A much greater proportion of the bacterial than the higher eukaryotic DNA is used directly for coding functional proteins. The DNA of higher eukaryotes contains non-coding or junk DNA. The frog in particular contains an enormous excess of what is apparently junk DNA.

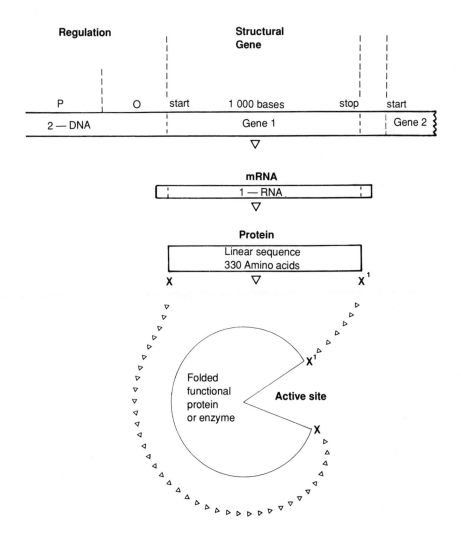

FIGURE 7.2   Arrangement of genes within the 2-DNA chromosome and the production of proteins. The regulatory upstream region contains the promoter (p) and operator (o) sites. The structural gene is transcribed into a messenger RNA (mRNA) carrier molecule which takes part in the production of the protein. The protein is initially formed as a linear sequence of amino acids which then folds to form its active configuration.

Gene structure and regulation

In prokaryotes, structural genes are usually arranged in groups or functional units called operons which contain all the genes specifying the enzymes of a particular biochemical pathway. Since it would be very inefficient in terms of energy for all genes of an organism to be functioning or expressed when not required, gene expression is extremely well controlled by regulatory regions or genes upstream of the structural genes of an operon (Miller and Reznikoff 1978) (Figure 7.2). Immediately upstream of the structural genes are the promoter (p) and operator (o) sites. The p site regulates the efficiency of structural gene expression, and the o site is the on/off switch. Distal regulator genes produce regulatory proteins which interact with the p and o sites (Ptashne 1986).

Although the structural genes for common proteins of prokaryotes and eukaryotes often show close homology, there are specific differences in the upstream regulatory sites of prokaryotes and eukaryotes. Therefore, in transferring genes between bacteria and animals or plants it is important that the structural genes be preceded by the correct regulatory region for gene expression in the particular host. Genes without an upstream regulatory region are not expressed and remain silent. Linking a regulatory region to a silent gene results in its expression. Therefore, in assessing genetic diversity it is important to consider not only changes or movements of structural genes, but also the impact of changes or movements in regulatory regions. This is emphasized by the ability of regulatory regions to affect the expression of more than one gene, whereas changes in a structural gene affect only its own gene product.

Split genes

Another important general difference between the structure of prokaryote and eukaryote genes encoding proteins is that prokaryote genes are continuous. However, many eukaryote and archaebacteria genes are interrupted or split by intervening sequences (introns), and the coding regions of the expressed gene (exons) are scattered among these non-coding introns (Chambon 1981). The introns and exons are initially transcribed into an intact precursor messenger RNA (mRNA) molecule, which is processed by spligases to cut out the introns and join the exons to produce the mature mRNA. The regions at the borders of introns and exons are important, since they are accurately recognized and spliced by the spligases (Zaug et al 1986). Mutations in these regions or in the spligases could result in gross changes in the mature mRNA and the protein product.

Chromosomal, plasmid and viral DNA

The structure of the bacterial chromosome has been elucidated. It is a single, double-stranded (2-DNA), circular DNA molecule of approximately 4×10^6 bp. In addition to the chromosome, bacterial cells usually contain supernumerary self-replicating DNA molecules known as plasmids, which can vary in size from small (approximately 3 000 bp) to large (approximately 5×10^5 bp) (Figure 7.3). The majority of plasmids are circular 2-DNA molecules, but large and small linear 2-DNA plasmids have recently been isolated (Barbour and Garon 1987).

Although plasmids are supernumerary and are not essential for bacterial growth or viability, they may encode important properties, for example, antibiotic resistance. Plasmids replicate independently of the chromosome and are present either as low copy number plasmids (2 to 5 per cell) or high copy number plasmids (30 to 50 per cell). Sophisticated systems have evolved to control copy number and partition in order to ensure that both daughter cells inherit the plasmid after cell division. Nevertheless, a fundamental property of plasmids and cytoplasmic genes is that they segregate at a frequency higher than the mutation rate. Segregation of plasmid genes obviously affects genetic diversity. Neither plasmids nor cytoplasmic DNA are confined to bacteria but are present in plant and animal cells, either in organelles or free in the cytoplasm.

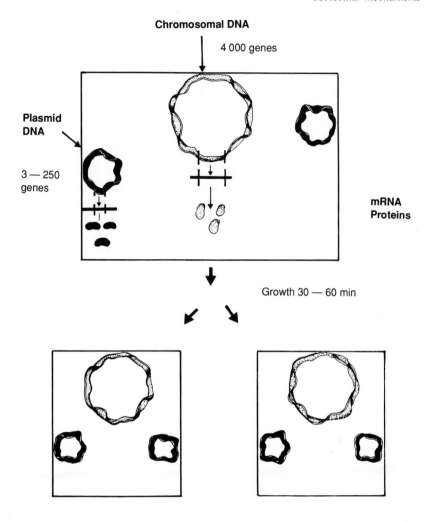

Chromosomal DNA

4 000 genes

**Plasmid
DNA**

3 — 250
genes

**mRNA
Proteins**

Growth 30 — 60 min

FIGURE 7.3   Simplified diagram of a bacterial cell containing the single circular 2-DNA chromosome and a low copy number 2-DNA circular plasmid

MOLECULAR MECHANISMS OF DIVERSITY

Mutation

Point mutations are spontaneous random events, unrelated to adaptation, which cause changes in the base sequence of DNA. An important point about mutation is that it causes changes in the genetic complement already present in the cell; it does not involve the acquisition of new genetic material. Mutations are caused by chemical or physical agents which directly affect or interact with the DNA molecule.

Homologous recombination

Genetic variation can also be brought about by homologous recombination. The process is characterized by pairing between relatively long (>25 bp) homologous regions of DNA, and reciprocal exchange of DNA.

Mutation and homologous recombination were considered to be the only important mechanisms for generating biological diversity. They seemed to account for the amount of diversity observed. The constraints of homologous recombination, which prevent the exchange of genetic information between unrelated organisms lacking extensive DNA sequence homology, appeared to be consistent with both a modest rate of biological evolution and the persistence of distinct species that retain their basic identity generation after generation.

Site-specific recombination

In contrast to homologous recombination, site-specific recombination involves relatively short sequences of homology (12 to 25 bp), on the plasmid and chromosome, for example. It takes place via a single cross-over event, and this requires that the recombined or integrated DNA sequence be a circular molecule prior to insertion (eg a plasmid). Site specific recombination normally occurs between chromosomal and plasmid or viral DNA so that the plasmid DNA, for instance, is incorporated into the chromosome.

The integration of plasmid or viral DNA into chromosomal DNA via site-specific recombination can affect genetic diversity in the following ways: by the addition of genes to the chromosomal complement; by the inactivation or disruption of a gene or genes (depending on the site of integration); and by regulating neighbouring genes. Site-specific rearrangements tend to be limited and conservative, and in higher organisms they are not often inherited because they occur only in somatic cells.

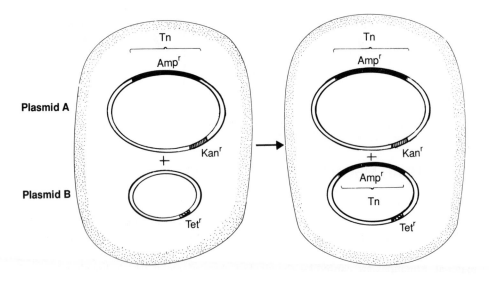

FIGURE 7.4   The transposition of a transposon (Tn) containing the gene for ampicillin resistance (Amp^r) from plasmid A, which also contains the gene for kanamycin resistance (Kan^r) to plasmid B, which contains the gene for tetracycline resistance (Tet^r). The plasmids A and B are contained within the same bacterial cell. Note that the transfer mechanism involves the replication of the Tn.

Transposition elements

The remarkable discovery that genomes of organisms were not as stable as originally thought was made by Barbara McClintock (1956) during her study of the rapid, random and renewable pigment changes of maize kernels (Frederoff 1984). She proposed the existence of 'controlling elements' which were responsible for the high frequency of genetic variation. These controlling elements seemed to be mobile and did not remain fixed at a particular site on the chromosome. As with many genetic phenomena, the molecular basis of controlling elements was investigated only when they were discovered some fifteen years later in bacteria and viruses (Figure 7.4) (Cohen and Shapiro 1980; Shapiro 1983).

McClintock's controlling elements are now known as transposition or transposable genetic elements (transposons) and are able to control their own movements. Transposition can also result in genetic rearrangements and deletions. Stress seems to induce transposons to become mobile and escape from apparently doomed genomes or cells. It is this property of survival that resulted in the name selfish DNA.

An important property of transposable elements which affects genetic diversity is that they often contain potent regulatory regions near their extremities. After insertion into a resident chromosome, these powerful regulatory regions will lie adjacent to resident genes, and their expression can be markedly affected. Examples of the regulation of resident genes and major alterations of appearance involve antigenic variation and expression of disease symptoms by a Spirochaete bacterium, *Borrelia* (Plasterk et al 1985), and a Trypanosome, *Trichomonas* (Donelson and Rice-Ficht 1985), which cause relapsing fever and sleeping sickness, respectively. These diseases are cyclic in nature and very difficult to cure, since the movement of the transposition elements results in the expression of new surface antigenic proteins. These render the organism resistant to the host immune defence system, and the disease breaks out again (Figure 7.5).

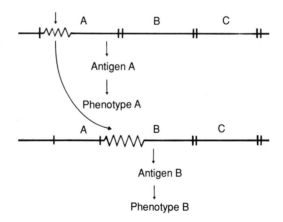

FIGURE 7.5 Diagram illustrating the regulation of antigenic variation by a transposable element containing a positive regulatory region. This is able to control after insertion the expression of individual antigenic structural genes (A – C) lacking their own regulatory regions. An analogous system involves a stationary regulatory region and transposable antigenic genes which can transpose themselves and insert next to the regulatory region where they will be expressed.

HORIZONTAL GENETICS

Bacterial plasmid and gene-transfer systems

Molecular genetics advanced rapidly after the discovery in 1946 by Lederburg and Tatum that bacteria were able to transfer plasmid and chromosomal DNA between adult cells (horizontal genetics) by a process termed conjugation. The advent and use of antibiotics after World War II resulted in the discovery of self-transmissible antibiotic resistance plasmids (R plasmids) which were capable of intraspecific and intergeneric transfer. The importance of R-plasmid transfer is well recognized by microbiologists and those in the medical field.

Transfer of genes from bacteria to plants

A significant landmark in molecular genetics was the discovery that Crown Gall tumours in plants are caused by an approximately 20-kb DNA region (T-region) of a large (200 kb) tumour-inducing (Ti) plasmid of the bacterium *Agrobacterium tumefaciens*. The bacterium attaches itself to the outside of plant cells exposed by wounding and, by a mechanism resembling a combination of transposition and conjugation, the T-region is excised from the Ti plasmid and transferred to the plant nucleus where it recombines with a chromosome and proceeds to cause cancer (Figure 7.6). The T-region carries genes coding for plant growth hormones which, after incorporation into the plant nucleus, causes unchecked cell division, in other words, a tumour.

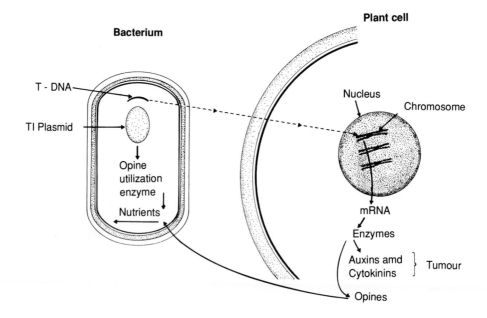

FIGURE 7.6   Transfer of tumour-inducing genes from the bacterium *A tumefaciens* to plant chromosomes which results in the over-production of plant growth hormones (auxins and cytokinins) and the formation of a tumour (Crown Gall). At the same time the transferred DNA produces opines which the plant cannot utilize. These opines are taken up by the bacterium which has a resident gene enabling it to utilize opines as its sole source of nutrients.

The Ti-T-DNA system is not a chance or haphazard system, but an example of a highly evolved type of parasitism at the DNA level. The plant growth hormone genes are not the only genes carried by the T-DNA; it also contains the genes for the synthesis of opines (amino acid analogues of arginine) which cannot be utilized by plants. Situated on a part of the Ti plasmid which is not transferred to plants is an opine oxidase gene which enables *A tumefaciens* to utilize opines as a sole source of carbon and nitrogen. Vigorous tumour growth fosters opine production which can be utilized only by the bacterium. The discovery of the Ti plasmid has not escaped the attention of biotechnologists, and the system has been exploited to enable the transfer of genes to plants (Chilton 1983; Goodman et al 1987; Vaeck et al 1987). Deletion of the plant growth hormone genes eliminates the development of tumours when plants are genetically engineered via the modified Ti and T-DNA systems. These natural and manipulated horizontal genetic systems between bacteria and plants have the potential of markedly enhancing genetic diversity.

A recent exciting discovery indicates that any transferable bacterial plasmid can facilitate the transfer of the T-DNA from *A tumefaciens* to plants (Buchanan-Wollaston et al 1987). Therefore the speculation that gene transfer from bacteria to eukaryotes (plants and animals) may be more general than the *A tumefaciens* system is not without foundation. There are numerous environments in which attachment occurs between bacteria and eukaryote cells (eg, intestine and rhizosphere), and many bacteria contain transferable plasmids. It is significant that two bacterial genera, *Rhizobium* and *Frankia*, which form root-nodules in plants, contain a glutamine synthetase enzyme, GSII, which seems to have originated in a plant and undergone gene transfer to the bacteria (Darrow and Knotts 1977; Edmands et al 1987). Furthermore, it is apparent that the gene for the enzyme superoxide dismutase in *Photobacterium leiognathi* was derived from bony fish, with which the bacterium has a close symbiotic relationship (Bannister and Parker 1985).

GENETIC ENGINEERING AND BIOTECHNOLOGY

Genetic engineering involves the isolation of specific genes, and their cloning on plasmid or viral vectors. This facilitates their amplification and transfer to a completely different organism via transformation, conjugation or transduction (Watson et al 1983). The discovery of a group of endonuclease enzymes which cut DNA to generate overlapping sticky ends (Figure 7.7), and enable the joining of any two DNA molecules cleaved with the same endonuclease, resulted in the birth of what is now known as genetic engineering. It is now possible to clone virtually any gene, and transfer and express it in bacterial, plant and animal cells (Figure 7.8). This achievement, and the ability of high copy number plasmids containing a particular gene to produce large amounts of the gene product, did not escape the attention of entrepreneurial industrialists, and in the mid-1970s the age of recombinant DNA biotechnology dawned.

Although the industrial, agricultural and medicinal potential of biotechnology, based on recombinant DNA technology, has been unrealistically boosted by the popular and scientific press, the practical realizations of a totally new technology in the space of only 15 years have been spectacular. To progress from a fundamental and novel discovery to the industrial production, authorization, sale and use of a medical product in a period of only 15 years is a remarkable achievement, and is an indication of the exciting prospects for industry, agriculture and medicine.

Initially, industries based on recombinant DNA technology focused on medical and high-value products. By the end of 1986 the following pharmaceutical products had been produced, licensed and marketed: human insulin (humulin), human growth hormone, (alpha)-interferon, and hepatitis-B vaccine. The markets for these products are relatively small compared with projected markets for pharmaceuticals such as tissue-type plasminogen activator (t-PA) or erythropoietin (EPO), which were awaiting approval by the USA Food and Drug Administration at the end of 1987.

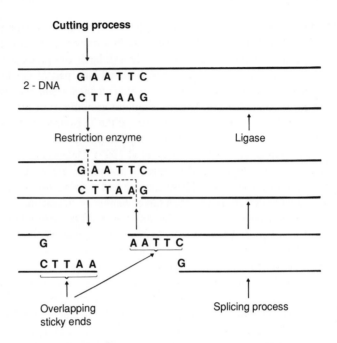

FIGURE 7.7   Cutting 2-DNA molecules with restriction endonucleases which recognize specific pallindromic base pair (bp) sequences, and splicing with ligase, which is able to join the DNA molecules after pairing of the overlapping sticky ends

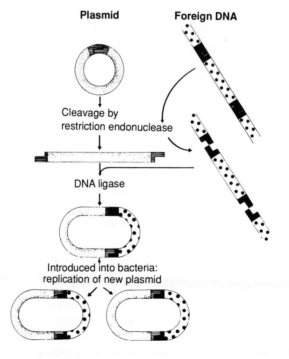

FIGURE 7.8   Cloning of foreign DNA into a bacterial plasmid vector, and its reintroduction into a bacterial cell (Source: Watson J D, Hopkins N H, Roberts J W, Steitz J A and Weiner A M. *Molecular Biology of the Gene, Volume 1.* © 1987. The Benjamin/Cummins Publishing Company.)

Many more products are on the way, and some of them will have annual sales amounting to hundreds of millions of dollars. Industrial companies have targeted nearly every major human disease, and are discovering proteins that have the potential to treat specific diseases.

Diagnostic products based on DNA probes (Klausner and Wilson 1983) and monoclonal antibodies are having a major impact on the diagnosis of disease and the identification of mutated genes responsible for genetic diseases and defects. The identification of these mutated genes and the cloning of their normal functional counterparts will eventually enable genetic engineering to be used in gene therapy. In preparation for the advent of gene therapy, national ethic committees are already considering the issues, and it is likely that within two to five years approved gene-therapy experiments will be carried out (Marx 1986; Robertson 1986).

In agriculture, genetic engineering and biotechnology have enormous potential. The cloning and transfer of growth hormone genes to animal embryos has been achieved (Palmiter et al 1982; Womack 1987). The transformed animals grew faster on the same amount of food as control animals, and their offspring also contained and expressed the additional growth hormone genes.

Insecticide and herbicide resistance have been transferred to plants, and the first field trials utilizing genetically engineered plants have been approved by the USA Environmental Protection Agency. Although gene transfer in plants at present involves mainly dicotyledonous plants, transfer to monocotyledons and cereal crops will be achieved in the near future. The goals of genetic engineering with regard to plants involve aspects such as nitrogen fixation and drought resistance. As recently as a few years ago these systems were considered too complex for manipulation by genetic engineering techniques. This illustrates the rate at which recombinant DNA technology and biotechnology have evolved.

The examples discussed so far involve natural genes which are being cloned, amplified and exploited. However, perhaps the most exciting area in the future of biotechnology is directed mutation and the planned alteration of genes and their products. Until the advent of recombinant DNA technology, mutagenesis was random and the isolation of improved products depended on efficient selection techniques. Directed mutagenesis is specific and allows the alteration of any base in a gene.

POTENTIAL AND APPLICATIONS OF GENETIC ENGINEERING AND BIOTECHNOLOGY FOR THE CONSERVATION OF BIOTIC DIVERSITY

Although the techniques enabling the genetic manipulation of cells and organisms are laboratory-based, they have important potential and applications for the conservation of the biotic diversity of populations. An application that has already been used successfully is the utilization of molecular biology techniques for assessing, monitoring and understanding genetic relationships (Klausner and Wilson 1983). Two interesting and contrasting examples are the discovery that almost indistinguishable animals, the northern and southern 'races' of white rhinoceros *Cerato-therium simum*, are genetically very distinct, whereas 'species' of *Leonthopithecus,* the lion tamarins, that were considered to be quite distinct, are genetically very similar (Forman et al 1986; Cherfas 1986). DNA-based techniques will contribute to molecular taxonomy and to understanding of the genetic and evolutionary relationships of organisms, which will have implications for breeding strategies. The techniques also provide precise methods of monitoring and tracing the offspring of breeding programmes. As well as contributing to the identification, characterization and preservation of endangered organisms, molecular genetics also enables the identification, monitoring and preservation of endangered genes.

Although conservationists tend to think in terms of schemes to preserve living endangered populations, an alternative and complementary approach is to store gene banks of endangered species. The entire genome of an organism can be stored deep-frozen at –70°C as a cosmid gene bank. This will not result in the resuscitation of a viable organism at a future date, but the genome and individual genes will be available for future characterization and distribution.

The plasmid storage of gene banks will eliminate the uncontrolled and risky reliance on random pieces of dried tissue. This was what was used in the characterization of the origins of the quagga, which has been extinct since 1883. Nevertheless, the power of recombinant DNA technology enabled the analysis of DNA sequences of the quagga *Equus quagga* which had been obtained from dried muscle and blood. The results showed that the DNA sequences were virtually identical to sequences from a living plains zebra *Equus burchelli*, and that the quagga was undoubtedly a sub-species of the plains zebra. A breeding programme involving selected plains zebra could therefore be initiated to re-create a quagga. This breeding programme has been initiated at Vrolijkheid under the auspices of the South African Museum in Cape Town. The aim of the programme is the retrieval of the quagga colour genes from the extant plains zebra population.

An alternative method of preserving genetic material involves tissue-culture techniques. Cultured cells of any species (plant or animal) are kept in liquid nitrogen long-term storage, and represent conservation of the entire intact genome in a living cell from which, it is conceivable, the plant (or animal) may be regenerated in the future.

As well as preserving the genetic material of endangered species as 'dead' gene banks, genetic engineering could also contribute to preserving living populations. For example, certain populations are threatened by disease. Resistance to many diseases is conferred by a single gene or a few genes, and the transfer of the resistant gene/s to the organism could be carried out by genetic engineering techniques. Subsequent breeding programmes and the release of the resistant organisms into the environment would enhance the spread of the resistant gene.

Recombinant DNA technology enables the transfer of novel and unrelated genes to organisms and genetic backgrounds, which would not have been possible via natural breeding mechanisms. This is an obvious method of enhancing genetic diversity and gene spread.

CONCLUSION

Although our understanding of molecular genetics has increased dramatically over the past four decades, the subject continues to be full of surprises and developments. Overlapping genetic codes, transposition elements, horizontal gene transfer between unrelated organisms, and nucleic acids as enzymes are some examples of the exciting discoveries that have been made in this field. Many of these discoveries are the result of recombinant DNA technology which is a very powerful scientific tool. This technology is being exploited successfully in industry, medicine and agriculture for the benefit of humanity. Perhaps the most exciting prospect for genetic engineering and directed mutagenesis is that it offers mankind the potential to change the world economy from one based on non-renewable resources to one based on renewable resources. This development will depend on the cloning and exploitation of various genes present in many different organisms and it enhances the importance of the preservation of genetic diversity. Molecular genetics and recombinant DNA technology are powerful new tools for assessing, monitoring, preserving and expanding genetic diversity.

REFERENCES

BANNISTER J V and PARKER M W (1985). The presence of a copper/zinc superoxide dismutase in the bacterium *Photobacterium leiofgnathi:* A likely case of gene transfer from eukaryotes to prokaryotes. *Proceedings of the National Academy of Science U S A* **82**, 149 – 152.
BARBOUR A G and GARON C F (1987). Linear plasmids of the bacterium *Borrelia burgdorferi* have covalently closed ends. *Science* **237**, 409 – 411.

BUCHANAN-WOLLASTON V, PASSIATORE J E and CANNON F (1987). The *mob* and *oriT* mobilization functions of a bacterial plasmid promote its transfer to plants. *Nature* **328**, 172 – 175.

CHAMBON P (1981). Split genes. *Scientific American* **244**, 48 – 59.

CHERFAS J (1986). Molecular biology is the key to the tamarin's survival. *New Scientist* **112**, 27.

CHILTON M D (1983) A vector for introducing new genes into plants. *Scientific American* **248**, 36 – 45.

COHEN S N and SHAPIRO J A (1980). Transposable genetic elements. *Scientific American* **242**, 36 – 45.

DARRROW R A and KNOTTS R R (1977). Two forms of glutamine synthetase in free-living root nodule bacteria. *Biochemical and Biophysical Research Communications* **78**, 554 – 559.

DONELSON J E and RICE-FICHT A C (1985). Molecular biology of Trypansome antigenic variation. *Microbiological Reviews* **49**, 107 – 125.

EDMANDS J, NORIDGE N A and BENSON D R (1987). The actinorhizal root-nodule symbiont *Frankia* sp. strain *CpII* has two glutamine synthetases. *Proceedings of the National Academy of Science U S A* **84**, 6126 – 6130.

FEDEROFF N V (1984). Transposable genetic elements in maize. *Scientific American* **250**, 65 – 74.

FORGMAN L, KLEIMAN D G, BUSH R M, DIETZ J M, BALLOU J D, PHILLIPS L G, COIMBRA-FILHO A F and O'BRIEN S J (1986). Genetic variation within and among lion tamarins. *American Journal of Physical Anthropology* **71**, 1 – 11.

FREIFELDER D 1987. *Microbial genetics.* Jones and Bartlett, Boston, Mass.

GOODMAN R M, HAUPTLI H, CROSSWAY A and KNAUF V C (1987). Gene transfer and crop improvement. *Science* **236**, 48 – 54.

HERSKOWITZ, I H (1973). *Principles of genetics.* MacMillan Company, New York, NY.

KLAUSNER A and WILSON T (1983). Gene detection technology opens doors for many industries. *BioTechnology* **1**, 471 – 479.

LEDERBERG J and TATUM E L (1946). Gene recombination in *Escherichia coli. Nature* **158**, 558.

MARX J L (1986). Gene therapy — so near yet so far away. *Science* **232**, 824 – 825.

MILLER J H and REZNIKOFF W S (1978). *The Operon.* Cold Spring Harbor Laboratory, Cold Spring Harbor, New York, NY.

PALMITER R D, BRINSTER R L, HAMMER R E, TRUMBAUER M E, ROSENFELD M G, BIRNBERG N C and EVANS R M (1982). Dramatic growth of mice that develop from eggs microinjected with metallothionein-growth hormone fusion genes. *Nature* **300**, 611 – 615.

PLASTERK R H A, SIMON M I and BARBOUR A G (1985). Transposition of structural genes to an expression sequence on a linear plasmid causes antigenic variation in the bacterium *Borrelia hermsii. Nature* **318**, 257 – 263.

PTASHNE M (1986). Gene regulation by proteins acting nearby and at a distance. *Nature* **322**, 697 – 701.

ROBERTSON M (1986). Gene therapy — desperate appliances. *Nature* **320**, 213 – 214.

SHAPIRO J A (1983). *Mobile genetic elements.* Academic Press, New York, NY.

VAECK M, REYNAERTS A, HOFTE H, JANSENS S, DE BEUCKELEER M, DEAN C, ZABEAU M, VAN MONTAGU M and LEEMANS J (1987). Transgenic plants protected from insect attack. *Nature* **328**, 33 – 37.

WATSON J D, HOPKINS N H, ROBERTS J W, STEITZ J A and WEINER A M (1987). *Molecular biology of the gene.* Benjamin Cummings Publishing Company Incorporated, Menlo Park, Ca.

WATSON J D, ROOZE J and KURTZ D T (1983). *Recombinant DNA: A short course.* Scientific American Incorporated, New York, NY.

WOMACK J E (1987). Genetic engineering in agriculture: Animal genetics and development. *Trends in Genetics* **3**, 65 – 68.

ZAUG A J, BEEN D M and CECH T R (1986). The *Tetrahymena* ribozyme acts like an RNA restriction endonuclease. *Nature* **324**, 429 – 434.

PART 3

The survey, evaluation and monitoring of biotic diversity

CHAPTER 8

The role of Red Data Books in conserving biodiversity

A A Ferrar

WHAT ARE RED DATA BOOKS?

In their most basic form Red Data Books (RDBs) are simply lists of more evident elements of our natural biota known to be rare or declining. In terms of biodiversity they set out to indicate existing or potential losses of species or populations and, where appropriate, to propose remedial conservation action. They also serve an important monitoring function with regard to time, setting out to answer the general question: 'How are we doing in conserving species?'

At present they answer this question rather inadequately and only at the species or population level of biodiversity. At the ecosystem or habitat level, and recently at the molecular genetic level (Higuchi et al 1984), changes in biodiversity are also being described. Lists of threatened habitats, protected areas, and taxa related to domesticated and economically valuable species all highlight trends of depletion and serve the same general purpose as RDBs, in both the scientific and popular literature (eg IUCN/UNEP 1987; Myers 1983). In a sense these lists will form the RDBs of the future at the ecosystem and genetic levels of the organizational hierarchy of biotic diversity.

Between 1976 and 1980 Ecosystem Programmes (CSIR) published the first series of Red Data Books for southern Africa, covering birds, mammals, reptiles, amphibians, fish and flowering plants (Siegfried et al 1976; Meester 1976; Skelton 1977; Skinner et al 1977; McLachlan 1978; Hall et al 1980). Their contents varied from the volume on plants, which was a plain check-list derived from existing and often inadequate data, to that on birds, a more systematically annotated list but explicitly provisional, containing many information gaps and assumptions.

A completely revised and expanded second volume on birds was published in 1984, and volumes on terrestrial mammals, fish, and reptiles and amphibians followed (Brooke 1984; Smithers 1986;

Skelton 1987; Branch 1988), denoting a time-lapse of roughly 10 years between revisions. A summary of the number of taxa in each category of each of these volumes is presented in Table 8.1.

TABLE 8.1   Summary of the formal status of the threatened animal species in southern Africa. The data sets are not fully comparable, since the volume on fish is the only one to include Namibia, and the other volumes do not consistently include Lesotho and Swaziland.

	*EX	E	V	R	I	OD	TOTAL	Total number of species
Birds (continental only)	2	4	17	42	35	1	102	±800
Terrestrial mammals	3	3	14	25	45	2	90	243
Fish	0	8	13	26	2	1	49	–
Reptiles and amphibians	1	6	15	(30)	(32)	0	84	488
Butterflies	2	2	7	91	40	–	142	632

* for definitions of the categories see Appendix 8.1.
() estimates made from Branch's (1988) categories: 'peripheral' and 'restricted'

A comprehensive revision of the volume on plants has not proved feasible since the original by Hall et al (1980). Sectional revisions have been published for the fynbos and karoo biomes (Hall and Veldhuis 1985; Hall 1987) and for the Transvaal (Fourie 1986), but they are for the most part annotated lists of poorly known species. The data gaps for Natal and its neighbouring eastern and coastal regions, together with the northern and eastern regions of the Orange River catchment, stand little chance of being filled in the foreseeable future. A great deal more work is needed to complete the taxonomic and survey work necessary to produce adequate lists for these regions. As an example, the succulent karoo biome, with probably the most species-rich succulent flora in the world, is particularly intractable for both taxonomists and conservation biologists. This is largely owing to the abundance there of groups such as the family Mesembryanthemaceae (approximately 2 600 species) which still requires basic taxonomic revision. A practical and comprehensive review RDB on plants for the whole of southern Africa therefore remains a high priority task.

THE PURPOSE OF RED DATA BOOKS

Their purposes are:

— to identify rare and declining species;
— to establish the nature and extent of such rarity or decline; and
— to suggest priorities for conservation, monitoring and research.

There is a further very general objective, which is to focus public attention on conservation issues by highlighting the plight of individual species. This political or public-awareness role of RDBs achieves its ends by means of environmental education, political pressure groups and fund raising. Global and regional RDBs have the same basic objectives but they differ in one important respect with regard to content. A global list does not have to consider local endemism or the problem of responsibility for peripheral populations of widely distributed species. This problem will be discussed later.

HOW WELL DO RED DATA BOOKS PERFORM?

On the positive side, the Red Data list concept, originated by Sir Peter Scott in the mid-1960s, has been remarkably successful in addressing both its general and specific goals. One can see that the concept, although not formally articulated, has been in existence for a long time, if one considers the number of protected areas that have been established to protect particular endangered species. In South Africa protected areas for bontebok *Damaliscus dorcas dorcas*, mountain zebra *Equus zebra zebra* and Addo elephants *Loxodonta africana* were proclaimed explicitly for the protection of those endangered species in the 1930s (Greyling and Huntley 1984). Since then many other reserves have been established for similarly specific management goals, both in South Africa and elsewhere. In addition, the management of many reserves has focused on the propagation of rare species as a newly defined and dominant function (eg Nylsvley, Percy Fyfe and Itala protected areas).

Since the purposes of Red Data lists have become more clearly defined, their use has become almost universal among the more developed nations. Many national, regional and global RDBs have been published, with many more countries in the process of compiling lists. These lists are most comprehensive for vertebrates, especially mammals and birds and, as might be expected, are least developed for invertebrates and plants.

In respect of the public

General awareness of the Red Data list concept is very widespread, 'red data species' being frequently quoted both in the popular conservation press and in the scientific literature. Despite their deficiencies, which will be elaborated later, these publications play an increasingly important role in policy-making, research stimulation, environmental education and public awareness. Their key attributes are:

— that they simplify complex conservation issues to focus on a single (or several) species; and
— that they develop a strong popular demand for information by appealing to our curiosity and the emotions of compassion, guilt and even pursuit of a lost cause.

These characteristics are simultaneously both strengths and weaknesses. Their strengths tend to be obvious, and overall they outweigh the weaknesses by a wide margin. However, some insights into the effectiveness of RDBs, in particular their negative attributes, will be considered later.

In research and monitoring

RDBs have contributed solidly to basic research and to our knowledge of many of our lesser-known biota. The publication of the first somewhat tentative RDB 'Aves' (Siegfried et al 1976) gave rise in seven years to more than 260 research papers and theses, (mention of an entry in the RDB was cited or implied as part of the motivation for the publication)(Brooke 1984). Response to other Red Data lists such as plants and even mammals, was probably considerably less, but it has not been quantified.

Highlighting species under threat has provided a useful public focus on shrinking or altered habitats that deserve attention. The impetus provided by the wattled crane *Grus carunculata* campaign for the protection of wetlands, and the more specific issue of the blue swallow *Hirundo atro caerulea* and its dependence on mid-altitude, moist grassland habitats are two recent examples. The predominance of raptors (18 species) and other vertebrate predators (17 species) in the current RDB Birds has contributed significantly to efforts to reduce the use of poisons in predator control, and has provided data to fuel the ongoing debate on the environmental impact of pesticides.

A key function of the second generation of RDBs is that they provide the first iteration of a monitoring cycle, setting out to compare the changed status of species after a time-lapse of 8 to 10 years. This has been only partially successful with most changes in the conservation status of species being due to more accurate information and more confident allocation of categories. As a result, in the current RDB Birds (Brooke 1984) about half the species included in the 1976 list (66 out of 137) were excluded from the 1984 list on the basis of improved information. Of those that appeared on both lists (53 species) only two threatened species are recorded as having a probable upward population trend at present. This is owing in both cases to human intervention (the pink-backed pelican *Pelecanus rufescens*, owing to protection of its single nesting site, and the lemon-breasted canary *Serinus citrinipectus*, owing to agricultural development). It has been found that 16 of the 53 species occurring in both lists deserve a higher category of conservation threat; in other words, they were assessed as declining in numbers. Of these, 10 species inhabit grasslands and wetlands, providing a clear indication of the changes to these ecosystem types that for several years have been repeatedly highlighted by ecologists.

In planning and administration

Because Red Data lists are based on consensus between leading biologists and conservation representatives, they provide the most authoritative reference work available on the subject. As such they provide the basis for provincial and national legislation controlling conservation, utilization and trade with regard to the species concerned. The regular review of these lists is therefore as much a regulatory as a scientific priority.

RDB information is also much in demand by landuse planners and consultants for the purpose of compiling Environmental Impact Assessments (not yet mandatory in any southern African state). Owing to the large number of species and the lack of detailed distribution records for many of them, southern Africa lags 15 to 20 years behind many northern hemisphere countries in this respect. We lack the high-resolution data necessary to make these lists relevant to most site-specific information needs.

Producing RDBs based on politically determined boundaries presents special problems for species with centres of distribution outside these limits. In the southern African RDB series this has been approached first by identifying endemicity to the region as a special attribute of rarity that enhances conservation priority. Secondly, the authors have all had to wrestle with the problem of how to deal with cosmopolitan rare species, and with peripheral rare species (species that have centres of distribution outside the region). In general RDBs have been inclusive, since these species contribute to local biodiversity and therefore to local conservation priorities, regardless of whether such conservation efforts contribute significantly to the global status of the species. The executive decision as to what conservation action to take is usually obvious. There are many peripheral and cosmopolitan species that require little or no action. However, there are others, such as the two rhino species, where one-time peripheral populations have been elevated to having global priority, while conservation efforts are collapsing elsewhere on the continent. The principle involved here is that peripheral and cosmopolitan rare species must be included, even if only to remind conservation authorities to monitor global trends in species as a basis for reviewing local priorities for action.

THE PRINCIPAL WEAKNESSES OF RED DATA BOOKS

Sparse and subjective data

Despite local acceptance of standardized IUCN definitions for all conservation categories (see Appendix 8.1), our information base is poor, owing primarily to the vast number of species involved. Even the more complete lists (birds and mammals) are subjectively compiled subsets of the real data set. I use the term 'subjective' here not to imply any scientific bias or negligence but in recognition of the predominance of information on the more evident and charismatic species. Lack of adequate information is a far greater problem with regard to plants and other groups that have not yet been seriously dealt with, such as invertebrates: RDB Butterflies is pending, however (Henning and Henning 1989). In these groups large numbers of species are undescribed or are known by little more than a description of one or two specimens. There are also many species or species-complexes that, although known to be scarce, are not yet clearly established as discrete species. These deficiencies point strongly to the need for more taxonomic and biotic survey work including, as a high priority, more information on population status and distribution. From a monitoring point of view the rigour of the data included in RDBs is inadequate for the purpose of detailed time-lapse comparisons. This is because subjective mechanisms are used in allocating species to the various categories.

In recognition of this problem, Brooke (1984) has developed a useful method for dealing with threat assessment for birds. The method involves allocating unweighted scores to each species according to seven population characteristics, namely:

— spatial distribution (breeding);
— numerical abundance (breeding);
— regional uniqueness (endemicity);
— taxonomic/genetic status (uniqueness);
— intrinsic rate of increase (potential);
— degree of stress (anthropogenic); and
— decrease in numbers (percentage based).

This ranking system has proved useful for conservation agencies and has already resulted in conservation action for many of the high-priority species. Hall et al (1984) also suggest a ranking procedure for plants, but this will require widespread testing with field data before it gains general acceptance.

Given that adequate data, even to a level comparable with that for European and North American biota, are not within reach in any of our lifetimes, we must look for innovative shortcuts. More pragmatic and effective approaches to understanding and classifying our diverse biota must be found. For example, concepts such as ecological indicator species, pivotal or key species, functional guilds, and even categories based on economic values must be more actively developed. Such approaches should enable us to identify and assess the conservation status of our more valuable species and thereby realign our conservation priorities accordingly. I shall return to the comparative values of species later.

Inadequate scale of resolution

For the primary users of biodiversity data at the species level, namely planners, managers and those responsible for the monitoring of our biota, the scale of resolution of current distribution data is usually inadequate. The existing information allows only for the most approximate of subregional analyses of the threats to biodiversity. In terms of the ecological insights required, such as the functional value of species, the analytical potential of the data is even lower. Data from the RDB Fish and other aquatic biota are more specific because of the spatially restricted nature of freshwater ecosystems.

Specialized species from many groups, particularly plants and invertebrates, have highly restricted distributions which can be pinpointed to areas of a few hectares. This level of detail is not usually presented in RDBs even if it is known, primarily to protect the confidentiality of the site. Adding the need for confidentiality to the existing plethora of species and site information only complicates the already difficult decision-making process for conservation managers. The following are examples of some real responses to the identification and limited publicizing of highly restricted, endemic or endangered species.

Negative scenarios:
— the landowner ploughs up the site to forestall an expected protection order or visitor invasion;
— the landowner inflates the selling price of the site in expectation of government expropriation;
— the species is placed under increased threat by collectors and visitors;
— conservation legislation bans all collection of the species by amateurs, thereby limiting a potential source of biological data and conservation knowledge; or
— the information is kept confidential in order to protect the site from any of the above actions, and a potentially sympathetic developer destroys it unwittingly.

Positive scenarios include the opposites of all the above responses, which may result in:
— creation of private or public protected areas;
— donation of land for conservation;
— registration of Natural Heritage Sites;
— raising of funds and public support for protection, education and research; and
— effective law enforcement which may reduce the threat.

The decision-maker's dilemma in such cases requires that virtually every species, and sometimes each site, must be handled on its individual merits. However, a policy of openness and public involvement must always be the ideal because, if there is a final solution, it can be achieved only through public awareness and education.

Excessive emphasis on rarity

THE ROAD TO EXTINCTION MUST PASS THROUGH RARITY

The consequence of this truism is that in considering biodiversity and extinction, there is a general tendency for an uncritical preoccupation with rarity. It has already been said that one of the principal disadvantages of this emphasis is that it shifts the attention of the man in the street from the more important causes to the more comprehensible effects. How much more readily will a member of the public be persuaded to donate money to a 'Save the blue swallow' campaign than to put real pressure on the timber industry to follow a more environmentally sensitive policy in allocating land for silviculture?

Another aspect of this preoccupation with rarity is encapsulated in a second truism.

RARITY IS AN ECOLOGICALLY RESPECTABLE CONDITION

In most natural or near-natural ecosystems the majority of species are relatively rare. Most of these species are, of course, overlooked in conservation management, being catered for by the general goals of managing for biodiversity. Real rarity in some of these species may be referred to as natural or static rarity, which is quite distinct from rarity induced by a decrease in numbers. In such cases, the reduction of rarity *per se* is not a biologically valid goal for conservation. The experience of translocating cheetah *Acinonyx jubatus* to the Kruger National Park, probably a case of overestimating the carrying capacity of the park (Anderson 1984), provides an appropriate field example to illustrate this lesson. In this context, rarity is often quite independent of population dynamics, and even of population trend in the short term.

THE RARITY PHENOMENON

Rarity is a complex attribute. Smith (1976) explores the issues thoroughly and contrasts biological rarity (whether real or assumed) with cultural or perceptual rarity, and the history of the management and administrative actions that have ensued in North America. However, even if one ignores the socio-economic aspects and sticks to biology it is conceptually simplistic to consider rarity as a unidimensional ordering of species from extinct to superabundant. To start with, rarity varies both in time and in space, whether in undisturbed or man-modified ecosystems (Harper 1981), and is acutely dependent on the effects of scale. Such attributes need to be taken carefully into account in designing and evaluating monitoring tools such as RDBs. For example, catering for the time dimension in monitoring demands one overriding condition: data collection and analysis must be repeatable so that subsequent, time-lapsed data sets are truly comparable.

Rarity as a central concept in conservation biology needs to be thoroughly examined in order to develop a sound basic philosophy. There is insufficient space here to do justice to the topic, and there are thorough treatments of it by several authors in synthesis volumes, for example Synge (1981) and Soule (1986).

Generalizations about rarity have been based mainly on the analysis of plant biogeographical data (eg Rabinowitz et al 1986). It is useful here to propose a variation (Figure 8.1) of Rabinowitz's eight-box conceptual model. It does not aspire to any scientific rigour, but it does help to illustrate the relationships between three primary aspects of biological rarity intrinsic to a species or population. The model does not account for external components of rarity such as threat assessment, nor can it address any time-dependent parameter such as population trend.

A conceptual model of rarity

Conservation biologists have noted that species, particularly plants, may be rare in terms of three basic attributes. These are:

1. Habitat specificity (specialization);
2. Geographic range (distribution); and
3. Local population size (abundance).

These attributes may be represented as a Venn diagram (Figure 8.1) which provides a conceptual model for ranking the levels of rarity, and deriving such potential rankings from an understanding of their interactive causes.

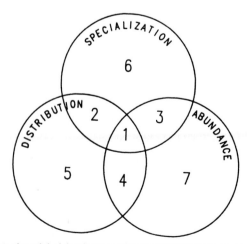

FIGURE 8.1   A conceptual model of the three predominant biological aspects of rarity

I have deliberately changed the order of these attributes from that used by Rabinowitz, although she attaches no purpose or significance to the order she uses (her sequence is 2,1,3). The significance of the order used here is based on the ability of conservation management to influence, *in situ*, the trait in question. My response is as follows:

— we have little chance of influencing the habitat specificity of a species that is highly complex and genetically determined;
— we have the capability to influence geographic range in many ways, but always within the limitations of the adaptability of the species and the extent and distribution of habitable areas; and
— we have an even greater capability to influence local population size by a variety of direct and indirect management techniques, potentially applicable to each local population.

If the logic behind this rank ordering holds, a more refined ordering of the different categories of rarity as indicated in Figure 8.1 is suggested (elaborated with some southern African examples).

Category 1 (most vulnerable to extinction pressure):
Highly specialized and restricted endemics, with markedly reduced abundance, eg some lycaenid butterflies and the riverine rabbit *Bunolagus monticularis*.

Category 2:
Highly specialized and restricted endemics, with local populations that are not critically reduced, eg Rudd's lark *Mirafra ruddi*, Ngoya red squirrel *Parax erus palliatus ornatus* and Lebombo cycad *Enchephalartos lebomboensis*.

Category 3:
Highly specialized species with reduced populations not restricted in their distribution, eg pangolin *Manis temminkii* and yellow-billed oxpecker *Buphagus africanus*.

Category 4:
Species that are less specialized, or non-specialized, but have very limited distributions and reduced abundance, eg Stanley's bustard *Neotis denhami*, and giant golden mole *Chrysospalex trevelyani*.

Category 5:
Localized endemics that are neither habitat specialists nor critically reduced in abundance, eg bontebok *Damaliscus dorcas dorcas* and jackass penguin *Spheniscus demersus*.

Category 6:
Specialized species with wide distribution, and populations that are not reduced to levels that pose any real threat, eg antbear *Orycteropus afer* and Cape vulture *Gyps coprotheres*.

Category 7:
Generalist species of wide distribution but with markedly reduced populations eg black rhino *Diceros bicornis* and bateleur eagle *Terathopius ecaudatus*.

Species falling outside some arbitrary level of these three traits that characterize rarity should be omitted from the list, or be included only on the basis of external threats. Subjectivity remains in the imprecise threshold levels selected for inclusion.

One case of an external threat is that of genetic 'contamination' by cross-breeding between taxa. An example of this is the African wildcat *Felis lybica* interbreeding freely with the domestic cat

Felis catus, probably throughout its range (Smithers 1983). The unsubstantiated but reputedly widespread production of fertile offspring from this cross raises the question of whether these two taxa warrant discrete species status.

A second category of rarity that does not fit the model is more correctly an external threat category, namely species that are commercially exploited and are linked to substantial human industries. The great whales, the African elephant *Loxodonta africana*, and various forest hard-woods may be included in this category. Often superabundant generalists, they generate such a level of economic dependence that their exploitation is frequently uncontrollable. When finally controlled by the limitation of cost-per-unit-effort, their reduced ecological role has changed their habitat to such an extent that recovery to former levels is not possible, even with complete protection.

THE COMPARATIVE VALUE OF SPECIES

A major criticism of Red Data lists by biologists is that they cannot avoid the bias imposed by limited inclusion criteria. First, this bias favours those species that are known, and those that are known well are favoured most. Secondly, it favours those species that are attractive or charismatic over those that appear insignificant or boring. This attitude gives rise to reference by the occasional cynical scientist to 'hugger clubs' — elephant huggers, tree huggers and even fynbos huggers. There is no escape from this particular problem because this high level of enthusiasm should not be discouraged. Our only approach is to be aware of it and counteract it even-handedly.

An extension of the 'attractive species attribute' is that of uniqueness and curiosity value. This is often associated in the public mind with genotypic uniqueness or genetic distance between species, a characteristic that undoubtedly bestows high conservation value. Theoretically, there-fore, the conservation manager striving for objectivity should have little trouble ranking the antbear (a monospecific order) above the pangolin (a monogeneric order comprising four species), if genetic criteria alone were applied. However, genetic uniqueness alone should not be allowed to be too predominant in our strategy to conserve biotic diversity.

Even if we are able to incorporate and balance the other major components of rarity and the value of species in ordering our priorities, there is one other attribute that is arguably more important than all the factors I have already mentioned. I refer here to the functional role of a species which may be crucial to the ecological processes that drive and define the limits of our life-giving ecosystems. This facet of biotic diversity conservation has been articulated clearly in two previous forums in southern Africa (Siegfried and Davies 1982; Siegfried 1984).

Despite these earlier exhortations to identify our functionally important species, little or no progress has been made. The problem is certainly complex but if the value of fundamental ecological research was ever in doubt, this surely provides a response. How can we possibly identify ecologically valuable species if we have not yet identified the key ecological processes in which they play a vital role?

A ROLE IN STIMULATING RESEARCH

The world and southern Africa are going to lose species in large numbers (Myers 1979) and as they disappear, research and management money will have to be spent on species that are more important for sustaining life. Researchers and managers will have to face increasingly tough decisions that require more objective and more practical information. If the Red Data list concept does not evolve into a reduced and more critically analysed data set, the volume of information will inhibit the development of viable plans of action.

The most obvious value of the current style of Red Data information is for monitoring the dynamics of threatened species and populations. To be effective, however, the ecological basis for the selection of species, and of the most appropriate attributes to monitor more than mere numbers, needs to be carefully thought out. Sampling design and large-scale analytical methods such as those used in biological atlassing (ie the Southern African Bird Atlas; Harrison this volume) will have to be skilfully designed to provide enough sound data, the collection of which must be repeatable over appropriate time scales. Such a task presupposes the existence of an agency that accepts the archival function of data storage and retrieval. The Department of Environment Affairs has made a strong start in this direction (the SA Plan for Nature Conservation, the SA Natural Heritage Programme, the National Atlas of Critical Environmental Components), but it is not enough. In keeping with the government's policy of privatization, and following the example of SAFRING (South African Bird Ringing Unit, which is government-funded but managed by the University of Cape Town), it is obvious that more teamwork is required between the public and private sectors.

Another role for Red Data-type information is to provide for more effective development of the key-species concept. Here we must think in terms of species with outstanding information value. Functionally pivotal species (which are not necessarily rare) and indicator species (which often are) both have the potential to provide a higher than usual return of information on the original investment in research.

Plant taxonomists, understaffed, underfunded, and locked into a painfully time-consuming set of procedures, stand little chance of ever reaching their goal of a complete Regional Flora at the present rate of progress. With species already becoming extinct before they have been properly described, the need is urgent. The Red Data lists and other criteria for ascribing values to species, provide at least one decision-making aid to the ordering of taxonomic research priorities. If such an aid can be developed, it will have to be soon; otherwise it will serve only to describe how far the horse has bolted by the time we plan to have the stable door closed.

Finally, information on rare, threatened and valuable species provides a check-list of priority research subjects for the improvement of the theoretical basis of biology and ecology, and the development of new theory. The new focus being developed in North America and Britain, referred to as Conservation Biology (Soule 1986), is a hybrid of many disciplines which aspires to be as relevant to the field manager as to the mathematical theorist. It addresses a wide range of issues including extinction and speciation, minimum viable population size, and superabundance phenomena. The common thread among these issues is that biology will be the final arbiter of the success or failure of conservation (Ehrenfeld 1987).

ACKNOWLEDGEMENTS

Some of the ideas in this paper originated in discussions with Peter Frost and indirectly with each of the authors of the current South African Red Data Book series. Richard Brooke, Tony Hall, Peter Lloyd, and an anonymous referee provided useful comments on an earlier draft. This paper is dedicated to the memory of Reay Smithers who passed away not long after the publication of his *RDB Terrestrial Mammals*. The wise, warm and infallibly willing doyen of mammal research in southern Africa will be remembered with fondness and respect by all who knew him.

APPENDIX 8.1

IUCN categories of threat

('Species' is used as a synonym for 'taxa'. Definitions are from Wells et al 1983 with minor adaptions.)

Extinct
Species not definitely located in the wild in the past 50 years, despite deliberate searches in most likely places. (This includes species that are extinct in the wild but surviving in cultivation or captivity.)

Endangered
Species in danger of extinction and whose survival is unlikely if the causal factors continue operating. (Included are taxa whose numbers have been reduced to a critical level or whose habitats have been so drastically reduced that the taxa are deemed to be in immediate danger of extinction. This includes species with populations so low as to be vulnerable to breeding collapse due to lack of genetic diversity. Also included are taxa that are possibly already extinct but have definitely been seen in the wild in the past 50 years.)

Vulnerable
Species believed likely to move into the 'endangered' category in the near future if the causal factors continue operating. (Included are taxa of which most or all the populations are decreasing because of over-exploitation, extensive destruction of habitat, or other environmental disturbance; taxa with populations that have been seriously depleted and whose ultimate security has not been assured; and taxa with populations that are still abundant but are under threat from severe adverse factors throughout their range.)

Rare
Species with small world populations that are not at present 'endangered' or 'vulnerable', but are at risk. (These taxa are usually localized within restricted geographical areas or habitats, or are thinly scattered over a more extensive range.)

Indeterminate
Species known to be 'endangered', 'vulnerable', or 'rare' but where there is not enough information to say which of the three categories is appropriate. (Threatened is a general term to denote species that are 'endangered', 'vulnerable,' 'rare' or 'indeterminate'.)

Out of danger
Species formerly included in one of the above categories, but now considered relatively secure because effective conservation measures have been taken or the previous threat to their survival has been removed.

Insufficiently known
Species that are suspected but not definitely known to belong to any of the above categories, because of lack of information. (In practice, 'endangered' and 'vulnerable' categories may include, temporarily, taxa whose populations are beginning to recover as a result of remedial action, but whose recovery is insufficient to justify their transfer to another category.).

Commercially threatened
Species not currently threatened with extinction but most or all of whose populations are threatened as a sustainable commercial resource, or will become so unless their exploitation is regulated. (Generally taxa with large populations, eg commercial marine species.)

Threatened community
A group of ecologically linked taxa occurring within a defined area, which are all under the same threat and require similar conservation measures.

Threatened phenomenon

Aggregates or populations of organisms that together constitute major biological phenomena, endangered but not as taxa.

REFERENCES

ANDERSON J L (1984). A strategy for cheetah conservation in Africa. In *Proceedings of an international symposium on The Extinction Alternative*. (ed Mundy P) Endangered Wildlife Trust, Johannesburg. pp 127 – 135

BRANCH W R (1988). South African Red Data Book: Reptiles and Amphibians. *South African National Scientific Programmes Report* **151**, CSIR, Pretoria. 240 pp.

BROOKE R K (1984). South African Red Data Book: Birds. *South African National Scientific Programmes Report* **97**, CSIR, Pretoria. 213 pp.

EHRENFELD D (1987). Editorial. *Conservation Biology* **1(1)**, 6 – 7.

FOURIE S P (1986). The Transvaal, South Africa. Threatened plants programme. *Biological Conservation* **37(1)**, 23 – 43.

GREYLING T and HUNTLEY B J (eds)(1984). Directory of southern African conservation areas. *South African National Scientific Programmes Report* **98**, CSIR, Pretoria. 311 pp.

HALL A V (1987). Threatened plants in the fynbos and karoo biomes, South Africa. *Biological Conservation* **40(1)**, 29 – 53.

HALL A V, DE WINTER B, FOURIE S P and ARNOLD T H (1984). Threatened plants in southern Africa. *Biological Conservation* **28(1)**, 5 – 21.

HALL A V, DE WINTER B and VAN OOSTERHOUT S A M (1980). Threatened Plants of Southern Africa. *South African National Scientific Programmes Report* **45**, CSIR, Pretoria. 244 pp.

HALL A V and VELDHUIS H A (1985). South African Red Data Book: Plants — fynbos and karoo biomes. *South African National Scientific Programmes Report* **117**, CSIR, Pretoria. 160 pp.

HARPER J L (1981). The meanings of rarity. In *The biological aspects of rare plant conservation*. (ed Synge H 1981) John Wiley and Sons, London. pp 189 – 204.

HENNING S F and HENNING G A (1989). South African Red Data Book: Butterflies *South African National Scientific Programmes Report* **158**, CSIR, Pretoria.

HIGUCHI R, BOWMAN B, FREIBERGER M, RYDER O A and WILSON A C (1984). DNA sequences from the quagga, an extinct member of the horse family. *Nature* **321, 5991**. 282 – 284.

IUCN/UNEP (1987). *The IUCN Directory of Afrotropical Protected Areas*. IUCN, Gland, Switzerland and Cambridge. 1034 pp.

MCLACHLAN G R (1978). South African Red Data Book: Reptiles and Amphibians. *South African National Scientific Programmes Report* **23**, CSIR, Pretoria. 53 pp.

MEESTER J A J (1976). South African Red Data Book: Small Mammals. *South African National Scientific Programmes Report* **11**, CSIR, Pretoria. 59 pp.

MYERS N (1979). *The sinking ark. A new look at the problem of disappearing species*. Pergamon Press, London. 307 pp.

MYERS N (1983). *A wealth of wild species. Storehouse for human welfare*. Westview, Colo. 274 pp.

RABINOWITZ D, CAIRNS S and DILLON T (1986). Seven forms of rarity and their frequency in the flora of the British Isles. In *Conservation Biology — The science of scarcity and diversity*. (ed Soule M E) Sinauer, Sunderland, Mass. pp 182 – 205.

SIEGFRIED W R (1984). Red Data Books: Bibles for protectionists, Aunt Sallies for conservationists. In *Proceedings of an international symposium on The Extinction Alternative*. *(ed Mundy P)* Endangered Wildlife Trust, Johannesburg. pp 119 – 126.

SIEGFRIED W R and DAVIES B R (1982). Conservation of Ecosystems: Theory and Practice. *South African National Scientific Programmes Report* **61**, CSIR, Pretoria. 97 pp.

SIEGFRIED W R, FROST P G H, COOPER J and KEMP A C (1976). South African Red Data Book: Aves. *South African National Scientific Programmes Report* **7**, CSIR, Pretoria. 108 pp.

SKELTON P H (1977). South African Red Data Book: Fishes. *South African National Scientific Programmes Report* **14**, CSIR, Pretoria. 39 pp.

SKELTON P H (1987). South African Red Data Book: Fishes. *South African National Scientific Programmes Report* **137**, CSIR, Pretoria. 199 pp.

SKINNER J D, FAIRALL N and BOTHMA J DU P (1977). South African Red Data Book: Large mammals. *South African National Scientific Programmes Report* **18**, CSIR, Pretoria. 29 pp.

SMITH R L (1976). Ecological genesis of endangered species: the philosophy of preservation. *Annual Review of Ecology and Systematics* **7**, 33 – 55.

SMITHERS R H N (1983). *The mammals of the southern African subregion*. University of Pretoria, Pretoria.

SMITHERS R H N (1986). The South African Red Data Book: Terrestrial mammals. *South African National Scientific Programmes Report* **125**, CSIR, Pretoria. 216 pp.

SOULE M E (1986). *Conservation Biology — The science of scarcity and diversity*. Sinauer, Sunderland, Mass. 584 pp.

SYNGE H (1981) (ed). *The biological aspects of rare plant conservation*. John Wiley and Sons, London. 558 pp.

WELLS S M, PYLE R M and COLLINS N M (1983). *The IUCN invertebrate red data book*. IUCN, Gland, Switzerland. 632 pp.

CHAPTER 9

Rare plant surveys and atlases

A V Hall

INTRODUCTION

The Foundation for Research Development has a project to determine the conservation status of threatened plant and animal species in southern Africa and to publish Red Data Books (RDBs) about them. The programme for threatened plants began in 1974. The only available data sources were herbaria, the literature, and general field knowledge. Records of rare species were often thin, giving many taxa an uncertain conservation status until more detailed studies could be made in the field (Hall et al 1980). Some such studies have been made, mainly of the rarer and more threatened taxa. These have been guided by sight records from advisers and input from the planning professions on likely threats.

The aim was first to publish a preliminary RDB for southern Africa, and then produce detailed accounts for each major region. The preliminary RDB, covering South Africa, Namibia, Botswana, and all included territories, appeared in 1980 (Hall et al 1980). Some progress has been made with regional programmes (S P Fourie, C Hilton-Taylor, D F Laidler and R Pool pers comm; Hall and Veldhuis 1985). Computer methods have been widely adopted, allowing storage of confidential data on exact localities, and selective retrieval of information for various uses, including the preparation of RDBs.

The surveys have to work chiefly at species level, although it is the evolutionary and ecological health of their breeding populations that will determine their long-term future. The general condition of populations may be described and a census carried out, but this barely touches on important issues such as their genetic diversity, evolutionary significance and ecological dynamics.

Survey methods for RDBs are described and critically reviewed in this chapter, along with developments in geographical information systems and atlassing. The aim of the surveys is to gather data about rare plant diversity, mainly to help in restoring selected species in ecosystems and evosystems. The term evosystem is used here for the evolutionary processes that alter gene pools in space and time.

RED DATA BOOKS FOR PLANTS

The first aim in a species conservation programme is often to gather rare plant information into an RDB. This can be used for assessing the situation and planning further action.

The International Union for the Conservation of Nature and Natural Resources (IUCN) published an RDB consisting of case studies of 250 threatened plant species from various parts of the world (Lucas and Synge 1978). The accounts for each species give the status, distribution, habitat and ecology; the conservation measures taken and proposed; notes on biology and potential value; cultivation and other ex situ measures; and a short description with references.

Most regional RDBs give much less information than this. Several consist of lists of threatened species grouped according to status or provincial area, with a general text on rare plant conservation (Kartesz and Kartesz 1977; Ayensu and DeFilipps 1978; Lucas and Walters 1976). Examples of more detailed regional treatments are those published for each province in Canada. The atlas of rare plants in Ontario (Argus et al 1982-7) gives information on each species as follows: distribution maps for the province and for North America; habitat data; notes on biology, conservation status and any protection measures; and literature references. The provincial map has spot records accurate to an area of about 15 km in diameter. Each spot is marked to show roughly when the record was made.

Amended figures from the first RDB published for southern Africa give 39 species and infra-specific taxa of vascular plants as Extinct, 110 Endangered, 223 Vulnerable, 700 Critically Rare, 393 Indeterminate and 908 Uncertain, giving a total of 2 373, about 10% of the flora (Hall et al 1980; Hall et al 1984).

An example of a regional RDB is that for the fynbos and karoo biomes (west of 26 degrees East, and south of the Orange River: Hall and Veldhuis 1985). In this work 1 808 species were listed with brief statements on status, habit, flowering time, and coarse distribution data, given as 1/4 by 1/4-degree geographical grid areas. General accounts of population status and threats based on recent field work were given for about 250 of the species.

An example of a general account is the entry for *Sorocephalus tenuifolius* (Proteaceae), which became extinct in the wild during the survey:

Extinct; shrublet; [flowering in] January–February; [geographical grid code] 3418BB. Population status: formerly existed in a single population extending over 2 ha which the landowner intended to set aside as a nature reserve. In 1980 there were 700 plants; all but 15 of these were destroyed in a fire the next year. By 1982 there appeared to have been good seedling regeneration with many hundreds of plants. The surroundings were then ploughed for apple orchards and a windbreak of Pines was planted through the population. In 1985 the habitat was destroyed by ploughing. Cultivation: 7 plants in the nursery of Provincial Nature Conservation were transferred to Kirstenbosch in 1985. The plant is of scientific interest and would be decorative in horticulture. Urgency for conservation: maximum priority.

A supporting text in this RDB gives the methods used in compiling the lists, compares the results with those from other areas, describes the causes of decline, and suggests the kinds of conservation measures that could be used. Maps and statistics show where the highest numbers of threatened species occur, as in Table 9.1. The region as a whole has a large proportion of thinly researched taxa, as shown by the figure of 702 (39%) in the Uncertain category.

TABLE 9.1   Statistics on the occurrence of threatened plants in the major phytogeographical regions in the area west of 26° E and south of the Orange River. The categories of threatened plants are abbreviated as: U – Uncertain; I – Indeterminate; R – Critically Rare; V – Vulnerable and Declining; E – Endangered; X – Extinct. The average density of threatened plants in each region is given by a species-area ratio, scaled up for clarity by a factor of 1 000.

Region	U	I	R	V	E	X	Total	Area (km²)	Spp/Area x 1 000
Cape Province W of 26° E and S of the Orange River	702	281	495	183	118	29	1 808	437 143	4
Fynbos biome (major part)	472	184	389	152	103	26	1 326	77 393	17
Karoo biome (major part)	261	104	115	36	20	3	539	359 750	1
W coastal lowland region	91	77	83	69	58	8	385	14 700	26
Western mountain region	262	80	202	52	36	11	643	17 223	37
Southern mountain region	178	56	142	45	16	5	442	31 230	14
S coastal lowland region	118	43	91	57	30	5	344	14 240	24

Another local RDB for the 50 km-long Cape Peninsula gives a list of 174 threatened species (Hall and Ashton 1983). Of these, 5 are given as Extinct, 25 Endangered, 28 Vulnerable, 50 Critically Rare, 40 Indeterminate and 26 (15%) Uncertain. The lower percentage of Uncertain taxa shows that this flora, close to Cape Town, is better known than the regions given in Table 9.1. Localities are given approximately in terms of six topographic regions. To help stimulate landowners' interest in conservation, the text contains short articles on the richness and scientific importance of the flora, and guides for the rescue and long-term management of threatened plants.

ALTERNATIVES TO THE RED DATA BOOK

Halting the extinction of species and their populations is the ultimate target of threatened plant programmes. RDBs can help raise awareness of this among the public and such key persons as educators, environmental planners, administrators, landuse advisers, agriculturists and land-owners.

However, for laymen they have at best a very limited appeal; at worst they are an unhelpful compilation of obscure taxonomic names. They have given factual support to appeals for conservation through the media and in society lectures, but a popular text such as that by Fisher (1987) for the British Isles would perhaps be a better tool for raising public awareness.

Another problem is that the RDB has its information fixed the moment it appears in print. It can record only a snapshot view of a dynamic situation. It may mislead by omitting the most recent impacts which may destroy the last remnants of an endangered species. As an example, the seven remaining plants of *Sorocephalus tenuifolius* (see p 149) died after their transfer to the Kirstenbosch Gardens. The professional conservationist prefers to obtain the latest information directly from rare plant surveyors or a regularly revised computer data bank.

In RDBs, localities have to be given approximately. This is to avoid increasing the danger to some species by guiding unscrupulous collectors to their exact sites. It also avoids alerting unco-operative landowners who have been known to destroy a threatened plant population rather than have some of their property set aside for conservation (Hall and Veldhuis 1985). Unfortunately, exact localities are essential data for managers, environmental planners and researchers.

This omission makes RDBs much less useful, but there appears to be no remedy. Once again computer data banks, which can carry confidential exact localities, are the only solution for the professional conservationist.

Computer data banks can be incorporated into large-scale geographical information systems. Two are being developed in southern Africa, one for the Cape Province and the other, which is in a planning stage, for South Africa. They are the Cape Inventory of Critical Environmental Components (CICEC) and the National Atlas of Critical Environmental Components (NACEC).

CICEC (D F Laidler pers comm) is to include, among other data, results from rare plant surveys in the karoo and the western and eastern Cape. The karoo data of Hilton-Taylor have already been incorporated, and results from the eastern Cape are expected shortly.

CICEC is adapted from The Heritage Programs of the Nature Conservancy of the USA (Jenkins 1981). The inventory will cover all known threatened plant and animal species in the Cape Province. The main data set covers each species, habitat and community ('elements') to be conserved; a second carries data on where each element is found; a third comprises data on each conserved area; and a final data set includes a computer index to the system and reference texts on the elements. It is planned to map all known localities of threatened species on a scale of 1:50 000. The identity of threatened plant species and their localities is being vouchered by herbarium specimens, indexed through a data bank of information gathered from their labels.

CICEC could be of outstanding value for practical conservation planning and action. A variety of reports such as RDBs could be extracted from it when needed. As the aim of CICEC is to record all critical elements of natural diversity, it could be of general value to planners for tasks such as choosing the least sensitive sites for development. For this purpose, it is planned to send the CICEC data to NACEC. NACEC is an important planner's tool which maps the sites of all environmentally sensitive elements in each region.

Both CICEC and NACEC are major geographical projects requiring large-scale compilations of data. The data are often thin and more field research is usually needed. In the USA the similar Heritage Program has multimillion dollar funding (L E Morse pers comm). It is of outstanding value in deflecting impacts from threatened plants and habitats. In southern Africa, the very small staffing of CICEC (2) and NACEC (nil, under review) means that it could be many years before they are complete, and able to be dependably used by planners. By that time much more natural diversity will have been lost as a result of major changes caused by high rates of population growth.

A latecomer to the plant survey programmes is the proposed atlassing project for rare Proteaceae, conceived along the same lines as those for birds in southern Africa and the genus *Banksia* in Australia (Rebelo et al 1987). The field surveys will be done by amateurs, supervised and co-ordinated by a team of botanists. A special field guide will aid identification of rare species in the context of the rich floras where Proteaceae are often found.

SURVEY DATA FOR ASSESSING CONSERVATION PRIORITIES

In most RDBs the urgency for action is implied but seldom specified, owing partly to lack of data. It is important to make even a rough ranking of priority where there are too many threatened species for all to be made safe because of limited funding and staff.

The urgency of rescue and conservation should be based on a mixture of rarity, rate of decline, and degree of artificial stress. Other factors are the scientific, evolutionary and utilization potentials of the species, and the logistics of getting conservers and equipment to the site of the threatened species and making the plant safe. This information is inadequately summarized by the IUCN status categories so it is necessary that a suitable scheme be devised.

A scoring scheme which may in the future be useful in southern Africa is given in Table 9.2 (Powell 1974; Hall 1987). Codes for priority ranking for conservation vary from 1 (mild) to 3 (severe) for each of five criteria. A species rated with 3s for all criteria would merit the most urgent action. It would be extremely rare, in a state of severe decline, heavily threatened, of major importance, and easily accessible for logistical support for conservation.

TABLE 9.2   Criteria for priority ranking of threatened species for conservation, based on Powell's system (1974). The multi-digit priority codes should show where data are missing for example, the codes 32122 would be 321-2 if the species' importance were unknown.

| Digit no | Ranking Factor | Criteria for codes | | |
		Code 1	Code 2	Code 3
1	Rarity:	Rare	Very rare	Extremely rare
2	Evidence of decline:	Mild decline	Medium decline	Severe decline
3	Artificial stresses:	Mild stresses	Moderate stresses	Severe stresses
4	Species importance:	Mild importance	Medium importance	Major importance
5	Logistical support:	Severely difficult	Somewhat difficult	Easily provided

These criteria are uncorrelated variables. Each should be considered separately in the ranking process. Totals may sometimes be compared, but it must be borne in mind that they do not disclose the state of individual factors. For example, a zero score for any criterion would mean that a species would not rate at all for conservation. In the absence of any artificial (human-caused) threats, decline to extinction might be accepted as a natural process not meriting interference, unless the species were of value to humanity.

A preliminary study of this ranking system suggests that the information most in need of refinement is on rates of decline, which shows a need for regular monitoring of census changes in field populations. Species importance often proves to be difficult to assess where so little is known about the threatened taxa. There is clearly much else that could be incorporated in an ideal situation with plenty of research, as in the case of population viability analysis of Gilpin and Soule (1986). A simple priority ranking scheme such as the above should be incorporated in future RDBs in southern Africa, even if scores for some of the criteria have to be left unassessed for lack of data.

RARE PLANT SURVEYS IN SOUTHERN AFRICA

In a study of known threatened plant species distributions in Africa, Hedberg (1979) showed that by far the highest concentration appears to lie in the western Cape. This was supported by the finding that 56% (1 326) of southern Africa's total of 2 373 threatened species and infra-specific taxa appear to be concentrated in the relatively small area of the fynbos biome (Hall et al 1984; Hall and Veldhuis 1985). These data are approximate. They are artificially biased by the fact that the flora of the western part of the fynbos biome has been more intensively collected than that in other regions such as the karoo. The karoo region is seriously undercollected and receives heavy impacts from pressures such as overgrazing, weed invasion, flooding and soil erosion.

Similar biases exist elsewhere in southern Africa. The only remedy for this problem would be to increase the size of herbarium specimen collections as much as fivefold to give more even coverage. The uneven coverage by past collectors would have to be taken into account (Gibbs Russell 1985). Collectors would have to avoid any personal bias and collect all plant groups in all kinds of habitat. The high cost of extra staffing and the herbarium facilities needed for this would be difficult to motivate unless there were a major cultural shift in favour of conservation. This, however, is at present unlikely.

The three principal southern African regions in which field surveys are under way are the Cape, Natal and the Transvaal.

Cape Province

The Red Data lists for the winter-rainfall karoo (the western Cape karroid domain) are being revised and all the information is being stored in a computer-based data bank (C Hilton-Taylor pers comm). Material will be extracted for an RDB, but the computer-based data bank will be the main source of specialized information for conservationists. This will help prevent sensitive data, such as degree of rarity and localities of succulent species, from getting into the hands of unscrupulous collectors.

It has been estimated that in the eastern Cape (24–29° E, 31–35° S) one species has recently become Extinct, 3 are Endangered, 15 Vulnerable, 41 Rare, 117 Indeterminate, and 662 Uncertain (Lubke et al 1986). The lists of eastern Cape Red Data species are not published; they are stored in a computer-based data bank for use in research or conservation.

Natal

Work on Red Data species in Natal is much less advanced than in the Cape Province. Hilliard (1983) suggested that some 30% of the province's flora may be endemic, based on a geographical analysis of Natal Asteraceae (Hilliard 1978). A dozen centres of endemism were pointed out, among them the Natal Drakensberg, the coastal plain north of Durban, and areas overlying Table Mountain sandstone. Some areas such as the coastal forests and grasslands had been widely replaced by croplands and many extinctions of species and populations had been recorded there. Hilliard (1983) listed many probably threatened species. For the majority of the areas, threatened species and habitats cannot be fully assessed until detailed field surveys by qualified botanists have been carried out. These surveys will have to be supported by an efficient team of herbarium taxonomists.

Transvaal

A herbarium and field survey of Red Data species was started in the Transvaal in 1976. Planned and staffed by the Transvaal Provincial Administration, it has an active programme of field trips and, for remote areas and larger plant forms, aerial surveys have been made by helicopter. Data from the surveys are being accumulated in files for each of the currently listed 246 threatened plant species. When the field surveys are completed, status reports will be drawn up on each species for use in rescue and management. Special reports will be made for areas with locally high concentrations of threatened plants, for use in planning nature reserves (Fourie 1986).

A computer-based data bank has been made of the threatened species data from the Transvaal. Printouts can be made from this for environmental planners and nature conservation agencies. The data bank covers the minimal information of species and family name; region, locality grid, and farm name and number; and codes for endemism and IUCN conservation status category. Retrievals are made in selected fields. As in the Cape and Natal, the emphasis is primarily on a computer-based system rather than an RDB. Printouts and maps have been compiled for an atlas of threatened plants of the Transvaal: this is a confidential document for internal circulation to environmental planners and conservation units (S P Fourie pers comm).

A report on the status of the important genus *Encephalartos* (Zamiaceae) in the Transvaal was compiled from field records as an internal document (Fourie 1986). For nearly all the species the status proved more threatened than expected. All had been inadequately conserved.

The Transvaal programme is handled by a small team which receives taxonomic support from herbaria such as the Botanical Research Institute in Pretoria. In spite of advanced training in conservation at universities, this programme has suffered in recent years from a lack of recruits to its funded posts.

DISCUSSION: LIMITATIONS OF SURVEYS

There are limits to the kind of assessment that can be made from RDBs, mainly because the data are usually preliminary and incomplete. This problem must be borne in mind in conservation planning.

An example is the recent intensive field survey of the highly threatened genus *Encephalartos* (Zamiaceae) in the Transvaal (Fourie 1986). This showed that past records in the literature and herbaria were far too thin to show changes in conservation status in historical time. The records were so uneven that large disjunct populations of some species, for example *E lebomboensis*, had not been recorded at all. This is in spite of the species' prominence and scientific interest.

The reason for this thin coverage can be understood when it is realized that collections in herbaria were made to assess the broad systematic diversity of a region. Geographical distribution played only a minor role. Herbaria were never intended for use in conservation biology, which needs far more precise data, often down to the sizes and exact locations of populations. Early specimens seldom have detailed locality data, and they have no data at all on threats and declines, which makes a poor basis for comparing the past with the present. Modern documentation of herbarium specimens is far more complete.

Another problem is that geographical coverage of herbarium records is often very uneven in southern Africa (Gibbs Russell 1985). Extensive areas are severely under-represented in the large computer data bank of grid-referenced specimens in the National Herbarium, Pretoria. In the area covered by the fynbos and karoo survey (Hall and Veldhuis 1985), 14% of the 731 and 1/4 x 1/4-degree grid areas have no specimens in the data base, and 64% have only 50 specimens or fewer (Gibbs Russell 1985). Collections in regional herbaria are far from adequate to make up for this uneven coverage. Threatened status is therefore usually weakly signalled by the data in herbaria, but there is often nowhere else to start gathering evidence for a survey.

The species-level taxonomy of many genera still requires revision before their threatened species can be listed, other than in the IUCN Uncertain category which in the Cape holds far more species than any other (Hall and Veldhuis 1985). Whole families such as Mesembryanthemaceae still require more intensive and extensive collecting before taxonomic revisions can take place.

Searches over large areas of difficult terrain require great persistence, and the success rate varies widely. A major factor is the prominence of the plants in the field: the problems in finding large plants such as *Encephalartos* are greatly aggravated with small annuals or geophytes. Another problem may be the brevity and variability of flowering or other prominent phases. Known sites may be hard to reach and precise search patterns difficult to follow in rough terrain. There are often no logical links between distribution and ecology. The populations may be remnants from a random past factor mosaic, such as a pattern of fire histories, which gives no indication where the populations are likely to be found.

Some geophytes such as terrestrial Orchidaceae may not flower or even appear above ground for many years. Important seed-stores which may function as reservoirs of many potential new individuals may be overlooked, as in the case of the Marsh Rose, *Orothamnus zeyheri* (Boucher

1981). These and other cryptic aspects of plants, such as rare flowering of desert ephemerals, often create severe problems of assessment in the field.

Some groups are less prone to these artefacts of data collecting than others. Colourful, large Proteaceae and Ericaceae have been more attractive to collectors and are often (but not always) well represented in herbaria. Worst off are families such as Cyperaceae and Poaceae which not only lack distinctive search images but are taxonomically poorly known. Full account should be taken of these problems when attempting to compare distributions of families or life forms.

Identification is a severe problem in floristically rich areas in southern Africa. Identifications are often difficult especially when the species are unfamiliar, are seldom seen, and have close relatives. The literature search alone may take a trained taxonomist an hour or more to complete even in a good library. Some herbaria are establishing rapid-guide collections with limited specimens and keys for identification but this is of little help in the field, where rare species must be distinguished from look-alikes on semi-cryptic characters. It is generally only experts in each group that are capable of identifying rare species in unexpected places. Data-bank programs from which local field keys can be constructed, such as Dallwitz's Delta program (1980), may be a step in the right direction. In the meantime, many populations belonging to rare species will be extirpated unnoticed unless there is a realistic, major increase in taxonomic support for conservation in southern Africa.

CONCLUSIONS

Large amounts of data have been gathered since threatened plant surveys began in southern Africa some fifteen years ago. However, there is a severe shortage of data on many hundreds of possibly threatened species. These should be investigated soon to avoid a possibly serious loss of species diversity. About 55% of southern Africa's species on the preliminary threatened plant list are in the IUCN Indeterminate and Uncertain categories. Only about 14% of about 1 800 threatened species in the fynbos and karoo biomes are known in any detail. Some comments on the overall programme are as follows:

— Red Data Books serve as registers of the conservation status of species. They can be used to promote an interest in the problem of the loss of natural diversity. The main target groups are educators, environmental planners, administrators and interested members of the public.
— Computer-based data banks, such as those in the Cape, Transvaal and Natal, are more useful to conservationists than books because they can be kept more up to date. They also have the advantage of being able to hold confidential information on, for example, exact localities. The data banks can be incorporated into large-scale geographical information systems which can be used to plan development so that it has the least impact on biotic diversity and other natural resources.
— Future Red Data Books should evaluate priorities for conservation action, to provide assistance where staff and funds are too limited to cope with large numbers of threatened species.
— Taxonomy lends vital support to conservation biology and its present weaknesses in southern Africa should receive urgent attention. Herbarium collections need to be greatly enlarged for undercollected areas. Important groups still require a great deal of revision before they can be surveyed and conserved: at present there seem to be many 'pseudo-rare' taxa, which are undercollected or wrongly held to be distinct.
— The identification of threatened plants should be made much easier, both in the herbarium and in the field. Identification centres are needed, with regional or local keys made from computer data banks.

A major aim of this work is to restore threatened species to a safe state. This is best done in the natural habitats in which the species receive ecological support and may form part of long-term evosystems, in which some may evolve to take significant future roles.

The task is large and it is slow in getting under way. It seems that not one threatened plant species has yet been made safe enough to be removed from the southern African Red Data Book on plants. Leon et al (1985) found that the same held for the Mediterranean area. This they attributed to the need to gather enough data first and then to use it in the political arena to obtain funds before a start could be made on conservation. The same situation may be arising in some areas in southern Africa. For the most critical species it should be possible to bypass this lengthy system and start conserving them at once.

REFERENCES

ARGUS G W , PRYER K M, WHITE D J and KEDDY C J (eds)(1982-1987). *Atlas of the rare vascular plants of Ontario.* 4 Parts. National Museum of Natural Sciences, Ottawa. (Looseleaf).

AYENSU E S and DEFILIPPS R A (eds)(1978). *Endangered and threatened plants of the United States.* Smithsonian Institution and World Wildlife Fund, Washington DC. 301 pp.

BOUCHER C (1981). Autecological and population studies of *Orothamnus zeyheri* in the Cape of South Africa. In *The biological aspects of rare plant conservation.* (ed Synge H) John Wiley, Chichester. pp 343 – 53.

DALLWITZ M J (1980). A general system for coding taxonomic descriptions. *Taxon* **29**, 41 – 46.

FISHER J (1987). *Wild flowers in danger.* Victor Gollancz, London. 194 pp.

FOURIE S P (1986). The Transvaal, South Africa : Threatened Plants Programme. *Biological Conservation* **37**, 23 – 42.

GIBBS RUSSELL G E (1985). PRECIS: the National Herbarium's computerised information system. *South African Journal of Science* **81**, 62 – 5.

GILPIN M E and SOULE M E (1986). Minimum viable populations: processes of species extinction. In *Conservation Biology.* (ed Soule M E) Sinauer, Sunderland, Mass. pp 19 – 34.

HALL A V (1987). Threatened plants in the Fynbos and Karoo Biomes, South Africa. *Biological Conservation* **40**, 11 – 28.

HALL A V and ASHTON E R (1983). *Threatened plants of the Cape Peninsula.* Threatened-Plants Research Group, University of Cape Town, Cape Town.

HALL A V, DE WINTER M, DE WINTER B and VAN OOSTERHOUT S A M (1980). Threatened Plants of Southern Africa. *South African National Scientific Programmes Report* **45**, 241 pp.

HALL A V, DE WINTER B, FOURIE S P and ARNOLD T H (1984). Threatened plants of Southern Africa. *Biological Conservation* **28**, 5 – 20.

HALL A V and VELDHUIS H A (1985). South African Red Data Book: Plants — Fynbos and Karoo Biomes. *South African National Scientific Programmes Report* **117**, 157 pp.

HEDBERG I (1979). Possibilities and needs for conservation of plant species and vegetation in Africa. In *Systematic botany, plant utilization and biosphere conservation.* (ed Hedberg I) Almqvist and Wiksell International, Stockholm. pp 83 – 104.

HILLIARD O M (1978). The geographical distribution of Compositae native to Natal. *Notes from the Royal Botanic Garden, Edinburgh* **36**, 407 – 425.

HILLIARD O M (1983). Conservation of plant species in Natal with comments on the Natal Drakensberg. *Report to the Working Group for Threatened Plants.* CSIR, Pretoria. 26 pp.

JENKINS R E (1981). Rare plant conservation through elements-of-diversity information. In *Rare plant conservation: geographical data organization.* (eds Morse L E and Henifin M S) New York Botanical Garden, Bronx, NY. pp 33 – 40.

KARTESZ J T and KARTESZ R (1977). *The biota of North America.* Volume I: Rare plants. Part I: Vascular plants. International Council for Outdoor Education, Bonac, Pittsburgh, Penn. 361 pp.

LEON C, LUCAS G and SYNGE H (1985). The value of information in saving threatened Mediterranean plants. In *Plant conservation in the Mediterranean area.* (ed Gomez-Campo C) Junk, Dordrecht. pp 177 – 196.

LUBKE R A, EVERARD D A and JACKSON S (1986). The biomes of the eastern Cape with emphasis on their conservation. *Bothalia* **16**, 251 – 261.

LUCAS G and SYNGE H (eds)(1978). *The IUCN plant red data book.* Morges. IUCN. 540 pp.

LUCAS G and WALTERS S M (1976). *List of rare, threatened and endemic plants for the countries of Europe.* IUCN, Morges. 166 pp.

POWELL W R (1974). *Inventory of rare and endangered vascular plants of California.* California Native Plant Society Special Publication No 1. Berkeley, Ca. 56 pp.

REBELO A G, COWLING R M, GIBBS RUSSELL G E, HOCKEY P A R, JARMAN M L, BOUCHER C and HILTON-TAYLOR C (1987). Guidelines for the plant atlas of southern Africa: the outcome of a workshop held in Cape Town, 5 December 1985. *Ecosystems Programmes Occasional Report* **23**. CSIR, Pretoria. 29 pp.

CHAPTER 10

Atlassing as a tool in conservation, with special reference to the Southern African Bird Atlas Project

J A Harrison

INTRODUCTION

A prerequisite for the conservation of biotic diversity is a description of the existing diversity of organisms and their spatial and temporal distributions and abundance. With the land available for nature conservation becoming ever more scarce, and the average size of the patches of undisturbed natural habitat becoming ever smaller, the need for accurate and relatively fine-scale distribution maps increases. There is a parallel need for more detail to support the increased rigour of ecological and biogeographical research. Gathering the amount of data required for detailed mapping has been made possible by the growing public interest in environmental issues and wildlife conservation. The technical ability to analyse large data sets has been greatly facilitated by the development of computer technology. For these and various other reasons, the concept of biological atlassing[1] has come of age and is now receiving the attention of various biological disciplines in many parts of the world.

Maps in a modern biological atlas should be distinguished from other types of distribution maps. The distributions of species have traditionally been described by drawing lines between localities at the limits of their known ranges and filling in gaps with informed guesswork. The data used in

---

[1]'Atlas' has become an accepted biological term and simply refers to a collection of distribution maps for species. 'Atlassing' has come to mean the gathering of data for an atlas but does not refer to any specific methodology.

this large subjective process can be drawn from any and all time periods meaning that the hypothetical distributional picture obtained may differ dramatically from the current situation. As a quantitative and hence more objective technique, atlassing provides an accurate map of current distributions without involving guesswork or interpolation.

In the discussion which follows I attempt to provide information on the theory and practice of atlassing, particularly as it applies to birds. Readers should bear in mind that every group of organisms would impose its own peculiar set of constraints on an atlas project.

THE ELEMENTS OF THE ATLAS APPROACH TO SURVEYING

The only essential characteristic of the atlas approach to surveying is that observations are objectively and accurately linked to geographical location. Atlases can be categorized according to the degree to which they exceed this fundamental trait:

— those that record distribution regardless of temporal factors (Boshoff et al 1983);
— those that describe distributions at a particular point in time (Sharrock 1977; Bull et al 1985);
— those that document the seasonality of distributions (Cyrus and Robson 1980; Tarboton et al 1987; Earlé and Grobler 1987);
— those that measure the relative abundance of species through space and time (Blakers et al 1984); and
— those that measure the absolute abundance of species (the new atlas of breeding birds in Britain and Ireland, in progress).

Despite the fact that atlassing is currently in vogue, the objectives of an atlas project should be critically evaluated. The current state of knowledge of the group of organisms to be mapped should determine those objectives.

The Southern African Bird Atlas Project (SABAP), for example, would have difficulty in justifying its existence if it attempted only to establish the basic distributions of birds, since these are already broadly known. The SABAP has, therefore, ensured that in addition to relatively accurate distributional information, data relating to seasonality, relative abundance and breeding are also obtained (Hockey and Ferrar 1985; Harrison 1988). However, in planning an atlas project, the objectives of the project and degree of detail desired have to be weighed against the resources available to gather and process the data.

Geographic scale

The survey area is usually divided into uniform geographic units, often referred to as a 'grid', which allows for a degree of aggregation of data. The distributional picture is thus a mosaic of data sets from all the geographic units. The size and use of these geographic units should, to some extent, be determined by the nature of the biota being surveyed. Relatively large units are appropriate for animals which move about freely, birds for example, whereas a survey of more sedentary biota, or those with very specialized or restricted habitat requirements, would call for the use of smaller units or exact localities. An extreme case would be rare plants where it would be necessary to specify exact locality to allow individual plants to be located again in the future (Rebelo et al 1987; Taylor and Hopper 1984). However, it should be noted that in the absence of a grid system, it is difficult to define and monitor the coverage objectives for a project.

The size of the survey area combined with manpower limitations will also determine the degree of aggregation of data that is desirable. Thus the greater the total area and the smaller the work-force, the coarser the grid would need to be.

An 'optimism factor' should be built into the survey so that the level of geographic accuracy recorded is finer than the minimum requirements dictate. Data can always be lumped into larger geographical units if coverage of the smaller units is inadequate, but a project may be unexpectedly successful, enabling mapping on the finer scale.

Coverage in an atlas project is always patchy: certain areas have the potential to be excellently covered, whereas others will receive only minimal coverage. The potential for thorough coverage in some areas should not be wasted by adopting too coarse a grid. The procedures whereby the data are collected, processed and analysed should have built into them a degree of flexibility with regard to the scale of mapping to allow for the geographic variation in intensity of coverage.

Time scale

Unfortunately atlas projects must be given temporal boundaries because of the costs involved. By restricting atlassing to a particular period of time, the distribution patterns obtained become specific statements for that period and can be used as a basis for comparison with past or future data sets.

The atlassing period should be long enough to provide an 'average' picture which is not excessively distorted by short-term fluctuations. It should also be short enough to prevent data collected early in the project from becoming obsolete before completion of the project. At the present rate of environmental change due to human activity, the period of an atlas project should probably not exceed 10 years.

Whereas atlas data can be recorded and computerized without any date, the inclusion of a date enables both long-term trends and seasonal changes to be detected in distribution or in any ancillary data being collected, for instance breeding activity. As seasonality can have important implications for management practices, dating of data should always be seriously considered.

Dating each item of data carries with it complicating effects as does recording exact locality for each observation. Again it is desirable to employ a degree of amalgamation for the sake of simplicity and efficiency. Temporal units that have been used are, for example, seasons of three months, and calendar months (Blakers et al 1984; Cyrus and Robson 1980).

An important consideration in determining the length of time units is the multiplicative effect these units have on the sampling rate of the survey. Clearly, monthly sampling would entail three times the number of samples and three times the volume of data that would be required by quarterly (three-monthly) units. This has obvious implications for the effort required not only in data collection but also for data collation and analysis.

Scope of data gathering

The scope of atlas data can easily be expanded beyond simple presence/absence information. Provision should be made at least for breeding data. Plant specimens can be examined in detail at leisure, and plant atlases therefore have greater potential for the collection of ancillary data.

A major constraint on the choice of ancillary information is the ability to record such information in the form of simple numerical codes. Whereas the technology exists to process lengthy non-numerical data, the cost of capturing such data becomes prohibitive when dealing with large numbers of field cards.

Computerization

Not only does computerization facilitate the compilation of large quantities of data, it also allows for the emergence of more subtle information based on statistical analyses of the data. This point is expanded in the section 'What can a bird atlas show?'.

Another great advantage of computerization lies in the relative ease with which interim analyses of data can be obtained. These are invaluable as a means of providing feedback to the work-force. Regular feedback on the progress of a project is vital to sustaining motivation and directing efforts towards areas of greatest need.

Purpose-written atlas programs are preferable to commercially available packages. Ideally they should be written by someone who is closely associated with the atlas project and who will be available to maintain and modify the programs during the course of the project. All programs should be written and thoroughly tested on data from a small-scale pilot project prior to the commencement of a major atlas project.

Computerization has important implications for the design of the data reporting form. If the data submitted by observers have to be manually transposed onto coding sheets prior to computerized data capture, the work-load involved in data capture is at least doubled. It is therefore highly desirable that the data be captured directly from the observer's report form. While being 'computer compatible', the form should nevertheless be carefully designed to be 'user friendly'. The SABAP appears to have been uniquely successful in this regard (Hockey and Ferrar 1985) (Figure 10.1).

Data capture is a process that inevitably introduces errors into the data bank. This problem can be largely overcome by using a series of programmed checks for 'impossible' and inconsistent information. The 'check digit' appended to each species code is an example of such a system employed by the SABAP (Hockey and Ferrar 1985).

Choice of observers

A crucial question in the planning of an atlas project is whom to use as observers in the field. The choice is essentially between using professionals only or using volunteers drawn from the general public. The relative merits and demerits of using professionals and volunteer amateurs as observers are outlined in Table 10.1.

TABLE 10.1   Professional versus amateur observers

Professionals	Amateur volunteers
Few participants therefore little data — more time required for adequate coverage	Many participants therefore many data — less time required for adequate coverage
Data relatively reliable — vetting of data less problematic	Data relatively unreliable — vetting of data vital
More sophisticated and subtle data can be recorded	Data collection needs to be kept as simple and as undemanding as possible
Small quantities of data restrict statistical validity of analyses	Large data sets lend themselves to statistical analysis which could lead to additional and perhaps unexpected information.
Administrative functions relatively simple and small scale	Sophisticated full-time administration for recruitment of and communication with volunteers, and for coping with large volume of data
Costs relatively modest if data gathering part of normal duties — costs prohibitive if professionals are specially employed	Costs relatively great in view of greater administrative complexity and greater volume of data
Project is relatively isolated from the community at large	Wide involvement with attendant benefits for environmental education
Option of long-term or ongoing data collection is more feasible	Suited to intensive short-term data collection in the region of three to five years
More suitable for little known, esoteric biota	Will respond best to well-known – 'popular' biota

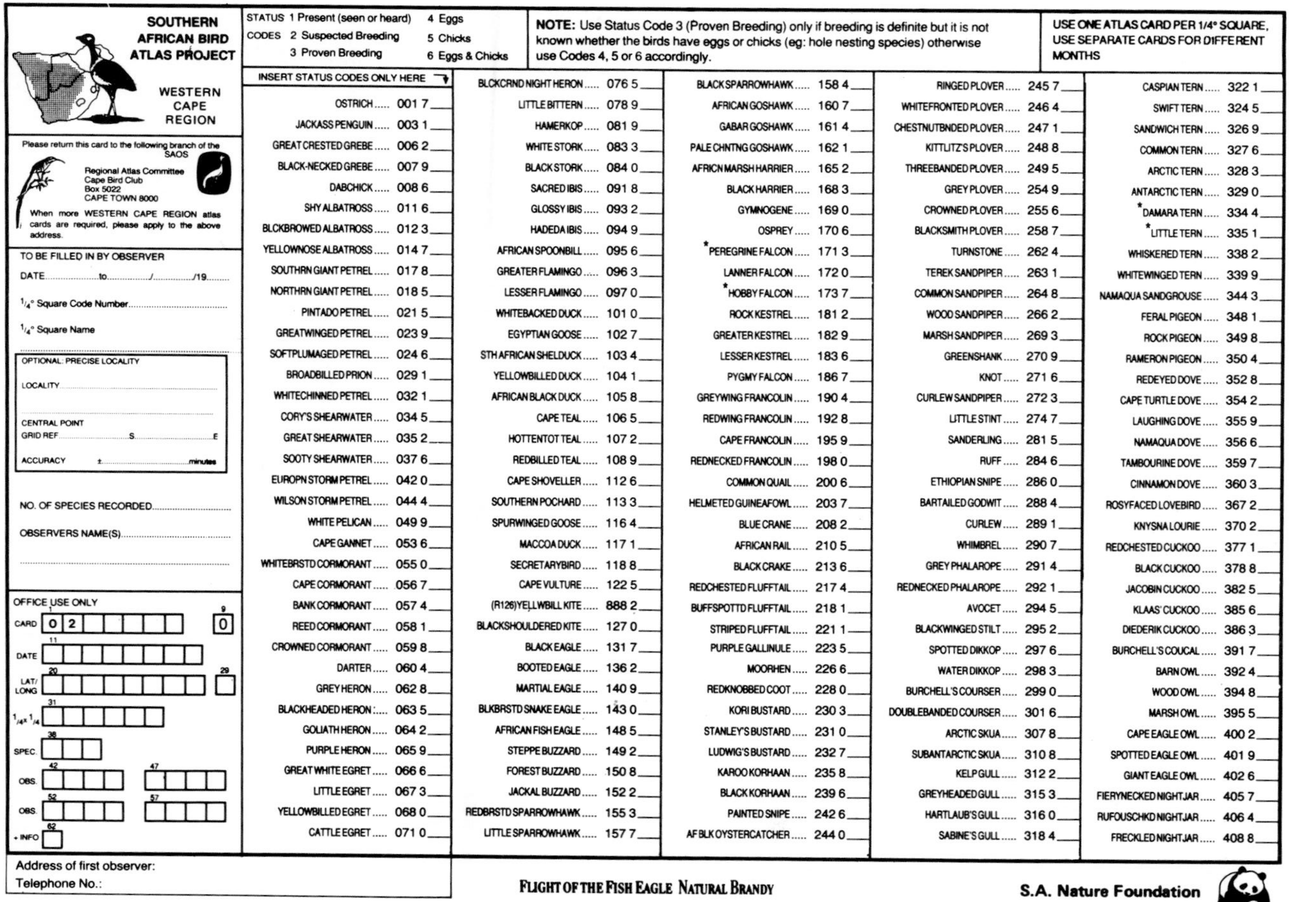

SOUTHERN AFRICAN BIRD ATLAS PROJECT

WESTERN CAPE REGION

Please return this card to the following branch of the SAOS

Regional Atlas Committee
Cape Bird Club
Box 5022
CAPE TOWN 8000

When more WESTERN CAPE REGION atlas cards are required, please apply to the above address.

TO BE FILLED IN BY OBSERVER

DATE .... to ..../........../19........

$^1/_4$° Square Code Number ....

$^1/_4$° Square Name

OPTIONAL: PRECISE LOCALITY

LOCALITY ....

CENTRAL POINT
GRID REF .... S .... E

ACCURACY ± .... minutes

NO. OF SPECIES RECORDED ....

OBSERVERS NAME(S) ....

OFFICE USE ONLY

CARD 0 2 (1) ... (9) 0

DATE (11)

LAT/LONG (20) ... (29)

$^1/_4$ x $^1/_4$ (31)

SPEC (38)

OBS (42) ... (47)

OBS (52) ... (57)

+INFO (62)

Address of first observer:
Telephone No.:

STATUS CODES: 1 Present (seen or heard); 2 Suspected Breeding; 3 Proven Breeding; 4 Eggs; 5 Chicks; 6 Eggs & Chicks

NOTE: Use Status Code 3 (Proven Breeding) only if breeding is definite but it is not known whether the birds have eggs or chicks (eg: hole nesting species) otherwise use Codes 4, 5 or 6 accordingly.

USE ONE ATLAS CARD PER 1/4° SQUARE, USE SEPARATE CARDS FOR DIFFERENT MONTHS

INSERT STATUS CODES ONLY HERE		BLCKCRND NIGHT HERON	076 5	BLACK SPARROWHAWK	158 4	RINGED PLOVER	245 7	CASPIAN TERN	322 1
OSTRICH	001 7	LITTLE BITTERN	078 9	AFRICAN GOSHAWK	160 7	WHITEFRONTED PLOVER	246 4	SWIFT TERN	324 5
JACKASS PENGUIN	003 1	HAMERKOP	081 9	GABAR GOSHAWK	161 4	CHESTNUTBNDED PLOVER	247 1	SANDWICH TERN	326 9
GREAT CRESTED GREBE	006 2	WHITE STORK	083 3	PALE CHNTNG GOSHAWK	162 1	KITTLITZ'S PLOVER	248 8	COMMON TERN	327 6
BLACK-NECKED GREBE	007 9	BLACK STORK	084 0	AFRICN MARSH HARRIER	165 2	THREEBANDED PLOVER	249 5	ARCTIC TERN	328 3
DABCHICK	008 6	SACRED IBIS	091 8	BLACK HARRIER	168 3	GREY PLOVER	254 9	ANTARCTIC TERN	329 0
SHY ALBATROSS	011 6	GLOSSY IBIS	093 2	GYMNOGENE	169 0	CROWNED PLOVER	255 6	*DAMARA TERN	334 4
BLCKBROWED ALBATROSS	012 3	HADEDA IBIS	094 9	OSPREY	170 6	BLACKSMITH PLOVER	258 7	*LITTLE TERN	335 1
YELLOWNOSE ALBATROSS	014 7	AFRICAN SPOONBILL	095 6	*PEREGRINE FALCON	171 3	TURNSTONE	262 4	WHISKERED TERN	338 2
SOUTHRN GIANT PETREL	017 8	GREATER FLAMINGO	096 3	LANNER FALCON	172 0	TEREK SANDPIPER	263 1	WHITEWINGED TERN	339 9
NORTHRN GIANT PETREL	018 5	LESSER FLAMINGO	097 0	*HOBBY FALCON	173 7	COMMON SANDPIPER	264 8	NAMAQUA SANDGROUSE	344 3
PINTADO PETREL	021 5	WHITEBACKED DUCK	101 0	ROCK KESTREL	181 2	WOOD SANDPIPER	266 2	FERAL PIGEON	348 1
GREATWINGED PETREL	023 9	EGYPTIAN GOOSE	102 7	GREATER KESTREL	182 9	MARSH SANDPIPER	269 3	ROCK PIGEON	349 8
SOFTPLUMAGED PETREL	024 6	STH AFRICAN SHELDUCK	103 4	LESSER KESTREL	183 6	GREENSHANK	270 9	RAMERON PIGEON	350 4
BROADBILLED PRION	029 1	YELLOWBILLED DUCK	104 1	PYGMY FALCON	186 7	KNOT	271 6	REDEYED DOVE	352 8
WHITECHINNED PETREL	032 1	AFRICAN BLACK DUCK	105 8	GREYWING FRANCOLIN	190 4	CURLEW SANDPIPER	272 3	CAPE TURTLE DOVE	354 2
CORY'S SHEARWATER	034 5	CAPE TEAL	106 5	REDWING FRANCOLIN	192 8	LITTLE STINT	274 7	LAUGHING DOVE	355 9
GREAT SHEARWATER	035 2	HOTTENTOT TEAL	107 2	CAPE FRANCOLIN	195 9	SANDERLING	281 5	NAMAQUA DOVE	356 6
SOOTY SHEARWATER	037 6	REDBILLED TEAL	108 9	REDNECKED FRANCOLIN	198 0	RUFF	284 6	TAMBOURINE DOVE	359 7
EUROPN STORM PETREL	042 0	CAPE SHOVELLER	112 6	COMMON QUAIL	200 6	ETHIOPIAN SNIPE	286 0	CINNAMON DOVE	360 3
WILSON STORM PETREL	044 4	SOUTHERN POCHARD	113 3	HELMETED GUINEAFOWL	203 7	BARTAILED GODWIT	288 4	ROSYFACED LOVEBIRD	367 2
WHITE PELICAN	049 9	SPURWINGED GOOSE	116 4	BLUE CRANE	208 2	CURLEW	289 1	KNYSNA LOURIE	370 2
CAPE GANNET	053 6	MACCOA DUCK	117 1	AFRICAN RAIL	210 5	WHIMBREL	290 7	REDCHESTED CUCKOO	377 1
WHITEBRSTD CORMORANT	055 0	SECRETARYBIRD	118 8	BLACK CRAKE	213 6	GREY PHALAROPE	291 4	BLACK CUCKOO	378 8
CAPE CORMORANT	056 7	CAPE VULTURE	122 5	REDCHESTED FLUFFTAIL	217 4	REDNECKED PHALAROPE	292 1	JACOBIN CUCKOO	382 5
BANK CORMORANT	057 4	(R126)YELLWBILL KITE	888 2	BUFFSPOTTD FLUFFTAIL	218 1	AVOCET	294 5	KLAAS' CUCKOO	385 6
REED CORMORANT	058 1	BLACKSHOULDERED KITE	127 0	STRIPED FLUFFTAIL	221 1	BLACKWINGED STILT	295 2	DIEDERIK CUCKOO	386 3
CROWNED CORMORANT	059 8	BLACK EAGLE	131 7	PURPLE GALLINULE	223 5	SPOTTED DIKKOP	297 6	BURCHELL'S COUCAL	391 7
DARTER	060 4	BOOTED EAGLE	136 2	MOORHEN	226 6	WATER DIKKOP	298 3	BARN OWL	392 4
GREY HERON	062 8	MARTIAL EAGLE	140 9	REDKNOBBED COOT	228 0	BURCHELL'S COURSER	299 0	WOOD OWL	394 8
BLACKHEADED HERON	063 5	BLKBRSTD SNAKE EAGLE	143 0	KORI BUSTARD	230 3	DOUBLEBANDED COURSER	301 6	MARSH OWL	395 5
GOLIATH HERON	064 2	AFRICAN FISH EAGLE	148 5	STANLEY'S BUSTARD	231 0	ARCTIC SKUA	307 8	CAPE EAGLE OWL	400 2
PURPLE HERON	065 9	STEPPE BUZZARD	149 2	LUDWIG'S BUSTARD	232 7	SUBANTARCTIC SKUA	310 8	SPOTTED EAGLE OWL	401 9
GREAT WHITE EGRET	066 6	FOREST BUZZARD	150 8	KAROO KORHAAN	235 8	KELP GULL	312 2	GIANT EAGLE OWL	402 6
LITTLE EGRET	067 3	JACKAL BUZZARD	152 2	BLACK KORHAAN	239 6	GREYHEADED GULL	315 3	FIERYNECKED NIGHTJAR	405 7
YELLOWBILLED EGRET	068 0	REDBRSTD SPARROWHAWK	155 3	PAINTED SNIPE	242 6	HARTLAUB'S GULL	316 0	RUFOUSCHKD NIGHTJAR	406 4
CATTLE EGRET	071 0	LITTLE SPARROWHAWK	157 7	AF BLK OYSTERCATCHER	244 0	SABINE'S GULL	318 4	FRECKLED NIGHTJAR	408 8

FLIGHT OF THE FISH EAGLE NATURAL BRANDY

S.A. Nature Foundation

FIGURE 10.1 An example of a field card used by the SABAP. Note the minimum of coding required (bottom left-hand corner) before actual data capture, which is carried out directly from the field card. The fourth digit in the species code is a computer check digit.

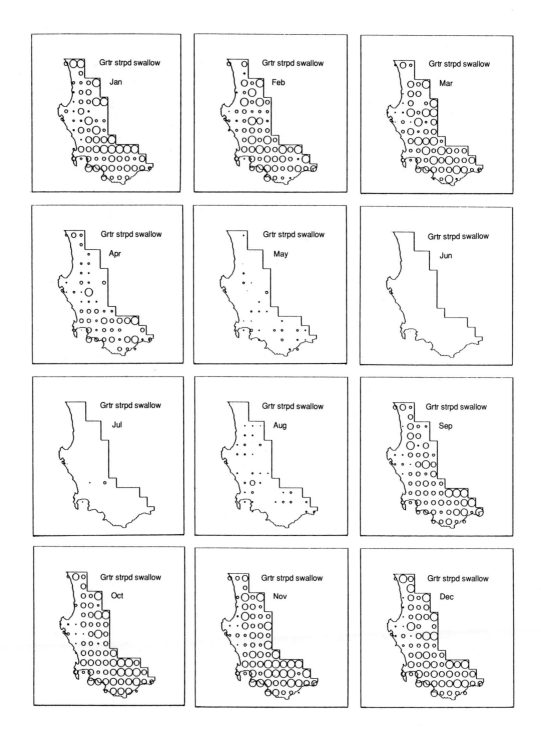

FIGURE 10.2 Monthly distribution maps for the greater striped swallow, an intra-African migrant, in the south-western Cape. Note the relatively brief period of absence from May to August. (The diameter of the circle is proportional to the reporting rate for each quarter-degree square.)

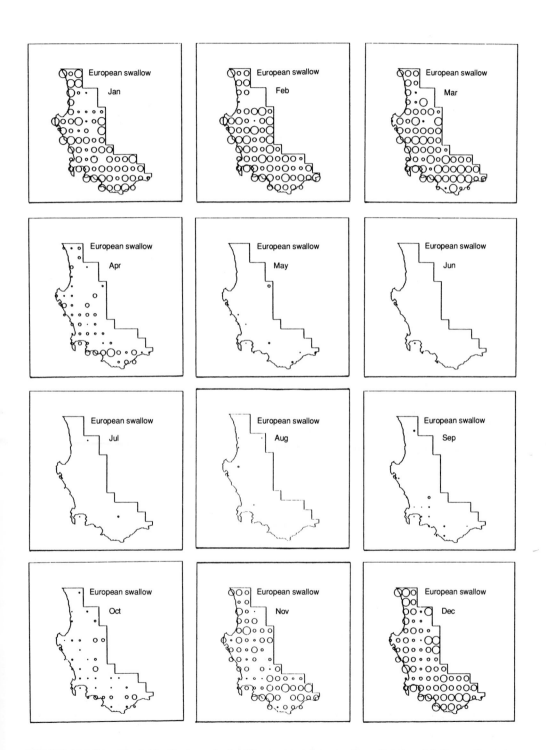

FIGURE 10.3 Monthly distribution maps for the European swallow, a palaearctic migrant, in the south-western Cape. Note the relatively long period of absence from April to October.

Because of the large geographic scale of most atlas projects there is, in practice, often no alternative to using volunteer amateurs to collect data. As can be seen from Table 10.1, this has important implications for the organization of an atlas project and it also determines many of the strengths and weaknesses of the atlas approach.

When is an atlas project indicated?

Although atlassing may appear to be the answer wherever baseline distributional data are required, there are certain fundamental limitations in the applicability of this approach.

Any group within which there are many unknown or inadequately described species will not lend itself to atlassing. Furthermore, there should be convenient and readily available field guides for the group to facilitate identification in the field.

If it is decided to use volunteer assistance with field work one is further restricted to those organisms that are relatively accessible and appealing to members of the public. In this regard birds are ideal but reptiles, for example, are not. A large, well-organized amateur organization can assist greatly with the recruitment and organization of volunteers. In the case of the SABAP, the Southern African Ornithological Society (SAOS) has played a pivotal role (Harrison 1987).

The scale and ambitiousness of a project obviously holds immense implications for the cost of the project, the most important factor being staffing. If the project is big enough to require the employment of special staff, a great increase in expense can be expected.

ATLASSING AND LONG-TERM MONITORING

Atlas projects generally attempt to cover a large geographic area with the help of a large volunteer amateur work-force. Such projects are usually relatively short-term and intensive with well-defined goals, so that costs can be kept down, fund raising motivated, and volunteers kept enthusiastic. Atlas projects *per se* are thus not equivalent to long-term monitoring; they attempt rather to provide baseline information. However, there are ways in which atlassing can contribute significantly towards long-term monitoring.

First, atlassing methods of data collection can be used for monitoring relatively small areas, particularly where a professional work-force is permanently available, as in forestry and nature conservation agencies.

As atlas projects span several years, trends can often be detected within the period (Underhill and Hockey 1988). These observations can provide the motivation for further, more specialized monitoring subsequent to the atlas project.

Objective methods make repeatability a feature of atlas projects. By repeating a project at intervals, the data series becomes the equivalent of 'pulsed' monitoring. For very large areas this type of monitoring may be more effective than continuous, low-intensity monitoring.

An atlas project such as the SABAP is, among other things, an ambitious public relations exercise. The possibility exists that sufficient dedicated volunteers will be recruited during the course of the project to allow data gathering to continue on a less intense but more structured basis after the initial atlassing period. Other options for long-term monitoring, namely ongoing data collection versus periodic repetition of atlas projects, should be investigated before the end of the SABAP's data -gathering period in 1991. The feasibility of a continuous monitoring attempt will depend largely on continuity of administration, which is envisaged in the form of a Bird Populations Data Unit (Ledger 1985 a, b; Prys-Jones 1984).

WHAT CAN A BIRD ATLAS SHOW?

The *raison d'être* of an atlas project is to describe distributions and this it can certainly do well if the desired standards of coverage are attained.

Beyond distribution *per se*, date-specific atlas data provide pictures of seasonal distributions for resident species, and times of arrival and departure for migratory species. This type of information enhances the potential for correlating atlas data with other environmental variables thus elucidating ecological mechanisms. See, for example, a comparison of seasonality of the greater striped swallow (an intra-African migrant) and the European swallow (a Palaearctic migrant) in the south-western Cape (Hockey et al in prep) (Figures 10.2 and 10.3).

This potential is greater still if breeding records have been included in the atlas data bank. Animals, particularly birds, often range well beyond their breeding grounds. Comparison of breeding and non-breeding distributions and pinpointing the timing of breeding can give further insight into ecological relationships and into the conservation priorities for species.

With regard to breeding it should be noted that where breeding activity is cryptic, as it is in many avian species, breeding records tend to be incidental and subsidiary to basic presence/absence information. This weakness can be overcome by the introduction of a formal search methodology, but this is not always practicable when using a semi-skilled work-force of amateur volunteers.

Although this is not one of the primary aims of an atlas project, it may be possible to detect changes from year to year. Such changes may be due to temporary exceptional circumstances such as droughts or floods, or they may be part of a cyclical pattern or ongoing trend. Despite the relatively short period of an atlas project, irruptions and range expansions and contractions can often be detected. Trends such as these can be further elucidated by reference to historical data (Blakers et al 1984). The hadeda ibis, for example, showed a marked westward expansion during the course of the Cape Bird Club's atlas project (Macdonald et al 1986; Hockey et al in prep) (Figure 10.4).

An important dimension of population variability is population size. Most atlas projects have not attempted to census species in order to avoid complicating atlassing techniques to a prohibitive degree. Nevertheless, an index of relative abundance can be obtained from the reporting rate (Blakers et al 1984). The diversity, ubiquity and conspicuousness of birds in particular make this type of analysis possible.

The technique depends on multiple sampling of the same geographic area, and calculating the proportion of field cards on which a particular species is recorded. For a given species, a difference in reporting rate between two areas is taken to indicate a difference in abundance.

It is important to appreciate that the calculation of the reporting rate assumes unbiased reporting of all species seen. It also assumes a reasonable minimum effort in the preparation of each field card. The field card is the basic sample unit and should list as many species as possible. All the data must be captured (computerized) in field card units in spite of the apparent redundancy of many of the data. The reporting and/or computerization of only certain 'special' species would make the calculation of reporting rates statistically meaningless.

Spatial variations in abundance of a species can indicate the 'core' range of the species as opposed to peripheral extensions of its range. Comparison of reporting rates for different species can be undertaken but only with great circumspection, since reporting rate is a function of conspicuousness and identifiability as much as it is a function of abundance. Where species are very similar and equally conspicuous, comparisons can be made, however. These are particularly interesting where the species are sympatric over parts of their ranges or where their ranges abut.

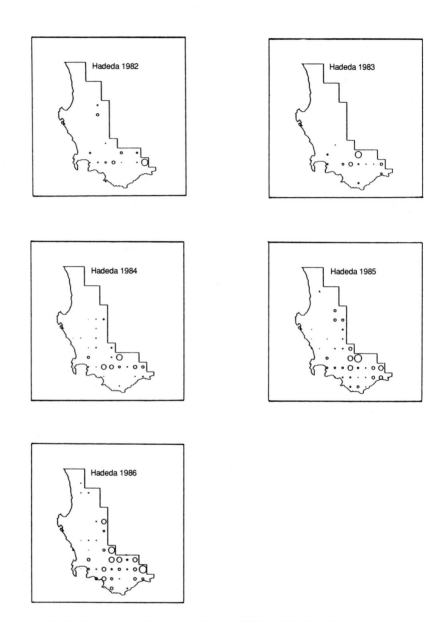

FIGURE 10.4 Annual distribution maps for the hadeda ibis from 1982 to 1986. The diameter of the circles is proportional to the reporting rate in the respective quarter-degree squares. The increasing number of squares in which the species was seen, together with an increased reporting rate in some of those squares, strongly suggests rapid expansion of the species' range into the region.

The replacement of one species by another in space invites analysis of habitat preferences and of the mechanisms maintaining allopatry or permitting sympatry.

It should be emphasized that far subtler insight into these issues can be gained from analyses of relative abundance than from absolute presence/absence patterns alone. The pied and black crows of the south-western Cape are a good case in point. Based on presence/absence data alone the two species appear sympatric in the south-western Cape, but when data are presented in terms of relative abundance, the distributions are found to be almost allopatric (Hockey 1987) (Figure 10.5).

Relative abundance over time can indicate temporal fluctuations in population size. Again these fluctuations can be related to environmental variables. For example, the irruptions of blackheaded canaries in the south-western Cape in 1982 and 1984 were probably the result of the severe drought conditions prevailing in their normal range in the northern Cape (Underhill and Hockey 1988).

A further example of the use of reporting rates concerns the fluctuations in population size indicated for certain species of migratory waders (Summers and Underhill 1987). These cyclical fluctuations were first detected using the recapture rate of ringed birds at Langebaan. The fluctuations were correlated with cycles of predation pressure, mainly by Arctic foxes, in their northern hemisphere breeding grounds. These cycles were in turn linked to the demographic cycle of lemming. Reporting rates in the south-western Cape provided independent corroboration of cyclical fluctuations in wader population sizes. The reporting rates for waders not subject to the same predation pressures showed no comparable fluctuations (Underhill and Hockey 1988).

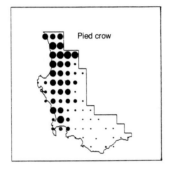 

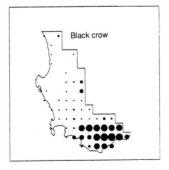

FIGURE 10.5 The distributions of pied crow, black crow and whitenecked raven in the south-western Cape. Whereas absolute presence/absence patterns suggest sympatry, relative abundance of the black crow and pied crow, as indicated by the reporting rates (circle diameters), suggest that these two species tend towards allopatry in the region. The whitenecked raven appears to favour the more mountainous parts.

The above examples should demonstrate that an atlas project can go well beyond the drawing of basic distribution maps provided that appropriate data have been collected and the necessary analytical programs are available.

THE BENEFITS OF A BIRD ATLAS TO CONSERVATION

Research required by nature conservation agencies can be categorized, in ascending order of complexity, as follows:

1) cataloguing of species within and outside conservation areas;
2) description of the distribution and status of threatened species;
3) evaluation of potential sites for nature reserves on the basis of features such as diversity and the extent of endemism;
4) monitoring of the status of species, particularly exploited species, and species that may be used as 'indicators' of more general biological conditions (eg the encroachment of alien vegetation) (Macdonald 1986);
5) determining appropriate management strategies for conservation areas, depending on management objectives (ie for diversity or for particular species); and
6) development of theory relating to the size, distribution and biological viability of nature reserves.

It is clear that the SABAP or any other atlas project will make a significant direct contribution to categories (1) and (2). Categories (3), (4) and (5) would also benefit greatly from the use of atlas data as part of broader data-gathering exercises.

Surveys of conservation areas usually result in the publication of 'check-lists' which frequently lack any reference to the status of individual species. Thus the inclusion of a species on a check-list does not necessarily mean that the species is being conserved in that area. It may be merely an occasional vagrant to the area or it may be on the decline. The quantification of relative abundance achieved by the SABAP means that a crude but useful measure of the status of species is obtained. Changes in relative abundance through space and time provide information relevant to conservation. Underhill and Hockey (1988) have developed a statistical technique with the potential to detect significant change in relative abundance.

Indications from the computerized south-western Cape Bird Atlas data are that several species will emerge as useful indicator species which will provide a key to the evaluation of habitat type and quality (P A R Hockey pers comm). This has a direct bearing on categories (4) and (5), in as much as atlas data will provide baseline information against which future changes may be measured.

The effective conservation of biotic diversity will entail the identification of centres of species diversity and of endemism. Evidence that centres of diversity and endemism tend to be congruent (Crowe and Crowe 1982) suggests that conservation could be especially advanced by identifying and setting aside land in such centres. Although atlas information is usually presented in the form of distribution maps for individual species, there is no obstacle to using geographic units as the focus of analysis. In this way species lists for localities can be compiled and compared in order to identify areas of particular species richness.

Given the necessary computer program, we could use our present knowledge of the composition of bird communities to map particular types of habitat. The computer would locate those areas in which a particular assemblage of species occurs, thus helping to pinpoint patches of habitat. The validity of this approach could be tested using habitat such as forest, and then the approach expanded to other habitats whose distribution is less easily detected by other standard techniques. This could be 'fine-tuned' to select those geographic units containing patches that are not yet degraded and are therefore potential candidates for conservation action. Conversely, the technique could be used to indicate areas of known habitat which are being degraded and therefore require further monitoring action.

Category (5) is important to professional conservationists, since it relates directly to the achievement of their aims and the execution of their duties. If the value of atlas data to categories (1) to (4) is demonstrated, it is reasonable to assume that they will be used to contribute towards the drawing up of management plans for the conservation of birds and, concomitantly, of biotic diversity.

Category (6) concerns the scientific basis for conservation action. Most studies of this nature would need to refer to a distributional data base as a foundation for the development of theory. As the SABAP promises to produce the most comprehensive and detailed distributional data base for a whole group of organisms yet available in southern Africa, it may prove essential to the development of this type of advanced theory in South Africa.

ACKNOWLEDGEMENTS

I gratefully acknowledge the constructive criticisms of Prof W R Siegfried, Prof L G Underhill, and Dr P A R Hockey during preparation of the text. Permission from the Cape Bird Club to use data from the Southwestern Cape Bird Atlas Project is acknowledged with thanks.

REFERENCES

BLAKERS M, DAVIES S J J F and REILLY P N (1984). *The atlas of Australian birds.* Melbourne University Press, Melbourne.

BOSHOFF A F, VERNON C J and BROOKE R K (1983). Historical atlas of the diurnal raptors of the Cape Province (Aves: Falconiformes). *Annals of the Cape Provincial Museums (Natural History)* **14**, 173 – 297.

BULL P C, GAZE P D and ROBERTSON C J R (1985). *The atlas of bird distribution in New Zealand.* The Ornithological Society of New Zealand, Wellington.

CROWE T M and CROWE A A (1982). Patterns of distribution and endemism in Afrotropical birds. *Journal of the Zoological Society of London* **198**, 417 – 442.

CYRUS D and ROBSON N (1980). *Bird atlas of Natal.* University of Natal Press, Pietermaritzburg.

EARLÉ R A and GROBLER N J (1987). *First atlas of bird distribution in the Orange Free State.* National Museum, Bloemfontein.

HARRISON J A (1987). The Southern African Bird Atlas Project. *South African Journal of Science* **83(7)**, 400 – 401.

HARRISON J A (1988). *SABAP Instructions Booklet.* Southern African Bird Atlas Project, University of Cape Town, Cape Town.

HOCKEY P A R (1987). Atlassing in the southwestern Cape: the first five years. *SABAP News* **(2)**, 7 – 8.

HOCKEY P A R and FERRAR A A (eds) (1985). Guidelines for the bird atlas of southern Africa. *Ecosystem Programmes Occasional Report* **2**, CSIR, Pretoria. 1 – 55.

HOCKEY P A R, UNDERHILL L G and NEATHERWAY M (in prep.) *Atlas of birds of the southwestern Cape.* Cape Bird Club, Cape Town.

LEDGER J A (1985a). Working towards a BPDB for South Africa. *Bokmakierie* **37(1)**, 21 – 23.

LEDGER J A (1985b). Progress report on BPDB activities in 1985. *Bokmakierie* **37(3)**, 67 – 69.

MACDONALD I A W (1986). Range expansion in the Pied Barbet and the spread of alien tree species in southern Africa. *Ostrich* **57**, 75 – 94.

MACDONALD I A W, RICHARDSON D M and POWRIE F J (1986). Range expansion of the Hadeda Ibis *Bostrychia hagedash* in southern Africa. *South African Journal of Zoology* **21**, 331 – 342.

PRŶS-JONES D (1984). A bird populations data bank for South Africa? *Bokmakierie* **36(4)**, 99 – 102.

REBELO A G, COWLING R M, GIBBS RUSSELL G E, HOCKEY P A R, JARMAN M L, BOUCHER C and HILTON-TAYLOR C (1987). Guidelines for the plant atlas of southern Africa. *Ecosystem Programmes Occasional Report* **23**, CSIR, Pretoria. 1 – 29.

SHARROCK J T R (1977). *The atlas of breeding birds in Britain and Ireland.* T & A D Poyser, Berkamsted, for the British Trust for Ornithology and the Irish Wild Bird Conservancy.

SUMMERS R W and UNDERHILL L G (1987). Factors related to breeding production of Brent Geese *Branta b. bernicla* and waders (Charadrii) on the Taimyr Peninsula. *Bird Study* **34**, 161 – 171.

TARBOTON W R, KEMP M I and KEMP A C (1987). *Birds of the Transvaal.* Transvaal Museum, Pretoria.

TAYLOR A and HOPPER S D (1984). *Banksia atlas instruction booklet and supplementary field guide.* Government Printer, Western Australia.

UNDERHILL L G and HOCKEY P A R (1988). The potential of the Southern African Bird Atlas Project for long-term population monitoring. In Long-term data series relating to southern Africa's renewable natural resources. (eds Macdonald I A W and Crawford R J M). *National Scientific Programmes Report* **157**, CSIR, Pretoria. 468 – 476.

CHAPTER 11

Assessment of animal diversity using molecular methods

W S Grant

INTRODUCTION

The conservation of biotic diversity ultimately depends upon the preservation of genetic information encoded in DNA (deoxyribonucleic acid). Species are not entities that necessarily act as
cohesive units: they consist of numerous populations which may act individually to some extent,
depending upon their degrees of reproductive isolation from one another. Population genetic
parameters, such as the magnitude of migration between populations and population size, have
important influences on the preservation of genetic variation. Events at the population level may
ultimately exert the greatest influence on the survival of a species. In small populations inbreeding
may lower selective fitness and reduce the long-term potential for species survival. With increasing
amounts of habitat disturbance by man, animal populations are being subdivided, and some
wildlife management practices may lead to the mixing of subspecies or genetically unique
populations.

Molecular methods are available which can be used to detect genetic variation at the level of
the gene, and this has enabled biologists to estimate levels of genetic variation within a species.
The results of numerous studies show that there is a wealth of molecular variation at the level of
the gene. A study of the geographic distributions of allele frequencies can reveal how populations
are structured genetically. Molecular methods can also be used to monitor the genetic effects of
manipulating populations through culling or by translocating individuals. The purpose of this paper
is briefly to outline the molecular basis of genetic techniques used for studying natural populations,
and to highlight the results of studies in population genetics and systematics which demonstrate
the application of this approach to understanding biotic diversity.

MEASURING GENETIC DIVERSITY

There are several molecular methods currently in use in population genetics and systematics. One group of techniques takes advantage of immune reactions to foreign proteins in higher animals. These methods include blood-group typing, which detects the presence of surface antigens on red and white blood cells, and microcomplement fixation, which measures the intensity of an antibody reaction to a specific protein. Another method, DNA-DNA hybridization, measures nucleotide sequence divergence between taxa by the degree of hybridization between the DNAs of the taxa. There are two techniques, however, that lend themselves most to studying genetic variation in natural populations: these are protein electrophoresis, and restriction-enzyme analysis of DNA.

Protein electrophoresis is used to detect mutations (nucleotide substitutions) in DNA that are reflected in the structures of the encoded proteins. Some nucleotide substitutions in DNA result in amino acid replacements in the encoded protein which change the overall electrostatic charge of the protein. These charge differences are inherited in a simple genetic fashion and can be viewed as mobility differences on an electrophoretic gel (Figure 11.1). The distributions of these variants can be used to infer the genetic stuctures of populations. For instance, the proportion of heterozygous genotypes in a sample can be used to measure the amount of genetic variation within a population, and the geographic distributions of alleles can be used to estimate the degree of reproductive isolation among populations.

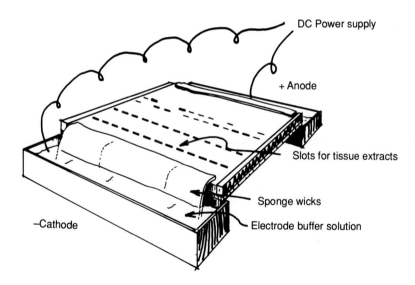

FIGURE 11.1 Diagram of a protein electrophoretic apparatus

The number of locations and individuals sampled and the number of proteins examined in a study depend upon the hypothesis being tested. When the object of a study is to infer the genetic structure of a group of populations, it is usual for 50 to 100 individuals to be sampled at several locations over the area of interest so that small differences in allelic frequencies can be detected. In taxonomic studies, on the other hand, fewer individuals of each taxa need to be analysed, but a large number of protein loci are required to produce accurate estimates of genetic distance between taxa. Computer simulations show that at least 30 loci should be examined in order to provide adequate phylogenetic resolution (Nei et al 1983). Although protein electrophoresis was first applied to the study of natural populations in 1966 (Lewontin and Hubby 1966), it is still widely used today because of its simplicity and ease of application.

Advances in biotechnology in the 1970s generated another method for studying molecular variation in which a class of bacterial enzymes called restriction enzymes could be used to cleave DNA at specific nucleotide sites. These enzymes recognize short DNA sequences, usually from four to six nucleotides in length, and cleave the DNA at specific points in the sequence. The resulting fragments of DNA are separated by agarose or acrylamide electrophoresis, and the number of restriction sites for a particular enzyme (ie the number of restriction fragments) can be determined. In most studies 10 to 15 restriction enzymes are used and usually 5 to 10 individuals per location are examined if the object of the study is to infer geographic population structure. The profile of restriction fragments or restriction-site data can be used in several ways. If the relationships of the genotypes reflect simple changes in restriction sites, parsimony networks may be constructed which summarize the history of genotypic transitions (Avise et al 1987). Alternatively, the presence or absence of shared restriction fragments may be analysed cladistically (DeBry and Slade 1985). The proportion of fragments shared by two individuals may also be used to estimate the number of nucleotide differences between them (Nei and Li 1979).

It is difficult, however, to isolate the same regions of nuclear DNA from different individuals for routine surveys of natural populations. Therefore, another form of DNA found in small organelles, called mitochondria, is used in most studies of natural populations of animals. Mitochondrial DNA (mtDNA) can be separated from nuclear DNA by differential centrifugation of the intact organelles, or by ultracentrifugation in a caesium chloride density gradient (Figure 11.2). Mitochondrial DNA is a covalently closed circular molecule and is therefore more dense than nuclear DNA. It is about 17 000 nucleotides long in most higher animals. Since mitochondria are inherited only through the female parent, this kind of analysis yields different kinds of information about natural populations than does the examination of nuclear genes by protein electrophoresis (Wilson et al 1985). It is therefore possible to infer maternal lineages among populations (Avise et al 1987) and to infer the direction of introgressive hybridization between species (eg Gyllensten and Wilson 1987) with the analysis of mtDNA.

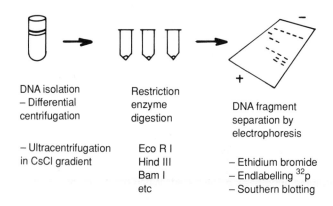

DNA isolation	Restriction	
– Differential centrifugation	enzyme digestion	DNA fragment separation by electrophoresis
– Ultracentrifugation in CsCl gradient	Eco R I Hind III Bam I etc	– Ethidium bromide – Endlabelling ^{32}p – Southern blotting

FIGURE 11.2 Diagram of procedures in the restriction enzyme analysis of mtDNA

One problem that sometimes arises in the study of rare or endangered species is that it is difficult to obtain samples for genetic analysis. Apart from tissue culture of biopsies from living individuals, most molecular techniques require samples of tissues such as liver, heart or muscle which normally leads to the death of an individual. A recently developed technique called polymerase chain reaction (PCR) (Saiki et al 1986) is being used to amplify small amounts of DNA for study. Theoretically a single strand of DNA can be amplified by cycles of heating, in which double-stranded DNA is 'melted' into single strands, and cooling, in which the single-stranded DNA is used as a template to produce double-stranded DNA. The amount of DNA doubles each cycle and very large quantities of DNA can be produced in about 30 cycles. Because primers at either end of the target DNA are required, DNA that has been sequenced previously is required for primer construction. Mitochondrial DNA may be ideal for this method because it has been sequenced for several species of mammal.

GENETIC VARIATION WITHIN POPULATIONS

Until the mid-1960s, little was known about the levels of inherited molecular variation in natural populations. Early genetic research using fruit flies led to the hypothesis that there was very little genetic variation in natural populations, because most morphological mutants were rare and harmful. Most natural populations were therefore thought to consist of genetically uniform, 'wild' type individuals. One of the first results of protein electrophoresis was to show that natural populations actually harboured a considerable amount of genetic variation at the level of the gene (Lewontin and Hubby 1966).

Measuring genetic variation

One convenient way of measuring the amount of genetic variation in a species or population is to take advantage of the fact that most plants and animals have two sets of chromosomes. Each set, or genome, is inherited from one of the two parents. There are therefore two bits of information for every gene. The two genes for a given trait may be identical, in which case the individual is said to be homozygous for that trait, or the genes may be different. When there is a difference in the nucleic acid sequence between two homologous genes, the individual is said to be heterozygous. One way of estimating the amount of variation in a population is to estimate from a sample of individuals the average proportion of heterozygotes for a collection of genes.

Usually we assume that the sample of individuals was taken from a population with random mating so that the proportions of homozygotes and heterozygotes fit a Hardy-Weinberg equilibrium. Expected heterozygosities for each locus (h) may be calculated from gene frequencies by subtracting the expected proportions of homozygotes from one:

$$h = 1 - \Sigma p_i{}^2$$

This yields the proportion of heterozygous genotypes and, when averaged over loci (H), is a measure of the amount of genetic variation in a population. Gorman and Renzi (1979) have shown that if a large number of loci have been examined, reasonably accurate estimates of H may be calculated from the examination of only a few individuals.

Genetic variation in major taxonomic groups

The amount of genetic variation in several major groups of animals is summarized in Figure 11.3. The solid bars are the mean heterozygosities for the groups indicated, and the open bars are heterozygosity values of particular species. There tends to be less genetic variation in organisms that are more complex with regard to their evolution. Invertebrates, such as molluscs and insects for example, tend to have more genetic variation than vertebrates. Among vertebrates, the more advanced taxonomic groups such as mammals generally have less genetic variation than more primitive groups. The primary reason for this is that more primitive groups have larger population sizes and are less subject to the loss of genetic variation through genetic drift and inbreeding. More advanced organisms tend to show a greater degree of social organization which has the effect of subdividing species into smaller populations.

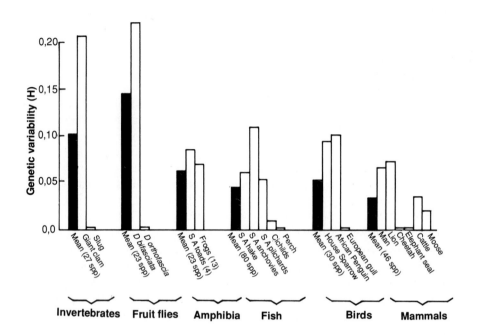

FIGURE 11.3 Distribution of heterozygosities (*H*) among major taxonomic groups. Overall averages for the different groups were taken from Nevo (1978) except for amphibians (Guttman 1985) and birds (Barrowclough 1983). Specific examples: slug, *Rumina* (Selander and Kaufman 1973); giant clam, *Tridacna* (Ayala et al 1973); *Drosophila orthofascia* Hawaii (Nevo 1978); *Drosophila bifasciata* Scandinavia (Saura 1974); hake, *Merluccius capensis* (Grant et al 1987); anchovy, *Engraulis japonicus* (Grant 1985b); pilchard, *Sardinops ocellata* (Grant 1985); cichlid (Van der Bank et al 1988); perch, *Perca fluviatilis* (Gyllensten et al 1985); toads, *Bufo* (Cherry and Grant in prep); house sparrow, *Passer domesticus* (Parkin and Cole 1985); European gull, *Larus argentatus* (Johnson 1985); man, *Homo sapiens* (Harris and Hopkinson 1972); lion, *Panthera leo* (Newman et al 1985); cheetah, *Acinonyx jubatus* (O'Brien et al 1983); elephant seal, *Mirounga angustirostris* (Bonnell and Selander 1974); cattle, *Bos taurus* (Wurster and Benirsch-ke 1968); moose, *Alces alces* (Ryman et al 1980).

The second trend that can be seen in Figure 11.3 is that within each major group there is a large amount of variation in heterozygosity values. The primary sources of this variation are population genetic effects brought about, for instance, by variation in population size, inbreeding, and the amount of migration between populations. Examples of a reduction in genetic variation brought about by founder events or historical reductions in population size (population bottlenecks) can be seen in invertebrate species such as a slug that was introduced to North America, and a species of fruit fly *Drosophila* that was most likely established on one of the Hawaiian islands by a very small founding population. Among mammals, elephant seals and, apparently, cheetahs are examples of species that have experienced recent reductions in population size and as a result have lost nearly all of their genetic variation.

GENETIC VARIATION AMONG POPULATIONS

Species are not single, cohesive entities in which all individuals are genetically homogeneous. Instead, they consist of mosaics of genetically differentiated populations knitted together to some degree by gene flow. The degree of reproductive isolation, and hence the degree of genetic differentiation, among populations is determined largely by the amount of migration between populations. Populations themselves may be further subdivided into colonies, herds, flocks or prides, or by some other form of social behaviour.

Theoretical considerations

We can measure the degree of subdivision from the gene-frequency data, generated by protein electrophoresis, for instance, by partitioning the total average heterozygosity, H_T into its within- and between-population components (Nei 1973):

$$H_T = H_S + D_{SC} + D_{ST}$$

The variable H_S is the mean value over populations of the amount of genetic variation (H) within populations (Figure 11.4). The distribution of this variation among major taxonomic groups was discussed in the previous section. The greatest amount of genetic variation in a species is contained at this level of population organization. Another component is the variation that is due to genetic differences among populations. The frequencies of some genetic traits may differ from one population to the next, or some traits may exist in some populations but not in others. This component is measured by the D statistic which itself may be decomposed into additional levels of population subdivision. In our equation, for example, D_{ST} may be a measure of variation among subpopulations, and D_{SC} a measure of variation among colonies within subpopulations. The analysis of gene-frequency data using this approach can give us a valuable overview of the degree of differentiation at the subspecies and population levels of organization.

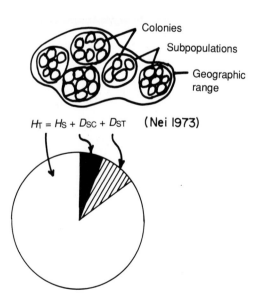

FIGURE 11.4 Analysis of genetic variation among populations. The components of the total amount of genetic variation in a species is represented by a pie diagram (Nei 1973).

Genetic differentiation among populations is determined largely by three interacting factors: random genetic drift due to finite population sizes; natural selection; and the magnitude of gene flow (migration). The amino acid substitutions detected by protein electrophoresis are thought to be neutral or nearly-neutral to selection so that the results of most genetic field studies can be interpreted in terms of random drift and gene flow (Lewontin 1974). Natural selection may, however, be reflected in molecular data to the extent that it affects gene flow. Although mutation is the ultimate source of genetic variation, recurrent mutation is not by itself an important force in bringing about evolutionary change.

Genetics of populations

We can see the effects of gene flow on population structure by comparing genetic data of species with differing life-history patterns and with differing amounts of migration among populations. For example, marine invertebrates, which have pelagic larvae that can disperse long distances in oceanic currents, tend to show very few genetic differences between populations (Figure 11.5). In the limpet *Siphonaria* there were virtually no gene-frequency differences among populations distributed over a distance of 3 000 km along the coast of Western Australia (Johnson and Black 1985). Mussel (*Mytilus galloprovincialis*) populations along the coast of southern Africa also show very little genetic differentiation, presumably because of large amounts of larval drift (Grant unpublished data).

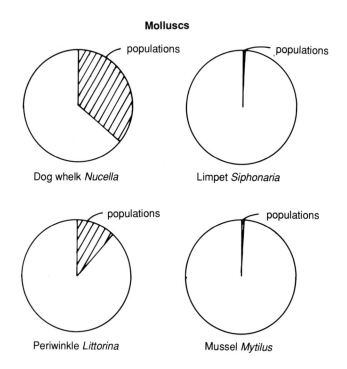

Molluscs

Dog whelk *Nucella*   Limpet *Siphonaria*

Periwinkle *Littorina*   Mussel *Mytilus*

FIGURE 11.5 The distributions of genetic variation within and among populations of molluscs with different life-history patterns. Sources of data: *Nucella* Grant and Utter (1988); *Siphonaria* Johnson and Black (1985); *Littorina* Janson (1987); *Mytilus* W S Grant (unpublished).

On the other hand, species that deposit egg capsules on rocks, such as the intertidal whelk *Nucella*, or species that brood their young, such as the periwinkle *Littorina*, have a much reduced potential for gene flow between populations. These species show a much larger degree of genetic fragmentation among populations. Populations of the whelk *Nucella lamellosa*, for instance, can have remarkably different shell forms and protein frequencies from one location to the next. About 35% of the total genetic variation in this species is due to population differences (Grant and Utter 1988). Populations of the Scandinavian periwinkle *Littorina saxatilis* show a remarkable degree of microdifferentiation among populations separated by only a few metres (Janson 1987).

Lower vertebrates, such as fish, also show the effects of migration on their genetic population structure. Along the coast of southern Africa fishes such as the anchovy *Engraulis japonicus*, and the hakes *Merluccius capensis* and *M paradoxus,* spawn in the open ocean, and larvae are capable of drifting long distances in gyres of the northward-flowing Benguela Current. For all of these species less than three per cent of the total genetic variation was due to regional differentiation among populations (Grant 1985a; Grant et al 1987). From a genetic point of view these species consist of a single population.

The potential for gene flow between populations of estuarine fishes, which are confined to river mouths and brackish waterways, is much reduced, and populations of these fishes show a correspondingly greater degree of genetic subdivision (Figure 11.6). The total amount of genetic variation in the Cape silverside *Atherina breviceps* and the estuarine minnow *Gilchristella*

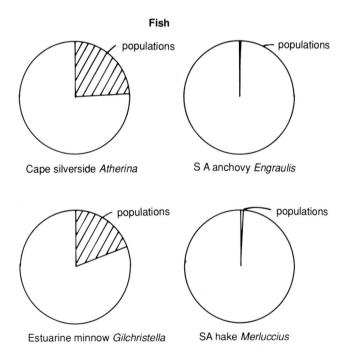

Fish

populations — Cape silverside *Atherina*

populations — S A anchovy *Engraulis*

populations — Estuarine minnow *Gilchristella*

populations — SA hake *Merluccius*

FIGURE 11.6 The distributions of genetic variation within and among populations of fishes with differing amounts of migration between locations. Sources of data: *Atherina* Grant et al (in prep); *Engraulis* Grant (1985a); *Gilchristella* Ratte et al (in prep); *Merluccius* Grant et al (1987).

aestuarius due to subdivision among estuaries is 24% and 18%, respectively (Grant et al in prep; Ratte et al in prep).

Mammals are affected by the same population genetic processes. The potential for long-distance migration and gene flow between populations has been much restricted because of the exploitation and destruction of natural habitats by man. Most populations of large mammals are far more fragmented now than they have been over evolutionary time. Nevertheless, mammalian populations are characterized by a large amount of genetic subdivision because of their high degree of social organization. For instance, populations of the springbok *Antidorcas marsupialis* show as much genetic fragmentation between herds within a region as nominal subspecies do between regions (Figure 11.7) (Robinson et al 1978). A similar degree of subdivision among local populations can be seen among populations of herds of elephant *Loxodonta africana* in the Kruger National Park (Osterhoff et al 1974). The results of a study of immunological blood groups in the olive baboon *Papio anubis* in Kenya also showed a high degree of genetic subdivision among localities. About 36% of the total variation in this species was due to gene-frequency differences among troops (Olivier et al 1986).

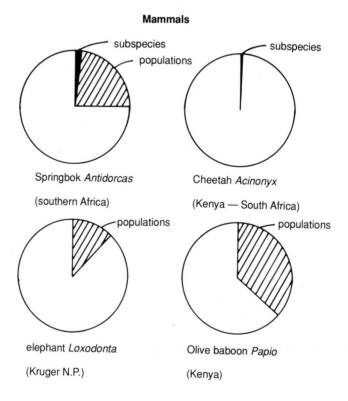

FIGURE 11.7 The distributions of genetic variation within and among populations of African mammals. Sources of data: *Antidorcas* Robinson et al (1978); *Acinonyx* O'Brien et al (1987); *Loxodonta* Osterhoff et al (1974); *Papio* Olivier et al (1986).

One apparent exception to the large amount of genetic subdivision among populations of mammals is the cheetah *Acinonyx jubatus*, which appears to show only a small degree of genetic differentiation between populations located as far apart as Kenya and South Africa (O'Brien et al 1987). The reason for this is that there is not enough genetic variation to reflect the degree of reproductive isolation that exists among populations.

Inbreeding

As populations of wildlife become more subdivided and smaller because of human population growth, the loss of genetic variation through random drift and inbreeding may adversely affect fitness and the chances of survival for some species. There is a large body of evidence from animal husbandry, zoo populations (Ralls et al 1979) and natural populations (Cothran et al 1983) to show that the loss of genetic variation can lead to the loss of disease resistance, developmental stability, fecundity and juvenile vigour. Over long periods, genetic variation is necessary for evolutionary change. Without variation on which natural selection can act, there can be no evolutionary change and populations may not be able to adapt to long-term environmental changes. Molecular methods are particularly useful for monitoring the levels of genetic variation in captive populations intended to restock or augment natural populations (eg Allendorf and Phelps 1980; Ryman and Stahl 1980), and for monitoring the genetic effects of management practices such as translocation and culling (eg Ryman et al 1981).

Preventive measures can be taken to reduce the degree of inbreeding in small populations. One possibility is to maintain large population sizes so that the loss of genetic variation through random drift is not significant. This is not always possible, however, because of the limits nature reserves impose on population growth and because some remnant populations, like the Knysna elephants or the Cape buffalo, are limited in size by slow growth rates. In many instances, however, artificial gene flow, through the transport of individuals from one locality to another, can greatly reduce the degree of inbreeding in each reserve. The successful introduction of only a single individual, into a breeding population, regardless of the size of the population, can theoretically reduce the degree of inbreeding at equilibrium from a value of 1,0 (totally inbred individuals) to 0,2 (Grant 1989).

SYSTEMATICS

Another very important use of molecular methods in understanding biological diversity is in the field of taxonomy. Unlike morphological analyses, molecular methods can provide quantitative estimates of genetic distance between taxa. Allele-frequency data in the case of protein electrophoresis, or the proportions of shared restriction fragments or sites in the case of mtDNA data, can be used to calculate a measure of genetic distance between two taxa. The results of a large number of studies show that there is a good correspondence between genetic distance and traditional classification (Thorpe 1982), so molecular genetic distance can be used to assist in problematic taxonomic classifications.

Inferring Phylogenies

Since the philosophy of modern systematics is that taxonomic classification should reflect evolutionary ancestry (phylogeny), much of the work in molecular systematics is concerned with the development of molecular and statistical methods for inferring phylogenies. Figure 11.8 shows an example of the use of protein electrophoresis in combination with a cladistic analysis to infer the phylogenetic relationships among species of cichlid fishes from the Caprivi Strip of Namibia (Grant et al in prep). In this approach the presence or absence of electrophoretic alleles was used to identify ancestral and share-derived character states among taxa. A parsimony computer algorithm (PAUP; Swofford 1985) was then used to produce the phylogenetic tree with the shortest total length of branches. This phylogenetic tree conforms to traditional taxonomic treatments based on the morphology, ecology and feeding habits of cichlid fishes.

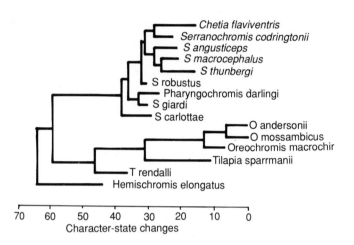

Cichlid fishes

FIGURE 11.8 A cladistic tree of phylogenetic relationships among 15 species of cichlid fishes from the Caprivi Strip, Namibia (Grant et al in prep). The phylogeny was based on the presence or absence of alleles in the various taxa and generated by a parsimony computer algorithm (PAUP; Swofford 1985).

Figure 11.9 shows an example of a phylogeny derived from the immunological analysis of albumins (Collier and O'Brien 1985). This approach is more suited to the study of distantly related species than is protein electrophoresis or the restriction-fragment analysis of mtDNA. These results resolved the large cats into three distinct lineages — the ocelot, domestic cat and panther lineages — which are also reflected in the chromosome numbers and morphologies of these groups.

Detection of morphologically cryptic species and hybrids

One of the more practical uses of molecular methods is the identification of species that have not been detected by morphological methods but which represent genetically and evolutionarily distinct lineages. Morphologically indistinct species of invertebrates — gastropods, bivalves and fishes — have been discovered during routine population surveys using molecular methods. In southern Africa, for instance, a previously unrecognized species of intertidal mussel on the west coast was detected using protein electrophoresis (Grant and Cherry 1985).

Hybrids can also be identified by molecular methods. A protein electrophoretic study of fishes in Hardap Dam, Namibia, showed that hybridization had occurred between species within two different genera, *Barbus* and *Labeo* (Van Vuuren et al in prep). The hybridizations might have occurred because of forced sympatry brought about by the construction of the dam.

When the analysis of nuclear genes is combined with the restriction-enzyme analysis of mtDNA, the sexes of the parental species contributing to the hybrid can be identified. The reason for this is that mtDNA is maternally inherited and remains unchanged from one generation to the next

Felids

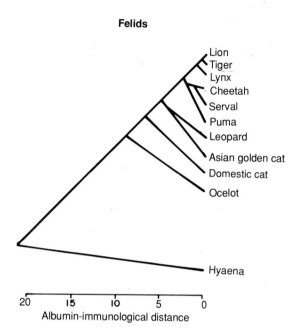

FIGURE 11.9 A molecular phylogeny of felids based on albumin-immunological distances between taxa (Collier and O'Brien 1985)

because of the general lack of recombination, at least in animals. An example of this was found in two species of European mouse, in the genus *Mus*, which form a hybrid zone extending from Denmark to Yugoslavia so that *Mus domesticus* occurs in western Europe and *Mus musculus* in eastern Europe. Although the mice of Scandinavia are morphologically and electrophoretically of the *Mus musculus* type, their mtDNA consists entirely of the *Mus domesticus* type (Gyllensten and Wilson 1987). The explanation for this is that the individuals that founded the Scandinavian populations most probably originated from a hybrid cross in which the female parent was of *Mus musculus* type. A maternal *Mus domesticus* mtDNA lineage became established in the other species, and individuals of this lineage colonized the whole of Scandinavia with the spread of grain culture about 4 000 years ago. Founder events established with hybrid individuals may also account for instances of cross-species transfer of mtDNA in other mammals (Tegelstrom 1987) and amphibians (Spolsky and Uzzell 1984).

CONCLUSION

Species are composed of individuals and populations displaying a dynamic array of genotypes, which are constantly changing in response to natural selection and genetic drift. Molecular tools can be used to assess genetic diversity at several hierarchical levels of population organization representing subspecies, populations and subpopulations, and herds, flocks, prides or other forms of social organization. There is no single ideal molecular method for the analysis of genetic diversity; the different methods are suited to addressing different questions. The analysis of mtDNA yields information about population events affecting maternal lineages but not about events affecting nuclear genes which undergo Mendelian segregation and recombination every generation. Although there are errors associated with estimates of molecular distance between taxa, estimates of genetic distance nevertheless provide quantitative answers to problems that might be intractable to traditional morphological methods. Molecular methods, however, cannot replace morphological, ecological or behavioural approaches to studying biological diversity, but should be viewed as complementary to these methods.

The discussion in this paper has been limited to the use of molecular methods for measuring genetic diversity and to population genetic mechanisms that may affect natural or captive populations. Although these processes can influence population viability, they can be overshadowed by other events such as random fluctuations in the demographic make-up of small populations, habitat changes, and natural catastrophes (Gilpin and Soule 1986). Any conservation effort must include consideration of all possible sources of population and species extinction.

REFERENCES

ALLENDORF F W and PHELPS S R (1980). Loss of genetic variation in a hatchery stock of cutthroat trout. *Transactions of the American Fishery Society* **109**, 537 – 543.

AVISE J C, ARNOLD J, BALL R M, BERMINGHAM E, LAMB T, NEIGEL J E, REEB C A and SAUNDERS N C (1987). Intraspecific phylogeography: the mitochondrial DNA bridge between population genetics and systematics. *Annual Review of Ecology and Systematics* **18**, 489 – 522.

AYALA F J, HEDGECOCK D, ZUMWALT G S and VALENTINE J W (1973). Genetic variation in *Tridacna maxima*, an ecological analog of some unsuccessful evolutionary lineages. *Evolution* **27**, 177 – 191.

BARROWCLOUGH G F (1983). Biochemical studies of microevolutionary processes. In *Perspective in Ornithology* (eds Bush A H and Clark G A Jr) Cambridge University Press, Cambridge. pp 223 – 261.

BONNELL M L and SELANDER R K (1974). Elephant seals: genetic variation and near extinction. *Science* **184**, 908 – 909.

CHERRY M I and GRANT W S (in prep). Evolution of the Bufonidae in southern Africa: a correlation of electrophoretic and behavioural evidence.

COLLIER G E and O'BRIEN S J (1985). A molecular phylogeny of the felidae: immunological distance. *Evolution* **39**, 473 – 487.

COTHRAN E G, CHESSER R K, SMITH M H and JOHNS P E (1983). Influences of genetic variability and maternal factors on fetal growth in white-tailed deer. *Evolution* **37**, 282 – 291.

DE BRY R W and SLADE N A (1985). Cladistic analysis of restriction endonuclease cleavage maps with a maximum-likelihood framework. *Systematic Zoology* **34**, 21 – 34.

GILPIN M E and SOULE M E (1986). Minimum viable populations: processes of species extinction. In *Conservation Biology*. (ed Soule M E) Sinauer, Sunderland, Mass. pp 19 – 34.

GORMAN G C and RENZI J JR (1979). Genetic distance and heterozygosity estimates in electrophoretic studies: effects of sample size. *Copeia 1979,* 242 – 249.

GRANT W S (1985a). Biochemical genetic stock structure of the southern African anchovy, *Engraulis capensis* Gilchrist. *Journal of Fish Biology* **27**, 23 – 29.

GRANT W S (1985b). Population genetics of the southern African Pilchard, *Sardinops ocellata*, in the Benguela upwelling system. In *Simposio Internacional sobre las areas de afloramiento mas Importantes del oeste Africano.* (eds Bas C, Margalef R, and Rubies P) Investigaciones Pesqueras, Barcelona. pp 551 – 562.

GRANT W S (1989). Genetic considerations in the management of small populations. *South African Journal of Wildlife Management.*

GRANT W S and CHERRY M I (1985). *Mytilus galloprovincialis* Lmk. in southern Africa. *Journal of Experimental Marine Biology and Ecology* **90**, 179 – 191.

GRANT W S, LESLIE R W and BECKER I I (1987). Genetic stock structure of the southern African hakes *Merluccius capensis* and *M paradoxus. Marine Ecology Progress Series* **41**, 9 – 20.

GRANT W S, RATTE T and LESLIE R W (in prep). Morphological and genetic differentiation of *Atherina breviceps* populations in southern Africa.

GRANT W S and UTTER F M (1988). Genetic heterogeneity on different geographic scales in *Nucella lamellosa* (Prosobranchia, Thaididae). *Malacologia* **1-2**, 275 – 287.

GRANT W S, VAN DER BANK H and FERREIRA J T (in prep). Molecular evolution of southern African cichlid fishes.

GUTTMAN S I (1985). Biochemical studies of Anuran evolution. *Copeia 1985,* 292 – 309.

GYLLENSTEN U, RYMAN N and STAHL G (1985). Monomorphism of allozymes in perch *(Perca fluviatilis L.) Hereditas* **102**, 57 – 61.

GYLLENSTEN U and WILSON A C (1987). Interspecific mitochondrial DNA transfer and the colonization of Scandinavia by mice. *Genetical Research, Cambridge* **49**, 25 – 29.

HARRIS H and HOPKINSON D A (1972). Average heterozygosity per locus in man: an estimate based on the incidence of enzyme polymorphisms. *Annals of Human Genetics* **36**, 9 – 20.

JANSON K (1987). Genetic drift in small and recently founded populations of the marine snail *Littorina saxatilis Heredity* **58**, 21 – 37.

JOHNSON C (1985). Biochemical genetic variation in populations of *Larus argentatus and Larus fuscus* in northwestern Europe. *Biological Journal of the Linnean Society* **24**, 349 – 363.

JOHNSON M S and BLACK R (1985). The Wahlund effect and the geographical scale of variation in an intertidal limpet, *Siphonaria* sp. *Marine Biology* **70**, 157 – 164.

LEWONTIN R C (1974). *The genetic basis of evolutionary change.* Columbia University Press, New York, NY.

LEWONTIN R C and HUBBY J L (1966). A molecular approach to the study of genetic heterozygosity in natural populations II. Amount of variation and degree of heterozygosity in natural populations. *Genetics* **54**, 595 – 609.

NEI M (1973). Analysis of gene diversity in subdivided populations. *Proceedings of the National Academy of Science, USA* **70**, 3321 – 3323.

NEI M and LI W-H (1979). Mathematical model for studying genetic variation in terms of restriction endonucleases. *Proceedings of the National Academy of Science, USA* **76**, 5269 – 5273.

NEI M, TAJIMA F and TATENO Y (1983). Accuracy of estimated phylogenetic trees from molecular data. *Journal of molecular Evolution* **19**, 153 – 170.

NEVO E (1978). Genetic variation in natural populations: patterns and theory. *Theoretical Population Biology* **13**, 121 – 177.

NEWMAN A, BUSH M, WILDT D E, VAN DAM D, FRANKENHUIS M R, SIMMONS L, PHILLIPS L and O'BREIN S J (1985). Biochemical genetic variation in eight endangered or threatened field species. *Journal of Mammalogy* **66**, 256 – 267.

O'BRIEN S J, WILDT D E, BUSH M, CARO T M, FITZGIBBON C, AGGUNDEY I and LEAKEY R E (1987). East African cheetahs: Evidence for two population bottlenecks? *Proceedings of the National Academy of Science, USA* **84**, 508 – 511.

O'BREIN S J, WILDT D E, GOLDMAN D, MERRIL C R and BUSH M (1983). The cheetah is depauperate in genetic variation. *Science* **221**, 459 – 462.

OLIVIER T J, COPPENHAVER D H and STEINBERG A G (1986). Distribution of immunoglobulin allotypes among local populations of Kenya olive baboons. *American Journal of Physical Anthropology* **70**, 28 – 38.

OSTERHOFF D R, SCHOEMAN S, OP'T HOF J and YOUNG E (1974). Genetic differentiation of the African elephant in the Kruger National Park. *South African Journal of Science* **70**, 245 – 247.

PARKIN D T and COLE S R (1985). Genetic differentiation and rates of evolution in some introduced populations of the House Sparrow, *Passer domesticus*, in Australia and New Zealand. *Heredity* **54**, 15 – 23.

RALLS K, BRUGGER K and BALLOU J (1979). Inbreeding and juvenile mortality in small populations of ungulates. *Science* **206**, 1101 – 1103.

RATTE T, GRANT W S and LESLIE R W (In prep). Morphological and genetic divergence among populations of the estuarine minnow (*Gilchristella aestuarius*).

ROBINSON T J, OP'T HOF J and GEERTHSEN J M P (1978). Polymorphic genetic markers in the springbok *Antidorcas marsupialis* (Bovidae). *South African Journal of Science* **74**, 84 – 86.

RYMAN N and STAHL G (1980). Genetic changes in hatchery stocks of brown trout (*Salmo trutta*). *Canadian Journal of Fishery and Aquatic Science* **37**, 82 – 87.

RYMAN N, BACCUS R, REUTERWALL C and SMITH M H (1981). Effective population size, generation interval, and the potential loss of genetic variability in game species under different hunting regimes. *Oikos* **36**, 257 – 266.

RYMAN N, REUTERWALL C, NYGREN K and NYGREN T (1980). Genetic variation and differentiation in Scandinavian moose (*Alces Alces*): Are large mammals monomorphic? *Evolution* **34**, 1037 – 1049.

SAIKI R K, GELFAND D H, STOFFEL S, SCHARF S J, HIGUCHI R, HORN G T, MULLIS K B and ERLICH H A (1986). Primer-directed enzymatic amplification of DNA with a thermostable DNA polymerase. *Science* **239**, 487 – 491.

SAURA A (1974). Genic variation in Scandinavian populations of *Drosophila bifasciata*. *Hereditas* **76**, 161 – 172.

SELANDER R K and KAUFMAN D W (1973). Self fertilization and genic population structure in a colonizing land snail. *Proceedings of the National Academy of Science, USA* **70**, 1186 – 1190.

SPOLSKY C and UZZELL T (1984). Natural interspecies transfer of mitochondrial DNA in amphibians. *Proceedings of the National Academy of Science, USA* **81**, 5802 – 5805.

SWOFFORD D L (1985). PAUP: phylogenetic analysis using parsimony. Illinois Natural History Survey, Champaign, Ill.

TEGELSTROM H (1987). Transfer of mitochondrial DNA from the northern red-backed vole (*Clethrionomys rutilus*) to the band vole (C. glareolus). *Journal of Molecular Evolution* **24**, 218 – 227.

THORPE J P (1982). The molecular clock hypothesis: biological evolution, genetic differentiation and systematics. *Annual Review of Ecology and Systematics* **13**, 139 – 168.

VAN DER BANK F H, GRANT W S and FERREIRA J T (1988). Electrophoretically detectable genetic variation in fifteen southern African cichlids. *Journal of Fish Biology* (In press).

VAN VUUREN N G, MULDER P F S, FERREIRA J T and VAN DER BANK F H (in prep). The identification of hybrids of *Barbus aeneus S B. kimberleyensis and Labeo capensis X L. umbratus* in Hardap Dam, S.W.A.–Namibia.

WILSON A C, CANN R L, CARR S M, GEORGE M, GYLLENSTEN U B, HELMBYCHOSCKI K M, HIGUCHI R G, PALUMBI S R, PRAGER EM, SAGE R D and STONEKING M (1985). Mitochondrial DNA and two perspectives on evolutionary genetics. *Biological Journal of the Linnean Society* **26**, 375 – 405.

WIRSTER D H and BENIRSCHKE K (1968). Chromosome studies in the superfamily Bovoidea. *Chromosoma* **25**, 152 – 171.

PART 4

Conservation status of terrestrial ecosystems and their biota

CHAPTER 12

Preservation of species in southern African nature reserves

W R SIEGFRIED

INTRODUCTION

Southern Africa contains a remarkable physiographic diversity and richness of plant and animal species (Werger 1978). Here southern Africa is taken to mean that part of the African continent that embraces the Republic of South Africa, the kingdoms of Swaziland and Lesotho, and the republics of Transkei, Ciskei, Venda and Bophuthatswana (Figure 12.1); the Walvis Bay enclave in Namibia, and all offshore and oceanic islands are excluded. In total, these territories cover about 1 268 600 km^2, accounting for about four per cent of Africa and about 0,8% of the total land area of the world.

Although relatively small, southern Africa contains some 20 300 species of vascular plants, or about eight per cent of the world's vascular flora (Table 12.1). The fynbos biome, in the south-western coastal strip of southern Africa, alone has some 7 300 vascular plant species. There are indications that the insect fauna of the fynbos biome, and perhaps other invertebrate groups as well, is also extraordinarily rich in species. However, the taxonomic and geographic data bases for most of southern Africa's invertebrate groups are still relatively poorly developed, and these taxa are not given attention in this review.

Among the vertebrates, southern Africa's amphibian, reptilian, avian and mammalian terrestrial faunas account for approximately two, six, seven and six per cent, respectively, of the world's total species of these groups. The relative richness of the southern African indigenous terrestrial flora and fauna, and their high level of endemism, is brought out more strikingly in a comparison involving numbers of species per unit area (Table 12.1). The survival of a significant part of this

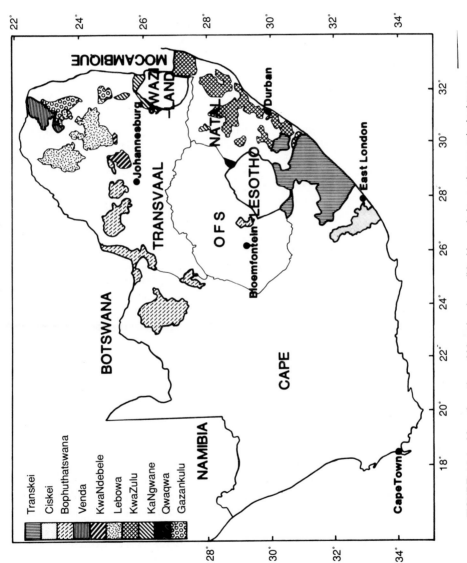

Transkei
Ciskei
Bophuthatswana
Venda
KwaNdebele
Lebowa
KwaZulu
KaNgwane
Qwaqwa
Gazankulu

FIGURE 12.1   Map of southern Africa showing political boundaries and some principal towns

species diversity (= species richness) is seriously threatened in southern Africa; for instance, at least some 1 000 vascular plant species alone are threatened (Hall this volume).

TABLE 12.1 Species richness of selected major biotic taxa reproducing in southern Africa, Africa and the world. Percentage of species endemic to southern Africa given in parentheses. The land areas of southern Africa, Africa and the world are 1 268 600, 30 264 000 and 148 354 000 km², respectively.

	No of species			No of species per 1 000 km²		
	Southern Africa	Africa	World	Southern Africa	Africa	World
Vascular plants	20 300[1] (36%)	45 000[1]	250 000[1]	16,00	1,49	1,69
Amphibians	84[2] (44%)	610[3]	>3 900[3]	0,07	0,02	0,03
Reptiles*	286[4,5,6] (31%)	1 010[7]	6 214[8,9]	0,23	0,03	0,04
Birds (breeding)	600[10,11] (6%)	1 655[12]	8 981[13]	0,47	0,05	0,06
Birds (non-breeding)+	118[10,11,14]	80[12]	—	0,10	0,01	—
Mammals*	227[15] (15%)	843[16]	3 927[16]	0,18	0,03	0,03

(1) G E Gibbs Russell and H P Linder (pers comm); (2) Passmore and Carruthers (1979); (3) Duellman and Trueb (1986); (4) Broadley (1981, 1983); (5) G R McLachlan (pers comm); (6) Greig and Burdett (1976); (7) Welch (1982); (8) Bertin and Burton (1980); (9) Pritchard (1979); (10) Maclean (1985); (11) Tarboton et al (1987); (12) Brown et al (1982); (13) Howard and Moore (1980); (14) Clancey et al (1987); (15) Smithers (1983); (16) Corbet and Hill (1980).
* Excludes marine forms which do not breed on land
+ Excludes purely marine visitors to South African waters

This paper attempts to answer the question: what proportions of southern Africa's indigenous terrestrial vascular plant flora and vertebrate fauna are contained in nature reserves? Subsidiary goals include assessments of the effectiveness of the system of nature reserves in representing the numbers of indigenous species of vascular plants, amphibians, reptiles, birds and mammals which reproduce in the five major terrestrial biomes found in southern Africa. The review does not include purely aquatic floras and faunas.

Although preliminary surveys have been made of the adequacy of a selection of South African nature reserves for the preservation of vegetation types (Edwards 1974) and certain rare and 'endangered' large-animal species (Von Richter 1974), no attempt has ever been made to assess comprehensively the extent to which the nature-reserve system actually includes representative floras and faunas in South Africa. Many of the reserves have been in existence for more than 50 years, and there is no obvious reason for this long-standing neglect.

THE NATURE RESERVE ESTATE

A 'nature reserve' is taken here to mean any protected terrestrial area that has the preservation of its native biota as one of its primary management objectives, and whose ownership and management are vested in a statutory authority. Hence, this study includes all publicly owned nature reserves in southern Africa, regardless of their different formal designations (eg national park, provincial nature reserve, wilderness area), but excludes privately owned areas such as those registered as nature reserves in the South African Natural Heritage Programme. In effect, the term 'nature reserve' is used here in much the same sense as the term 'protected area' is used by the International Union for the Conservation of Nature and Natural Resources (IUCN). Also excluded are purely marine areas. Consequently, the PFIAO catalogue (FitzPatrick Institute unpublished data) of nature reserves included in this report is far more comprehensive than Greyling and Huntley's (1984) directory of southern African conservation areas. The information in the current catalogue, which contains a total of 582 individual nature reserves, was verified by means of questionnaires sent in 1987 to all the statutory nature conservation agencies in southern Africa.

Distribution, number and size of reserves

Most of the total of 582 nature reserves in southern Africa are concentrated in a relatively narrow band of land between the major inland plateau and the marine coast, stretching from the north-east along the eastern and southern escarpments to the south-western part of the subcontinent (Figure 12.2). Taken together, the reserves cover some 7×10^6 ha (Table 12.2) and account for 5,8% of southern Africa. Most of these reserves were established during the last 25 years, after a period in which the rate of proclamation was relatively slow. Paradoxically, during this period more than 3×10^6 ha were acquired through the addition of the Kruger and Kalahari-Gemsbok national parks in 1925–1935 (Figure 12.3). It is encouraging to note that the increase in the rate of acquisition in both the number and area of publicly owned reserves has continued in recent years, as foreshadowed by Greyling and Huntley (1984). In addition, several large privately owned areas are today managed exclusively as nature reserves, effectively increasing the total proportion of reserved land to more than six per cent of southern Africa.

TABLE 12.2 The number and area of publicly owned nature reserves in southern Africa

	No of reserves	Reserved area (ha)
Republic of South Africa	518	6 810 740
National states (RSA)	35	237 105
Kingdom of Swaziland	5	60 209
Kingdom of Lesotho	1	6 805
Republic of Transkei	9	36 004
Republic of Ciskei	9	49 818
Republic of Venda	1	3 200
Republic of Bophuthatswana	4	66 380
	582	7 270 261

Most (70%) of the publicly owned nature reserves and most (83%) of the reserved areas occur in the Cape Province and the Transvaal, with the Cape Province containing about 44% of the total number of reserves, and 50% of the total reserved area (Table 12.3). Only five (<1%) of the 582 reserves are larger than 100 000 ha, and 413 (71%) are smaller than 5 000 ha (Figure 12.4). The total area of these relatively small (< 5 000 ha) reserves accounts for only seven per cent of

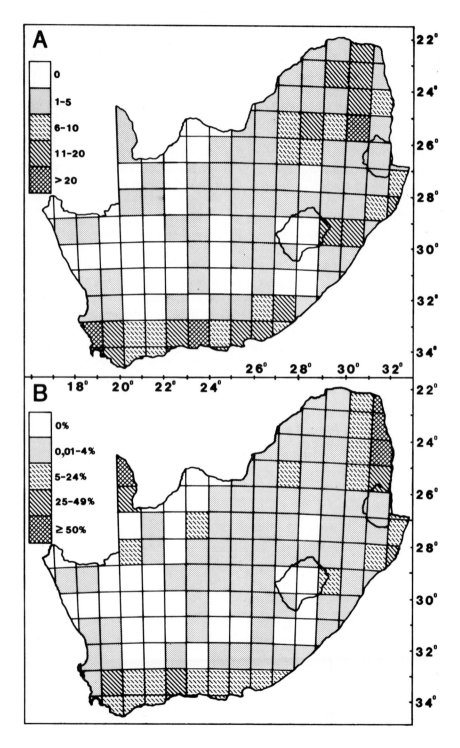

FIGURE 12.2 Number (A) and area (B) of 582 nature reserves according to one-degree grid squares in southern Africa

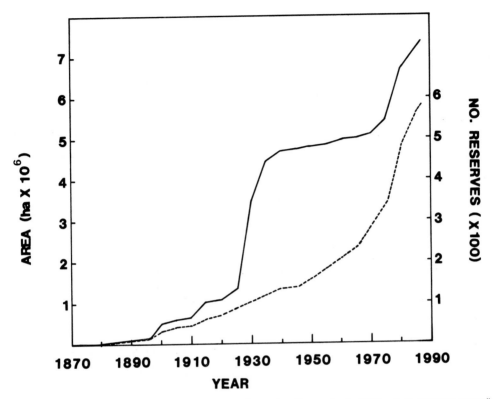

FIGURE 12.3 Cumulative number (dashed line) and area (continuous line) of 582 nature reserves according to year of proclamation in southern Africa

TABLE 12.3   The percentage number and area of 582 publicly owned nature reserves in southern Africa

	Percentage no of reserves	Reserved area
Republic of South Africa		
Cape Province	44,33	49,82
Natal	14,95	8,40
Orange Free State	3,78	2,17
Transvaal	25,95	33,29
National states	6,01	3,26
Other nations	4,98	3,06
	100,00	100,00

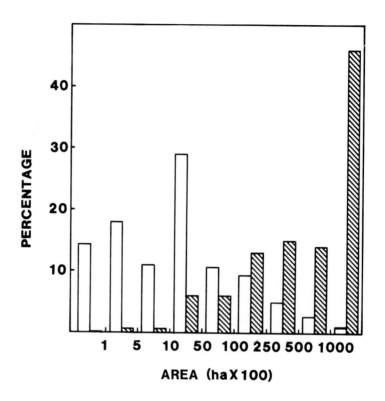

FIGURE 12.4 Percentage number (clear histograms) and area (shaded histograms) of 582 nature reserves in southern Africa

the total reserved area, about 46% of this area (7 x 10^6 ha) being covered by the five very large (>100 000 ha) reserves.

Most (64%) of the reserves are spatially isolated from one another (Figure 12.5). Moreover, most (>50%) of the land surrounding most (60%) of the reserves has been artificially transformed so as to militate against the exchange of floral and faunal elements between nearest-neighbour reserves supporting the same or similar habitats. Based on the results of a nearest-neighbour analysis (Clarke and Evans 1954), the distribution of southern African nature reserves is clumped (R = 0,59), with the mean distance between the nearest boundaries of reserves being 13,8 km (SE = 0,51). However, since 116 of the 582 reserves are individually named compartments of 26 larger reserves (mainly forestry areas), a re-analysis involving 466 (582 – 116) reserves yielded an R-value of 0,68 (reserves still clumped) and a mean nearest-neighbour distance of 17,8 km (SE = 0,63).

Representation of biomes

Five major terrestrial biomes are represented in southern Africa: fynbos, forest, karoo, grassland and savanna (Figure 12.6). For this report, the karoo is divided into the Nama-karoo and the succulent karoo (Rutherford and Westfall 1986), and the savanna into moist and arid components (Huntley 1984).

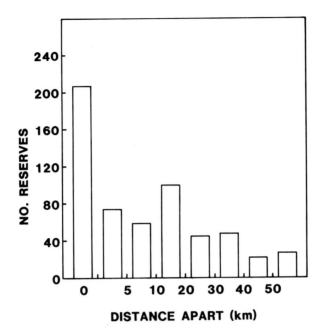

FIGURE 12.5 Nearest-neighbour distances (km) between 582 nature reserves in southern Africa

In accordance with the relative paucity of nature reserves in the central, and major, part of southern Africa (Figure 12.2), the karoo and grassland biomes are represented poorly in the reserve system (Table 12.4). This pattern differs very little from that presented by Greyling and Huntley (1984) who, in concluding that the neglect with regard to the establishment of reserves in the karoo and grassland had not changed much since its first reporting by Edwards (1974), made a strong recommendation for a better balance of nature reserves in South Africa. To date, however, this recommendation has not been implemented with any significant effect.

TABLE 12.4   The percentage number and area of 582 publicly owned nature reserves in relation to biomes in southern Africa

	Percentage no of reserves	Southern African area reserved (%)	Biome area (ha)	Biome area reserved (%)
Fynbos	22,85	24,24	6 987 500	26,26
Forest	4,81	3,29	309 000	77,36
Nama-karoo	4,64	3,96	34 610 700	0,83
Succulent karoo	1,89	0,89	8 190 800	0,79
Grassland	24,40	9,57	34 536 100	2,00
Moist savanna	24,05	8,69	12 669 100	4,99
Arid savanna	17,35	49,36	29 617 700	12,11
	100,00	100,00	126 942 800	

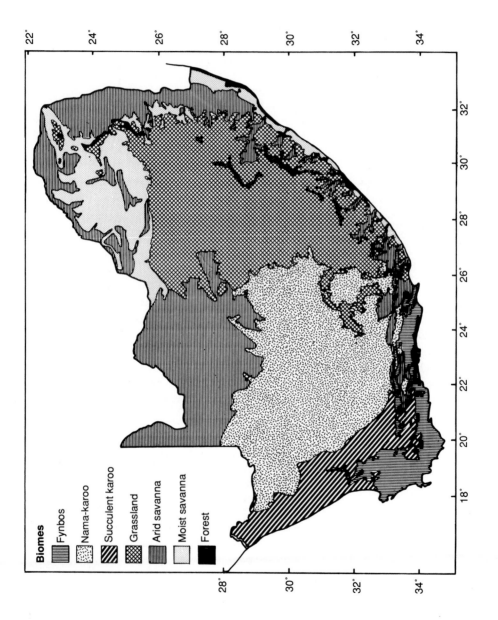

FIGURE 12.6 Distribution of biomes in southern Africa

In contrast, the very species-rich and generally heterogeneous fynbos biome, including the endemic Cape Floral Kingdom, appears to be represented adequately; about 30% of the biome's area being included in 133 nature reserves (Table 12.4). This is misleading, however, since most of the reserved area is covered by mountain fynbos vegetation; coastal fynbos and other lowland vegetation types are represented in about 10% of the reserves and account for about three per cent of the reserved area of the biome. Greyling and Huntley (1984) have drawn attention to the urgent need for better protection of these lowland vegetation types and ecosystems, since the survival of the biome's coastal renosterveld vegetation is critically endangered (Jarman 1986).

THE FLORA AND FAUNAS

Totals of 20 300 vascular plant, 84 amphibian, 286 reptilian, 600 avian and 227 mammalian species have indigenous breeding populations in southern Africa. Table 12.1 gives sources of the data which are stored in the FitzPatrick Institute's computerized data bank. It should be noted that these statistics reflect the modern situation, in that species that have become extinct during the historic period in southern Africa are excluded. As far as is known, the list of extinctions includes some 30 vascular plants (Hall 1987), two birds (Brooke 1984; Ledger 1985), and two mammals (Smithers 1983). It also should be noted that the statistics do not include those migratory species (eg > 100 avian spp) that visit southern Africa but do not breed there. Migratory species breeding in southern Africa are, however, included in the present treatment. All alien species are excluded from the statistics reported here.

TABLE 12.5   The number of species reproducing in biomes, based on totals of 20 300 vascular plants, 84 amphibians, 286 reptiles, 600 birds and 227 mammals in southern Africa. Number of species per 10 000 ha given in parentheses. (See Table 12.1 for sources of data.)

	No of species					
	Plants	Amphibians	Reptiles	Birds	Mammals	Totals
Fynbos	7 316 (10,47)	25 (0,04)	85 (0,12)	262 (0,37)	74 (0,10)	7 762 (11,12)
Forest	no data	13 (0,42)	21 (0,68)	313 (10,13)	37 (1,20)	no data
Nama-karoo	2 147 (0,62)	12 (0,01)	78 (0,02)	252 (0,07)	69 (0,02)	2 588 (0,75)
Succulent karoo	2 125 (2,59)	11 (0,01)	91 (0,11)	222 (0,27)	65 (0,08)	2 504 (3,06)
Grassland	3 378 (0,97)	33 (0,01)	104 (0,03)	417 (0,12)	94 (0,03)	4 430 (1,27)
Moist savanna	3 805 (3,00)	57 (0,04)	169 (0,13)	540 (0,43)	153 (0,12)	4 724 (3,73)
Arid savanna	3 380 (1,14)	52 (0,02)	177 (0,04)	519 (0,18)	171 (0,06)	4 299 (1,45)

Representation in biomes

Considering the taxa included in this report, the fynbos is the most species-rich biome in southern Africa, in both absolute and relative numbers, owing mainly to its remarkable diversity of plants (Table 12.5). The savanna features most prominently with regard to overall species richness of terrestrial vertebrate animals. It must be pointed out once again that the statistics concerned (reported in Table 12.5) reflect the modern status of species. For example, resident breeding populations of the African lion *Panthera leo* formerly occurred (300 years ago) in all of the South African biomes, whereas today such populations are restricted to the region's savanna biome (Smithers 1983). Consequently, this species is admitted exclusively to the savanna biome in the present treatment.

Representation in nature reserves

Either partial or complete check-lists of species for one or more of the five taxa considered here have been published in the open primary literature for only 88 (15%) of the 582 nature reserves in southern Africa. Published comprehensive suites of complete check-lists of species for all five taxa are available for only six reserves, four of these being national parks in the Republic of South Africa. Hence, questionnaires were sent to all the official agencies responsible for the management of nature reserves, requesting any unpublished lists they might have in their files.

Based on the response to this request, and the published information, comprehensive suites of lists of species for all five taxa could be assembled for only 28 individual reserves. Allegedly complete lists of species for four, three and two taxa were obtained for 84, 14 and 19 reserves, respectively. Twenty-six reserves had lists for only one taxon. Lists were not available for 300 (52%) of the reserves. Complete lists of avian species were most common, and complete lists of plant species were relatively rare (Table 12.6). Somewhat paradoxically, species lists were more often lacking for small reserves, especially in the Cape Province, than for large reserves (Figure 12.7). The poor showing with regard to plants reflects a traditional neglect of this taxon in the field of professional 'wildlife management' in southern Africa, and the relatively high incidence of animal taxa manifests a common bias in the knowledge and interest of reserve managers.

TABLE 12.6   The number of nature reserves for which either complete, partial or no lists of species for one or more of five taxa are available (published and unpublished) in southern Africa. Numbers in parentheses are nature reserves for which published lists exist.

	No of reserves		
	Complete	Partial	None
Plants	52 (20)	153 (17)	377
Amphibians	123 (12)	34 (1)	425
Reptiles	130 (13)	43 (1)	409
Birds	147 (23)	66 (4)	369
Mammals	128 (18)	105 (39)	349

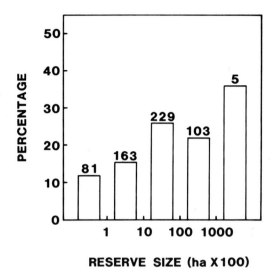

FIGURE 12.7   Number (at heads of histograms) of complete lists of species available for nature reserves and the area of the reserves in southern Africa

The paucity of species lists for nature reserves in southern Africa is remarkable, given that preservation of species was an important factor in motivating for the acquisition of many of the reserves, and that it is, ostensibly, a primary objective in their management and recurrent funding. Nevertheless, an assessment of the limited information made available in published reports and by the official agencies shows that 34% of southern Africa's vascular plant species, 92% of amphibian, 92% of reptilian, 97% of avian and 93% of mammalian species are represented in nature reserves. Moreover, some 50% of the animal species occur in more than 10 reserves (Table 12.7), and differences are small between the relatively high proportions of animal species represented in nature reserves in relation to the occurrence of the species in particular biomes (Table 12.8). This is a somewhat unexpected and satisfactory result, given that it is based on lists of species for fewer than half of the nature reserves in southern Africa. The relatively poor representation of karoo biome amphibians and reptiles should, however, be noted.

TABLE 12.7   The percentage number of reproducing species of five taxa occurring in one or more nature reserves, based on totals of 20 300 vascular plants, 84 amphibians, 256 reptiles, 600 birds and 227 mammals, and 52, 123, 130, 146 and 127 complete species lists, respectively (see Table 12.6), for nature reserves in southern Africa

No of nature reserves	% No of species				
	Plants	Amphibians	Reptiles	Birds	Mammals
1- 10	33,0	35,7	47,6	18,5	34,4
11- 25	1,0	20,2	17,2	24,3	16,3
26- 50	0,02	19,0	11,7	27,7	26,0
51-100	0	13,1	13,3	21,8	15,0
101-150	0	3,6	1,9	4,7	0,9

TABLE 12.8   The percentage number of reproducing species of five taxa reported occurring in nature reserves representative of biomes in southern Africa. Absolute numbers given in parentheses. (See Tables 12.1 and 12.5 for sources of data.)

| | % No of species | | | | | |
	Plants	Amphibians	Reptiles	Birds	Mammals	Totals
Fynbos	34,8 (2 540)	88,0 (22)	90,6 (77)	98,8 (259)	98,6 (73)	38,3 (2 971)
Forest	no data	100 (13)	100 (21)	99 (310)	100 (37)	no data
Nama-karoo	34,8 (748)	91,7 (11)	96,2 (75)	99,2 (250)	95,6 (66)	44,4 (1 150)
Succulent-karoo	40,1 (853)	72,7 (8)	75,8 (69)	98,6 (219)	93,8 (61)	48,3 (1 210)
Grassland	82,1 (2 775)	100 (33)	96,2 (100)	99,8 (416)	100 (94)	77,2 (3 418)
Moist savanna	53,4 (2 031)	100 (57)	94,1 (159)	98,1 (530)	96,7 (148)	61,9 (2 925)
Arid savanna	49,5 (1 673)	100 (52)	96,6 (171)	98,8 (513)	98,8 (169)	60,0 (2 578)

In an attempt to remedy the shortfall in the available lists of vascular plant species, the distributional ranges of all the species concerned, as mapped or otherwise recorded by PRECIS (Pretoria National Herbarium Computerized Information System), were digitized according to a standard quarter-degree grid of squares covering southern Africa (Gibbs Russell 1987). The dispersion of the southern African nature-reserve system was gridded similarly, facilitating matched overlays of species and reserves. This technique, although not free of shortfalls itself, at least allowed a more objective, complete and comprehensive assessment than the technique dependent entirely on the limited information provided by the management agencies of the reserves. The results of the grid-square analysis indicate that possibly some 74% of southern Africa's vascular plant species are represented in nature reserves (Table 12.9). However, the 99% of fynbos species apparently occurring in nature reserves (Table 12.9) is almost certainly too high. This probable bias could be due to several factors, the most important being that most of the fynbos reserves fall into grid squares that have been surveyed intensively because of their relatively high species richness. Once again, attention is drawn to the marked under-representation of the karoo biota in nature reserves (Table 12.9).

TABLE 12.9 The percentage number of vascular plant species possibly occurring in nature reserves representative of biomes in southern Africa based on quarter-degree grid congruency between dispersions of plants and nature reserves (see text). Absolute numbers given in parentheses.

	% No of species
Fynbos	99,4 (7 272)
Forest	no data
Nama-karoo	27,1 (581)
Succulent karoo	31,2 (664)
Grassland	78,2 (2 640)
Moist savanna	95,7 (3 642)
Arid savanna	61,6 (2 082)

DISCUSSION

The revelation that an extraordinarily high proportion of southern Africa's terrestrial vertebrate fauna, and possibly 74% of its vascular plant species, are represented by breeding populations in the region's nature reserves is remarkable, given that the nature-reserve system accounts for less than six per cent of the region and that it developed without any preconceived design for maximizing the preservation of biological diversity. It should be noted, however, that the term 'preservation' has been used deliberately and exclusively throughout this report. According to Frankel and Soule (1981), preservation 'provides for the maintenance of individuals or groups but not for evolutionary change'. This meaning is contrasted with the use by biologists of the term 'conservation', which denotes 'policies and programmes for the long-term retention of natural communities under conditions which provide for continuing evolution' (Frankel and Soule 1981). Here, the evolutionary response is seen as 'a condition of long-term survival' (Frankel and Soule 1981).

The potential of the computerized data base assembled for this study has not been fully realized here. Additional analyses could have been undertaken which would have allowed, for instance, the recognition of particular associations ('communities') of plants and animals in particular nature reserves in relation to particular biome types. The results of such analyses might be of considerable practical benefit in management for enhancing the long-term survival of a maximum amount of biological diversity in southern African nature reserves. On the other hand, and leaving aside potential socio-economic factors (Norgaard 1987), the absence of demographic information for nearly all the species considered here makes it virtually impossible to comment meaningfully on their individual prospects for long-term survival. However, it is possible to speculate in general terms about how the risks attending long-term survival of many species might be reduced, despite the current lack of lists of species for more than half of the nature reserves in southern Africa.

Given that a common principal goal of the nature conservation agencies in southern Africa is the long-term conservation of a maximum amount of biological diversity in nature reserves, it is not yet too late for devising and implementing an innovative strategic plan for realizing this goal. In this plan, the present nature-reserve system should not be regarded as sacrosanct. If and where necessary, parts of, or whole, reserves should be deproclaimed. The land could be either sold or exchanged in order to obtain new areas needed for conserving plant and animal associations that are under-represented in the current system, and for new groupings of, and links between, reserves. Such new networks should, for example, incorporate the steepest environmental gradients wherever possible. Of course, a rearranged southern African nature-reserve system, based on a strategic plan for conserving a maximum amount of biological diversity, is hardly a novel concept (Siegfried 1978).

CONCLUSIONS AND RECOMMENDATIONS

Most (70%) of southern Africa's 582 publicly owned terrestrial nature reserves are relatively small (< 5 000ha), but the combined area of the nature-reserve system accounts for almost six per cent of the region. The reserves tend to be spatially isolated from one another, being separated by artificially transformed land. Corridors of untransformed land and other links between nature reserves are few, militating against exchanges of floral and faunal elements.

All the major terrestrial biomes occurring in southern Africa are represented in the region's nature-reserve system. However, the karoo and grassland biomes are represented relatively poorly, in accordance with the general paucity of reserves in the central, and major part, of the region. Attention should be given to the establishment of more nature reserves in the karoo and grassland biomes and in the lowland area of the fynbos biome, to conserve as fully as possible examples of the different vegetation types of each biome.

Although the majority of South Africa's nature reserves came into being during the last 25 years, comprehensive suites of lists of species representing all five taxa (vascular plants, amphibians, reptiles, birds and mammals) are available for only 28 individual reserves. Lists of the species of one or more of these taxa were not available for more than 50% of the 582 reserves, despite the preservation of species having been an important factor in motivating for the proclamation of many of the reserves. The preservation of species is ostensibly a primary objective in the management and recurrent funding of these reserves. Consequently, complete lists of species, especially of plants, should be compiled for each reserve as a matter of urgency.

According to Scott et al (1987), the success of efforts to retain biological diversity will be judged in future on the number of surviving species, and not on whether the Californian condor *Gymogyps californianus,* for example, or some other currently endangered spectacular animal, is saved from extinction in the next decade or so. Scott et al (1987) advocate focusing on the protection of species-rich areas as offering 'the most efficient and cost-effective way to retain maximal biological diversity in the minimal area'. And, 'given the inevitability of further habitat loss, this strategy may be the only way to resolve conflicts between development and the preservation of genetic and species diversity.'

ACKNOWLEDGEMENTS

I dedicate this paper to Dr U de V Pienaar who compiled and published lists of the species of South Africa's national parks long before it became fashionable to do so. I thank the following persons for their advice and help: J Anderson, A Armstrong, A R Batchelor, P Bohnen, O Bourquin, R K Brooke, M Cohen, R F H Collinson, R A Conant, G de Graaff, P Dutton, G E Gibbs Russell, S K B Godschalk, C Kleinjan, J H Koen, C Kromhout, L A Lötter, I A W Macdonald, I MacFadyen, G R McLachlan, M M Mothepu, D B Müller, T Newby, I Newton, A Pulfrich, I C Sharp, M J Simpson, I W Terblanche, W R Thomson, C Tinley, M C Ward, L Wentzel, G Wright. The nature conservation organizations of southern Africa are thanked for their co-operation and assistance. Mesdames Ione and Helen Davies were responsible for computer-assisted computations and word processing. The study was supported financially by the University of Cape Town and the Foundation for Research Development of the CSIR.

REFERENCES

BERTIN L and BURTON M (1980). Reptiles. In *The new Larousse encyclopedia of animal life.* Hamlyn, London.
BROADLEY D G (1981). A review of the genus *Pelusios* Wagler in southern Africa (Pleurodira: Pelomedusidae). *Occasional Papers of the National Museums and Monuments of Zimbabwe. Ser B* **6**, 633 – 686.
BROADLEY D G (1983). *FitzSimons' snakes of southern Africa.* Delta Books, Johannesburg.
BROOKE R K (1984). South African Red Data Book — birds. *South African National Scientific Programmes Report* **97**, 213 pp.
BROWN L H, URBAN E K and NEWMAN K (1982). *The birds of Africa,* Vol 1. Academic Press, London.
CLANCEY P A, BROOKE R K, CROWE T M and MENDELSOHN J M (1987). *S A O S checklist of southern African birds (1980): first updating report.* Southern African Ornithological Society, Johannesburg.
CLARKE P J and EVANS F C (1954). Distance to nearest neighbour as a measure of spatial relationships in populations. *Ecology* **35**, 445 – 453.
CORBET G B and HILL J E (1980). *A world list of mammalian species.* British Museum (Natural History), London.
DUELLMAN W E and TRUEB L (1986). *Biology of amphibians.* McGraw-Hill, New York, NY.
EDWARDS D (1974). Survey to determine the adequacy of existing conserved areas in relation to vegetation types. A preliminary report. *Koedoe* **17**, 3 – 38.
FRANKEL O H and SOULE M E (1981). *Conservation and evolution.* Cambridge University Press, Cambridge.
GIBBS RUSSELL G E (1987). Preliminary floristic analysis of the major biomes in southern Africa. *Bothalia* **17**, 213 – 217.
GREIG J C and BURDETT P D (1976). Patterns in the distribution of southern African tortoises (Cryptodira: Testudinidae). *Zoologica Africana* **11**, 249 – 273.
GREYLING T and HUNTLEY B J (eds)(1984). Directory of southern African conservation areas. *South African National Scientific Programmes Report* **98**, 309 pp.
HALL A V (1987). Threatened plants in the fynbos and karoo biomes, South Africa. *Biological Conservation* **40**, 29 – 52.
HOWARD R and MOORE A (1980). *A complete checklist of the birds of the world.* Oxford University Press, Oxford.
HUNTLEY B J (1984). Characteristics of South African biomes. In Ecological effects of fire in South African ecosystems. (eds Booysen P de V and Tainton N M) *Ecological Studies* **48**. Springer, Berlin. pp 1 – 17.
JARMAN M L (ed)(1986). Conservation priorities in lowland regions of the fynbos biome. *South African National Scientific Programmes Report* **87**, 53 pp.
LEDGER J (1985). Egyptian extinction. *Quagga* **12**, 10 – 12.
MACLEAN G L (1985) *Roberts' birds of southern Africa.* Trustees of the John Voelcker Bird Book Fund, Cape Town.
NORGAARD R B (1987). Economics as mechanics and the demise of biological diversity. *Ecological Modelling* **38**, 107 – 121.
PASSMORE N I and CARRUTHERS V C (1979). *South African frogs.* Witwatersrand University Press, Johannesburg.
PRITCHARD P C H (1979). *Encyclopaedia of turtles.* T F H Publications, Neptune, NJ.
RUTHERFORD M C and WESTFALL R (1986). Southern African biomes. *Memoirs of the Botanical Survey of South Africa* **54**. Government Printer, Pretoria.
SCOTT J M, CSUTI B, JACOBI J D and ESTES J E (1987). Species richness: a geographic approach to protecting future biological diversity. *BioScience* **37**, 782 – 788.
SIEGFRIED W R (1978). Let the strandwolf fly. *African Wildlife* **32**, 10 – 14.
SMITHERS R H N (1983). *The mammals of the southern African subregion.* University of Pretoria, Pretoria.
TARBOTON W, KEMP M I and KEMP A C (1987). *Birds of the Transvaal.* Transvaal Museum. Pretoria.
VON RICHTER W (1974). Survey of the adequacy of existing conserved areas in relation to wild animal species. *Koedoe* **17**, 39 – 69.
WELCH K G R (1982). *Herpetology of Africa.* Robert E Krieger, Malabar, Fla.
WERGER M J A(ed)(1978). *Biogeography and ecology of southern Africa.* Junk, The Hague.

CHAPTER 13

Conservation status of the fynbos and karoo biomes

C Hilton-Taylor, A Le Roux

INTRODUCTION

Despite the development of global biogeographic and vegetation classification systems, no universally accepted system exists that is suitable for assessing the conservation status of habitats on a global scale (Huntley 1980). In South Africa the AETFAT nomenclature of vegetation types was used (for example Keay 1959) until Edwards (1974) proposed the use of Acocks's (1953) Veld Types. Subsequent reviewers have all used the Veld Types (eg Huntley 1978; Scheepers 1983; Cowan 1987). Huntley (1978) and Huntley and Ellis (1984), in subcontinental surveys, used the biome concept to delimit major biotic divisions with vegetation types and Veld Types serving as the basic units. While the concept of a Veld Type is useful in agro-ecological evaluations, and is adequate for broad assessments of conservation status on a national scale, it is too broad to reflect conservation priorities at the habitat level.

For this paper we have adopted the biome classification of Rutherford and Westfall (1986). According to this classification, the karoo biome (Huntley 1984) is subdivided into two biomes: the succulent karoo biome and the Nama-karoo biome. Although the boundaries of these two biomes extend into southern Namibia, this paper covers only the area within South Africa (Figure 13.1). For practical reasons we have excluded the Lesotho outlier of the Nama-karoo, and have included the savanna outlier in the Little Karoo and the Noorsveld.

Phytogeographically the succulent karoo and Nama-karoo biomes form part of the Karoo–Namib Region (Werger 1978a,b; White 1983). The fynbos biome is equated with the Cape Region (White 1983) which is also termed Capensis (Taylor 1978; Werger 1978a) but is perhaps best known as the Cape Floristic Kingdom (Takhtajan 1969; Good 1974). Kruger (1977), in proposing

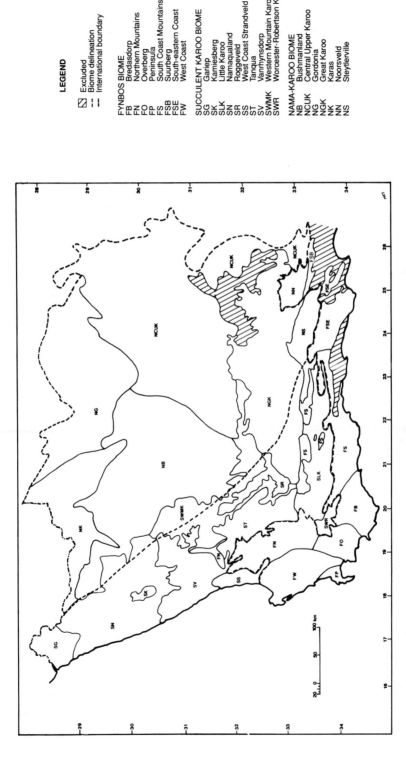

LEGEND

Excluded
Biome delineation
International boundary

FYNBOS BIOME
FB Bredasdorp
FN Northern Mountains
FO Overberg
FP Peninsula
FS South Coast Mountains
FSB Suurberg
FSE South-eastern Coast
FW West Coast

SUCCULENT KAROO BIOME
SG Gariep
SK Kamiesberg
SLK Little Karoo
SN Namaqualand
SR Roggeveld
SS West Coast Strandveld
ST Tanqua
SV Vanrhynsdorp
SWMK Western Mountain Karoo
SWR Worcester-Robertson Karoo

NAMA-KAROO BIOME
NB Bushmanland
NCUK Central Upper Karoo
NG Gordonia
NGK Great Karoo
NK Karas
NN Noorsveld
NS Steytlerville

FIGURE 13.1 Map showing the biogeographic areas of the fynbos, succulent karoo and Nama-karoo biomes. The boundaries are based on Weimarck (1941); Oliver et al (1983) and Hilton-Taylor (1987) with modifications based on Acocks (1953, 1975).

a strategy of ecological reserves for the fynbos biome, suggested the use of a biogeographic framework. He proposed a system of zones based on Acocks's (1953) Veld Types and Weimarck's (1941) biogeographic centres.

In this paper we relate conservation status to a framework of biogeographic areas (Figure 13.1). The areas used are those derived from Weimarck (1941), Oliver et al (1983) and Hilton-Taylor (1987), with modifications based on Acocks (1953, 1975). The boundaries for many of these areas, particularly in the succulent karoo and Nama-karoo, are speculative and need to be determined and tested quantitatively. This biogeographic approach places conservation in an evolutionary context. Many of the biogeographic patterns are not related to present-day environmental patterns, but are the result of events over the past 30 000 years. Unfortunately, most previous accounts of conservation status and the current South African Plan for Nature Conservation of the Department of Environment Affairs utilize Veld Types. We have therefore included Veld Types for comparative purposes. It should be noted that the vegetation map of the fynbos biome produced by Moll and Bossi (1984a) and as described in Moll et al (1984) is a far better basis for conservation assessment and planning. Reference is made to the vegetation types of Moll et al (1984) where appropriate.

The classifications used for this assessment of conservation status are all based on vegetation. This is because animal groups show little congruence with vegetation in their biogeographic patterns (Werger 1978a). While the fauna *per se* is not considered, it is assumed that if the habitat is conserved then so is the faunal component. This assumption appears to be valid for the vertebrates as shown by Siegfried in this volume.

It should be noted that throughout this paper we use the term 'conservation' in its popular sense, namely preservation of individuals or groups. Conservation in its biological sense allows for continuing evolution (Frankel and Soule 1981). Assessing conservation status in terms of evolutionary potential is beyond the scope of this paper.

In summary, we shall determine the status of conservation in relation to both Veld Types and biogeographic areas. We shall also focus on threatened taxa and to a certain extent on threatened habitats. Threatened taxa are often good indicators of threatened habitats, since they are more sensitive to changes in the environment (Tansley 1988a) than common, widespread species. They therefore serve as useful tools in conservation planning. The nature and extent of the threats to these taxa and habitats will be examined. From these data we shall establish conservation priorities which will be examined in the light of present conservation planning. These priorities are indicated by means of asterisks, where *** implies high priority, ** medium priority, * low priority, and no asterisk for areas considered to be adequately conserved. The allocation of these priorities was subjectively determined from the percentage area of the Veld Type and/or biogeographic area conserved, the number of threatened taxa per unit area and planned conservation areas. These priorities are not intended to prescribe any action but rather to help direct attention to certain areas.

The data presented here were obtained from the Chief Directorate Nature and Environmental Conservation (CDNEC), Cape Province. The data on the threatened taxa were extracted from the CDNEC's Critical Environmental Components (CICEC) data base.

FYNBOS BIOME

Description

The fynbos biome encompasses a broad category of evergreen sclerophyllous heathlands and shrublands (cf Moll and Jarman 1984) in which fine-leaved low shrubs and leafless tufted grasslike plants (Restionaceae) predominate. Trees and evergreen succulent shrubs are rare, while grasses form a small proportion of the biomass. Taylor (1978), Day et al (1979) and Rutherford and Westfall (1986) provide detailed descriptions of the biome.

The biome covers an area of approximately 70 000 km^2, some 5,72% of South Africa (Rutherford and Westfall 1986), and occurs almost exclusively in the south-western and southern Cape (Figure 13.1). It includes five Acocks (1953) Veld Types (abbreviated as VT): strandveld VT34; coastal rhenosterbosveld (now renosterveld) VT46; coastal macchia (coastal fynbos) VT47; macchia and false macchia (mountain fynbos) VT69 and 70. Mountain renosterveld (VT43) is variously included in the fynbos, succulent karoo and Nama-karoo biomes (Moll and Jarman 1984; Rutherford and Westfall 1986). This Veld Type can also not be allocated to any one biogeographic area. There are similar problems with strandveld. Here we have followed the divisions suggested by Boucher and Le Roux (in press). Strandveld south of the Olifants River Mouth, although included in the fynbos biome, does not form a distinct biogeographic unit. Moll et al (1984), although largely recognizing Acocks's (1953) Veld Types, further subdivided these to recognize fifteen vegetation categories which comprise the fynbos biome.

Moll and Bossi (1984b) compared the extent of natural vegetation as mapped by Acocks (1953) and that remaining as interpreted from 1981 Landsat satellite imagery. They concluded that 34% of the natural vegetation had been removed by intensive agriculture, urbanization and other human activities. The loss was most pronounced in coastal renosterveld (85% lost) and coastal fynbos (47% lost). This loss of natural vegetation is particularly important if one considers that the fynbos biome approximates the Cape Floristic Kingdom, the smallest of the world's six floristic kingdoms (Takhtajan 1969; Good 1974).

Biodiversity

Floristic diversity at both the beta and gamma levels, is particularly high (see Cowling et al this volume). Inventory diversity (*sensu* Cowling et al) in the fynbos biome is apparently higher than in any of the other biogeographic zones of the world, except perhaps the tropical rain forest, while the extremely high gamma diversities (*sensu* Cowling et al) are unmatched anywhere (Kruger and Taylor 1979). The flora consists of some 7 316 taxa (Gibbs Russell 1987). Many of these taxa are confined to the biome. There are seven endemic plant families, 210 endemic genera, and about 68% of the species are endemic (Goldblatt 1978; White 1983; Bond and Goldblatt 1984; Gibbs Russell 1987). Owing to the high levels of gamma diversity, many of the plant species are rare and hence threatened to some degree — 1 326 taxa (Hall and Veldhuis 1985; Hall 1987). (See section on threatened taxa.)

This floristic diversity is not paralleled by an equally rich fauna, but the biome nevertheless possesses a rich invertebrate fauna, with high levels of endemism. (See Jarvis (1979) for a review of the zoogeography of the biome.) The systematics and distributions of many of these invertebrate groups are relatively poorly known and hence these groups are not given any further attention in this paper. Fish are also excluded, since they are covered elsewhere in this volume.

The number of resident taxa in each major biotic group — plants, mammals, birds, reptiles and amphibians — and the numbers of endemic species are shown in Table 13.1.

Historical perspective

The Royal Society of South Africa, concerned about the destruction of Cape mountain vegetation appointed a committee to investigate the problem. The committee's report (Wicht 1945), which outlined the causes and extent of the destruction of mountain vegetation, proposed a number of management measures to preserve the vegetation. The report also suggested the establishment of a number of nature reserves in the mountain areas. These proposed reserves were the start of the State Forest Reserve and Mountain Catchment network.

TABLE 13.1   Total number of taxa in the five major biotic groups in the fynbos, succulent karoo and Nama-karoo biomes. Endemic taxa are indicated in [ ]. Data derived from: Gibbs Russell (1987); Hilton-Taylor (1987); Smithers (1983); Maclean (1985); Huntley (1978); Siegfried (this volume); E Baard and A de Villiers (pers comm).

Biome	Plants	Mammals	Birds	Amphibians	Reptiles	Total
Fynbos	7 316	74	262	25	85	7 762
	[4 872]	[7]	[6]	[9]	[20]	[4 914]
Succulent karoo		65	222	8	91	386
		[3]				[3]
Nama-karoo		69	252	12	108	441
		[4]		[1]	[8]	[13]
Combined karoo	±7 000					±7 000
	[±2 500]	[3]	[19]			[±2522]

Edwards (1974) estimated that 258 720 ha of mountain fynbos was conserved in State Forest Reserves by the early 1970s. In relation to other Veld Types Edwards (1974) did not consider any of the Veld Types of the fynbos biome to have a high priority conservation status. He did conclude, however, that additional reserves were required to conserve certain important local ecosystems and species.

Kruger (1977) indicated that conservation of fynbos was entering a critical phase in which problems would be solved only through an integrated approach including landuse planning and scientific management systems. He added that a prerequisite for long-term conservation success was an adequate network of ecological reserves. Kruger (1977) also discussed theoretical aspects of reserve location, design, size and number, and concluded that approximately 19 reserves of from 100 to 1 000 km^2 would be adequate for fynbos conservation. Huntley (1978) suggested that 'a network of small "witness stands" throughout the south-western Cape lowlands' was of higher priority than the creation of additional large mountain reserves.

The fynbos biome lowlands are discussed in detail under the section on conservation priorities. The national and international priority status of these lowlands has repeatedly been emphasized (Scheepers 1983; Huntley and Ellis 1984; Greyling and Huntley 1984).

Conservation status

The present conservation status of the fynbos biome in terms of area of Veld Types conserved is shown in Table 13.2 and in terms of biogeographic areas in Table 13.3. These tables are divided into state-owned conservation areas, semi-state conservation areas, mountain catchment areas and South African Defence Force areas. Not included are private nature reserves and natural heritage sites, since these have no secure long-term conservation status.

State-owned areas include National Parks, Provincial Nature Reserves and State Forest areas. Most of the latter have recently been transferred to the CDNEC. Semi-state conservation areas are nature reserves managed by local authorities and subsidized by the CDNEC, and the natural portions of National Botanic Gardens. Mountain catchment areas are privately owned, but are managed integrally with the adjacent state areas for purposes of water conservation. Management policy includes prescribed burning practices and the combatting of alien invasive plants; however, there are many problems with regard to control (Kruger 1982). If managed correctly, these areas, which account for some 536 449 ha (27%) of mountain fynbos, would serve a vital conservation function. No assessment of these areas has been published and they are therefore not included in the final totals of area conserved.

TABLE 13.2   Area of Veld Types conserved (excluding degraded areas) in the fynbos, succulent karoo and Nama-karoo biomes. Data from the CDNEC.

Veld Type number (Acocks)	Biome†	Total area Veld Type (S Africa) (ha)	State-owned conservation areas Area conserved (ha)	State-owned % of total Veld Type conserved	Semi-state conservation area Area conserved (ha)	Semi-state % of total Veld Type conserved	Mountain catchment and protected forest (private) Area conserved (ha)	Mountain catchment % of total Veld Type conserved	S A Defence Force Area conserved (ha)	S A Defence Force % of total Veld Type conserved	Permanently conserved areas (state and semi-state) Area conserved (ha)	Permanently % of total Veld Type conserved	Conservation priority rating
16	NK#	13 908 190	973 641	7,00	4 470	0,03	0	0,00	215 943	1,55	978 111	7,03	
17	NK#	1 804 570	22 696	1,26	700	0,04	0	0,00	0	0,00	23 396	1,30	
24	NK	274 350	0	0,00	0	0,00	0	0,00	0	0,00	0	0,00	***
25	SK NK	935 880	6 000	0,64	800	0,09	0	0,00	0	0,00	6 800	0,73	*
26	SK NK	3 378 590	16 511	0,49	3 089	0,09	0	0,00	12 211	0,36	19 600	0,58	*
27	NK	1 956 930	0	0,00	320	0,02	0	0,00	458	0,02	320	0,02	**
28	SK NK	2 135 760	2 074	0,10	1 846	0,09	0	0,00	0	0,00	3 920	0,18	*
29	NK	7 688 440	0	0,00	760	0,01	0	0,00	0	0,00	760	0,01	
30	NK	1 115 240	0	0,00	0	0,00	0	0,00	0	0,00	0	0,00	*
31	SK	3 714 910	200	0,01	40	0,00	0	0,00	15 705	0,42	240	0,01	***
32	NK#	3 714 190	12 195	0,33	70	0,00	0	0,00	21 078	0,57	12 265	0,33	**
33	SK NK	2 810 590	25 748	0,92	0	0,00	0	0,00	49 168	1,75	25 748	0,92	*
35	NK	2 783 870	0	0,00	1 300	0,00	0	0,00	2 053	0,07	0	0,00	***
36	NK	5 983 060	28 654	0,48	126	0,02	0	0,00	7 296	0,12	29 954	0,50	*
37	NK#	1 107 060	11 470	1,04	570	0,01	0	0,00	2 298	0,21	11 596	1,05	
38	NK	119 150	3 500	2,94	0	0,48	0	0,00	0	0,00	4 070	3,42	
39	SK NK	1 195 400	0	0,00	0	0,00	0	0,00	0	0,00	0	0,00	***
40	NK #	617 420	0	0,00	0	0,00	0	0,00	34 962	5,66	0	0,00	***
41	NK	121 750	0	0,00	0	0,00	0	0,00	0	0,00	0	0,00	***
42	SK NK	225 870	19 049	8,43	0	0,00	0	0,00	0	0,00	19 049	8,43	
34	F SK	733 540	5 040	0,69	400	0,05	0	0,00	2 407	0,33	5 440	0,74	**
43	F SK NK	621 280	11 060	1,78	3 175	0,51	0	0,00	3 500	0,56	14 235	2,29	
46	F	178 880	3 751	2,10	156	0,09	5 385	3,01	0	0,00	3 907	2,18	***
47	F	928 870	43 283	4,66	1 033	0,11	0	0,00	315	0,03	44 316	4,77	***
69	F	1 980 130	485 588	24,52	30 578	1,54	536 449	27,09	12 555	0,63	516 166	26,07	
70	F	1 747 290	270 926	15,51	652	0,04	72 103	4,13	875	0,05	271 578	15,54	
TOTAL AREA (HA)			1 941 386		50 085		613 937		380 824		1 991 471		

† F = Fynbos
SK = Succulent karoo
NK = Nama-karoo
# = Savanna (outside biomes under discussion)

TABLE 13.3 Area conserved of the biogeographic areas in the fynbos, succulent karoo and Nama-karoo biomes. Data from the CDNEC.

Biomes and biogeographical areas	Total area (ha)	State-owned conserved area (ha)	Semi-state conserved area (ha)	Total area conserved (ha)	% of total area conserved	Conservation priority rating
Fynbos biome	5 542 900	752 305	28 856	781 161	14,09	
West Coast	448 000	4 989	707	5 696	1,27	***
Northern Mountains	1 312 000	117 717	1 304	119 021	9,07	**
South Coast Mountains	2 308 900	243 154	2 004	245 158	10,62	*
South-eastern Coast	809 000	144 757	1 667	146 424	18,10	
Bredasdorp	302 000	87 051	2 306	89 357	29,59	
Peninsula	46 000	600	14 404	15 004	32,62	
Suurberg	43 000	20 788		20 788	48,34	
Overberg	274 000	133 249	6 464	139 713	50,99	
Succulent karoo biome	9 970 500	107 674	12 182	119 856	1,20	
Vanrhynsdorp	1 461 000			0	0,00	***
Roggeveld	553 000			0	0,00	***
Kamiesberg	82 000			0	0,00	***
Tanqua	1 215 000	27 064	11	27 075	2,23	***
Western Mountain Karoo	171 500	5 067	8 076	13 143	7,66	**
Namaqualand	4 591 000	14 864		14 864	0,32	*
West coast Strandveld	146 100	963		963	0,66	*
Worcester-Robertson Karoo	184 000	1 827	1 200	3 027	1,65	
Gariep	297 900	10 984		10 984	3,69	
Little Karoo	1 269 000	46 905	2 895	49 800	3,92	
Nama-karoo biome	26 000 250	117 062	11 679	128 741	0,50	
Noorsveld	274 350			0	0,00	***
Bushmanland	7 688 500		3 038	3 038	0,04	**
Gordonia	2 165 900		3 668	3 668	0,17	*
Central Upper Karoo	11 930 900	40 271	1 381	41 652	0,35	*
Karas	2 165 700	12 195		12 195	0,56	*
Steytlerville	634 000	17 514	2 988	20 502	3,23	
Great Karoo	1 140 900	47 082	604	47 686	4,18	

It is clear from Table 13.2 that the mountain fynbos Veld Types (VT69 and 70) have a large percentage area conserved (26,07% and 15,54%, respectively), whereas coastal fynbos (VT47) and coastal renosterveld (VT 46) have a relatively low percentage of their area conserved (4,77% and 2,18%, respectively). The difference between mountain and lowland areas conserved is more marked when the amount of natural vegetation remaining in each of these areas is considered (Moll and Bossi 1984b). Strandveld (VT34), which is essentially west-coast strandveld, is also poorly conserved. This will be discussed in more detail under the section on the succulent karoo biome. The status of mountain renosterveld (VT43) is difficult to evaluate, since it crosses three biome boundaries, but with only 2,29% of its area permanently conserved it is in need of attention. Much of the remaining area is threatened by expanding wheatlands.

Biogeographically (Table 13.3) the West Coast (1,27%), Northern Mountains (9,07%), South Coast Mountains (10,62%) and South-eastern Coast (18,10%) areas emerge as having the lowest percentage area conserved. Much of the northern area is private mountain catchment and is therefore better protected than the data in Table 13.3 suggest. These biogeographic areas include both mountains and lowlands, since floristically (Oliver et al 1983) and structurally (Cowling et al 1988) they cannot be separated. On closer examination it is clear that it is largely the lowland portion that is in need of conservation. These results correspond with those from the Veld Types.

Threatened taxa and habitats

The number of threatened taxa in the five major biotic groups are shown in Table 13.4 according to their evaluation in the South African Red Data Books, following IUCN threatened status criteria. From this it is clear that the fynbos biome contains a high number of threatened taxa. The total of 1 350 represents 17,4% of the total number of plant and vertebrate species (excluding fish) in the fynbos biome. The number of threatened plants (98% of all threatened taxa) is higher than in any other biome in South Africa (Hall et al 1984). Many of the species have become extinct only recently as a direct result of man's activities (Hall and Veldhuis 1985).

TABLE 13.4   Rare and threatened taxa in the fynbos, succulent karoo and Nama-karoo biomes. Data from the CDNEC's CICEC data base.

Biomes and biotic groups	Extinct	Endangered	Vulnerable	Rare	Indeterminate	Uncertain	Total
Fynbos	27	108	162	396	185	472	1 350
Plants	26	103	152	389	184	472	1 326
Mammals	1		5	5			11
Birds			3	2	1		6
Amphibians		3	1				4
Reptiles		2	1				3
Succulent karoo	4	15	43	199	90	235	586
Plants	4	15	33	182	89	235	558
Mammals			3	7			10
Birds			4	2	1		7
Amphibians				1			1
Reptiles			3	7			10
Nama-karoo	2	2	16	29	18	28	95
Plants		1	6	19	13	28	67
Mammals	1	1	5	7			14
Birds	7		5	2	5		13
Amphibians							0
Reptiles				1			1

Figure 13.2 (drawn from data in the CICEC data base) shows the relative concentrations of threatened plant taxa in the Cape. From this figure it is clear that the greatest concentrations occur in the south-western Cape. This may to some degree be a reflection of collecting intensity (Gibbs Russell et al 1984) but, based on the diversity patterns shown by Cowling et al (this volume), we feel this is an accurate reflection of the actual situation.

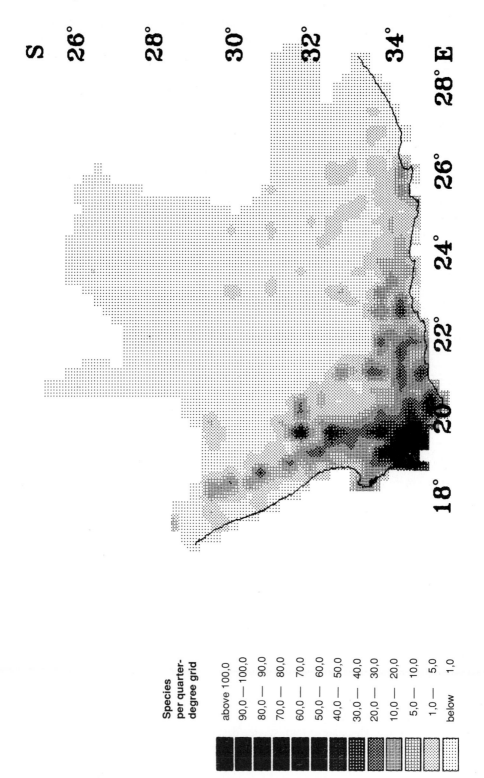

Species
per quarter-
degree grid

above 100,0
90,0 — 100,0
80,0 — 90,0
70,0 — 80,0
60,0 — 70,0
50,0 — 60,0
40,0 — 50,0
30,0 — 40,0
20,0 — 30,0
10,0 — 20,0
5,0 — 10,0
1,0 — 5,0
below   1,0

FIGURE 13.2   Distribution intensity map of rare and threatened plant taxa in the Cape Province. Data from the CDNEC's CICEC data base.

Table 13.5 shows the number of threatened taxa in relation to biogeographic areas. From the number of threatened taxa per unit area, we find that the greatest number of these taxa are in the Peninsula, Overberg, Bredasdorp and West Coast areas. The calculations are based on the area of each biogeographic unit, not the area of natural vegetation remaining. If the latter were used, the West Coast would have the highest priority. From the threatened taxa it is clear that West Coast renosterveld, south-coast renosterveld, limestone fynbos and Elim fynbos vegetation types are the most threatened.

TABLE 13.5   Number of threatened taxa within the biogeographic regions of the fynbos, succulent karoo and Nama-karoo biomes

Biomes and biogeographical areas	Taxa	Taxa/area (x1000)	Conservation priority rating
Fynbos biome			
Peninsula	362	7,87	***
Overberg	578	2,11	**
Bredasdorp	350	1,16	**
West Coast	319	0,71	*
Northern Mountains	388	0,30	*
Suurberg	11	0,26	*
South Coast Mountains	367	0,16	*
South-eastern Coast	92	0,11	*
Succulent karoo biome			
Western Mountain Karoo	187	1,09	***
West coast Strandveld	72	0,49	**
Kamiesberg	37	0,45	**
Worcester-Robertson Karoo	67	0,36	**
Little Karoo	293	0,23	*
Gariep	63	0,21	*
Vanrhynsdorp	166	0,11	*
Tanqua	58	0,05	*
Namaqualand	162	0,04	*
Roggeveld	2	0,00	*
Nama-karoo biome			
Great Karoo	82	0,07	***
Steytlerville	30	0,05	***
Noorsveld	4	0,01	**
Karas	17	0,008	*
Central upper Karoo	44	0,004	*
Gordonia	6	0,003	*
Bushmanland	19	0,002	*

West-coast renosterveld has been reduced by agriculture to three per cent of its former area (McDowell 1988). The once widespread renosterveld species are now restricted to a few remnants (eg *Protea odorata* McDowell 1986) and as a result many of these species are considered to be highly threatened. South-coast renosterveld, rich in geophytic endemics, is being replaced with cereal crops and planted pastures (Cowling et al 1986).

The limestone fynbos and Elim fynbos communities, because of their highly specialized and restricted nature, contain many threatened species. Elim fynbos is subject to medium grazing pressures, whereas limestone fynbos is utilized by flower pickers.

Coastal fynbos is often associated with wetter areas and contains many seasonal freshwater pans. The communities associated with these wetter areas and pans are severely threatened, since they are often disturbed and hence rapidly invaded by alien species. This is also true for riverine communities (Hall 1984), most of which are now infested with invasive alien plants, for example the Eersterivier at Stellenbosch.

Until recently, the west coast Strandveld (ie the area south of the Olifants River Mouth) was poorly conserved with only one nature reserve at Rocher Pan. This situation has subsequently changed with the proclamation of the Langebaan National Park which was recently expanded to include large areas of strandveld. An additional strandveld area that deserves attention is that around Bokbaai. Here, the dunes together with what is known as the 'sandveld' and the annual freshwater pans form a unique ecosystem which is under serious threat from invasive plants, agricultural practices and off-road vehicles.

Many of the dune-field areas, as on the Cape Flats, have a high number of associated threatened taxa. These areas have largely been stabilized through planting of alien invasive species (eg *Acacia cyclops),* which have outcompeted the indigenous flora to form almost mono-specific stands. Urban development has also destroyed many of these areas. The remaining dune areas should be high on the conservation priority list (Hall 1984; Low and McKenzie 1988).

Nature and extent of threats

The nature and extent of threats to both species and habitats in the fynbos biome have been extensively covered in the literature (eg Wicht 1945; Taylor 1978; Huntley 1978; Bigalke 1979; Kruger 1982; Hall 1984; Hall et al 1984; Macdonald and Jarman 1984; Hall and Veldhuis 1985; Tansley 1988b). The threats will therefore not be discussed in detail here.

The major threats are: fire — both controlled and uncontrolled (Kruger and Bigalke 1984); invasive alien organisms — plant and animal (Macdonald and Jarman 1984); agriculture — intensive, extensive, pesticides and herbicides; urbanization; flooding due to construction of dams; industrial/commercial development — including infrastructure of roads, railways, etc; quarrying which includes gravel borrow pits; and commercial and illegal wildflower picking. There are also many indirect threats as a result of any of the above eg trampling, soil erosion, and pathogens such as *Phytophthora cinnamomi* (Von Broembsen 1979).

Tansley (1988b) investigated threats to the Proteaceae and Restionaceae, two fynbos plant families that are dominant yet very divergent ecologically and structurally. She found that both were most severely threatened by fire (particularly fire interval), invasive alien organisms and agriculture (ploughing, trampling and grazing). Hall and Veldhuis (1985) analysed the threats to 95 threatened plant species in the area roughly corresponding to the Overberg. They found the five major threats to be invasive alien plants, especially acacias, hakeas and pines; agriculture; construction of houses and factories; uncontrolled fires; and presumed genetic decline in critically small populations. The last-mentioned threat could be as a result of either human-caused impacts or natural pressures. Hall and Veldhuis (1985) argue that it is important to emphasize the role of natural pressures on rare species. The floristic diversity of the fynbos biome is such that many of these species may well be palaeoendemics which are becoming extinct owing to climatic changes since the Pleistocene (Hall and Veldhuis 1985).

Threats to the animal component of the biome are not as well documented as for the plants. Generally the major threat is habitat destruction due to one of the factors mentioned. For example, the geometric tortoise *Psammobates geometricus* is severely threatened because of habitat destruction. A number of nature reserves have been proclaimed in order to protect this species from extinction (Greig 1984). Apart from habitat destruction, many reptile species, particularly lizards, are threatened by collectors — both amateurs and scientists (Baard 1988). Leopard *Panthera pardus* conservation has received much attention recently. This threatened species is considered a problem animal and is persecuted as vermin by farmers. The first of a number of Leopard Conservation Areas in the Cape mountains was recently proclaimed by the CDNEC (1988).

Conservation priorities

Following the recognition of the importance of the lowlands, a symposium was held at the University of the Western Cape to focus attention on the conservation status of the western Cape lowlands (Moll 1982). A resolution adopted at this symposium prompted the Minister of the then Department of Water Affairs, Forestry and Environmental Conservation, to establish a committee to investigate proposals for nature conservation areas. The report of this committee (Hall 1984) presented a general outline on the conservation status of the area and strategies to achieve the much-needed conservation objectives. The report, however, did not provide details for specific areas.

At the same time, the Fynbos Biome Project convened a number of working groups to prepare specific site reports for conservation-worthy areas. Brief reports and locality maps for each site were prepared and the sites ranked in terms of conservation, importance (Jarman 1986). The conservation significance ranking of the sites (see Jarman (1986) for details) was based purely on the biological attributes of the sites. The report did not provide specific recommendations regarding appropriate conservation mechanisms to conserve each of the proposed sites.

Through the auspices of the NAKOR National Plan for Nature Conservation a detailed conservation plan for the region was compiled (Burgers et al 1987). This plan includes comprehensive site reports, a conservation significance rating for each site, the degree of threat to the site with the sites ranked accordingly. Conservation action priorities are stated and mechanisms for the conservation of each site are listed.

The plan for nature conservation of lowland fynbos (Burgers et al 1987) together with the reports of Jarman (1986) and Hall (1984) form an invaluable 'conservation guideplan' for the coastal lowlands of the fynbos biome. If fully implemented, many of the conservation priorities outlined above will be achieved. In the light of current and future threats and pressures it is vital that the recommendations be implemented immediately.

It is necessary for any conservation guideplan to be constantly re-examined in the light of new information. This is clearly illustrated in the recent works of Tansley (1988a, b) and McDowell (1988). Tansley (1988a, b) recently completed a survey on threatened Proteaceae in the fynbos biome. This plant family is probably one of the most threatened, with three species recently extinct and less than 16% of the remainder protected in official nature reserves (Tansley 1988a). In examining the ecological characteristics of each species, Tansley (1988a) concluded that the small populations and restricted distributions of the rare and threatened Proteaceae presented special conservation problems. She suggested that a system of numerous reserves with areas from as small as 5 km^2 might be appropriate. The emphasis should also be on reserves representing unique habitats, in close proximity to one another, and with their specific sizes determined by the minimum population requirements of the species in that reserve.

In a survey on the influence of agriculture on the decline of the west-coast renosterveld, McDowell (1988) derived an 'agricultural threat' index. This index is a quantitative measure based on soil quality, gradient (slope) and rainfall measurements. Using this index McDowell (1988) found that 20 of the 40 west-coast renosterveld remnants had a high threat status. Three quarters of these still exist because they are either public lands (not necessarily set aside for long-term conservation) or the present owners are conservation orientated. The remainder have either been ploughed since the original survey was completed or have the same fate pending.

The situation regarding south-coast renosterveld is somewhat different. Cowling et al (1986) suggest a number of management practices which would return this vegetation to its original grassy condition. In this state pastoral utilization would be compatible with conservation.

McDowell (1988) argues that a 'socio-economic threat' index relating to agricultural, urban or industrial expansion, should also be used to supplement the attributes used in determining conservation priorities. He also referred to land that at present seems safe from agricultural exploitation because of steepness, lack of water, poor soil, etc, but which was not necessarily safe because of agro-technical (and other) innovations. McDowell's work underlines the fact that conservation plans need to be dynamic.

KAROO BIOME

In this section, the succulent karoo and Nama-karoo biomes will be discussed jointly. This is largely because most literature (pre-1986) regards the area as a single karoo biome. Also, Acocks's (1975) Veld Types cross the biome boundaries, making separate conservation assessments of the biomes difficult. Wherever possible differences between the two biomes will be highlighted.

Description: succulent karoo biome

The succulent karoo biome (Rutherford and Westfall 1986) covers an area of approximately 81 908 km^2, some 6,7% of South Africa (Figure 13.1). It is restricted to the year-round, winter and strong winter-rainfall areas with greatest summer aridity.

The vegetation is low to dwarf (usually less than one metre tall), open to sparse (15 to 50% canopy cover) succulent shrubland. This shrubland is dominated by stem and leaf succulents, fine-leaved evergreen shrubs and some obligately deciduous shrubs. Grasses are infrequent and are mainly annuals. The mass flowering displays of annuals (mainly Asteraceae) and geophytes (Liliaceae *sensu latu* and Iridaceae) in spring, particularly in disturbed areas, are characteristic. Low trees are common on rocky outcrops and along river courses where they form woodlands.

There are seven Veld Types (Acocks 1953, 1975) in the succulent karoo biome, namely karroid broken veld (VT26), Namaqualand broken veld (VT33), western mountain karoo (VT28), succulent karoo (VT31), strandveld (VT43), false succulent karoo (VT39) and mountain renosterveld (VT43). Some of these also occur in the Nama-karoo and fynbos biomes.

Description: Nama-karoo biome

The Nama-karoo biome (Rutherford and Westfall 1986) covers an area of about 346 107 km^2, some 28,35% of South Africa. This biome covers most of the vast central plateau region of the Cape Province, with smaller portions in the eastern Cape, eastern Orange Free State, Lesotho highlands and southern central parts of Namibia (Figure 13.1). The last two areas are not considered here. The biome can be subdivided in two, primarily on the basis of mean annual rainfall. A small portion with relatively high rainfall is known as the 'false' karoo. This corresponds to those areas formerly covered by grassland but now transformed through grazing pressures to a karroid shrubland. The vast remainder comprises the karoo proper.

The vegetation is a low to dwarf, open to sparse (see previous definitions), grassy shrubland. The shrubland is dominated by facultatively deciduous shrubs, some leaf succulents and perennial grasses. Grasses become more dominant from west to east. Scattered trees grow on rocky outcrops, low hills and along river courses, forming woodlands. On the northern fringes of the biome the dwarf shrubland often has an overstorey of subtropical shrubs and trees. There are 21 Veld Types (Acocks 1953, 1975) in the Nama-karoo, eight of which are classified as false karoo types.

For more detailed information on the vegetation and ecological characteristics of both the succulent karoo and Nama-karoo biomes, readers are referred to Acocks (1975); Werger (1978b); White (1983); Rutherford and Westfall (1986); Cowling (1986); Cowling et al (1986) and Cowling and Roux (1987).

Biodiversity

The succulent karoo biome is extremely rich in succulent plant species, particularly in the families Mesembryanthemaceae, Crassulaceae, Asclepiadaceae and Euphorbiaceae (Hilton-Taylor 1987; Van Jaarsveld 1987). This high succulent species diversity is unparalleled elsewhere in the world. This, together with the many geophytic and annual taxa, make the succulent karoo a unique biome of international importance.

The Nama-karoo biome does not appear to have a species-rich or unique flora. The area forms an ecotone between the Cape Flora to the south and the tropical savanna in the north. Many of the plant species of the Nama-karoo are shared with the savanna, grassland, succulent karoo and fynbos biomes (Gibbs Russell 1987).

The size of the floras of both karoo biomes is not accurately known. Gibbs Russell (1987) calculates the succulent karoo flora to have 2 125 taxa with 29% endemic, and the Nama-karoo to have 2 147 taxa, 18% of which are endemic. Le Roux (1988) estimates that there are 3 500 taxa in Namaqualand alone. For the whole Karoo-Namib Region White (1983) states that there are 3 500 species. Hilton-Taylor (1987), on the other hand, estimates the flora of the karoo biome to exceed 7 000 taxa with between 35% and 50% endemic. Hilton-Taylor's figures are used here (Table 13.1).

Very few data on the zoological components of either biome are available. The invertebrate fauna in the succulent karoo seems to be very rich with a large number of endemics. Some of the invertebrates have a significant impact on the vegetation of the Nama-karoo, notably the karoo caterpillar *Loxostege frustalis*, the brown locust *Locustana pardalina*, and the harvester termite *Hodotermes mossambicus* (Vorster and Roux 1983).

A distinct karoo fauna is not recognized by zoogeographers (Werger 1978a) although 19 bird species endemic to South Africa are largely confined to the karoo (Huntley 1984). The vast migratory herds of ungulates, including springbok *Antidorcas marsupialis*, black wildebeest *Connochaetes gnou*, blesbok *Damaliscus dorcas phillipsi* and the extinct quagga *Equus quagga*, have been replaced by domestic stock, particularly sheep and goats. The karoo in general has a rich rodent and reptile fauna (particularly tortoises), many of which are rare and threatened. (See Tables 13.1 and 13.4.)

Historical perspective

The poor conservation status of the karoo has long been a major cause for concern. By the time the national park concept gained momentum in South Africa, the once immense ungulate populations and attendant carnivores of the karoo had been so decimated that very little game remained to be conserved (Huntley 1978). Following the extermination of the ungulate herds, the karoo was fenced and stocked with sheep and goats. The deterioration of the veld as a result of inappropriate veld and stock management practices is well documented (eg Acocks 1953, 1964; Roux and Vorster 1983). This degradation included loss of vegetation cover, soil erosion, and replacement of highly palatable species with less palatable ones. This degradation was not restricted only to the karroid shrublands, but spread north-eastwards to the grasslands of the Orange Free State. Some 66 000 km² of former grassland is now occupied by karoo species (Acocks 1975). Huntley (1978) considered it a tragedy that no viable conservation areas existed within these rapidly changing ecosystems to serve as a basis for monitoring and comparison.

It is surprising that in the face of the desertification scenarios presented by Acocks (1953, 1964), and evidence presented by Edwards (1974), that government agencies have been so slow to react. Edwards (1974) stated that one of the major conservation deficiencies in South Africa lay in the karroid Veld Types. He stated that only 0,1% of the karoo and karroid bushveld types was conserved in provincial nature reserves or national parks, and only 0,3% of the false karoo types were conserved (Edwards 1974).

Despite these warnings, it was left to the initiative of voluntary conservation organizations such as the Southern African Nature Foundation to establish a karoo park. Only after considerable public pressure, the donation of an area of municipal commonage, and the raising of funds by school children did the government contribute towards the development of a karoo national park (Huntley 1978).

The proclamation of the Karoo National Park at Beaufort West, the Karoo Nature Reserve at Graaff-Reinet, and the Oviston and Rolfontein reserves did much to improve the conservation status of the Nama-karoo biome. However, Scheepers (1983) found the situation had changed very little from that reported by Edwards (1974). The karoo grass transition (ie false karoo types) had 1,03% conserved, the succulent karoo (*sensu* Scheepers 1983; a combination of Veld Types almost equivalent to the size of the biome) only 0,14%, and the central karoo (mostly Nama-karoo) 0,57%. These figures can also be regarded as generous, since they included all forms of conservation areas, not only publicly owned areas.

Conservation status

The present status of conservation in terms of area of the Veld Types conserved in the succulent karoo and Nama-karoo biomes is shown in Table 13.2. All the Veld Types excepting VT16, 38, 42 and 43 have less than one per cent conserved. Kalahari thornveld (VT16) is the highest with 7,06% conserved, although this is not strictly speaking a true karoo Veld Type. Of the remainder, VT 27, 29 and 31 have less than 0,1% conserved while VT24, 30, 35, 39, 40 and 41 have no portions conserved in state, semi-state or private conservation areas (see previous definitions). Some of these have some areas conserved, for example false Orange River broken veld (VT40), which has 34 962 ha in a South African Defence Force area.

The data for the succulent karoo Veld Type (VT31) exclude the newly acquired Tankwa Karoo National Park (27 063 ha). This is because the area is considered to be totally degraded (Department of Environment Affairs 1988). The park also does not include the Tanqua endemics and threatened taxa. In the short term the park has no conservation value. The siting of many reserves in the karoo has been purely opportunistic (eg around dams) or arbitrary. It is necessary to determine objective priorities.

From a biogeographic point of view (Table 13.3) we find that the Vanrhynsdorp centre, Roggeveld and Kamiesberg areas of the succulent karoo biome have no area conserved. The Western Mountain Karoo and Namaqualand, in terms of area, have little conserved. Namaqualand, apart from the Kommaggas and Spektakel areas, is adequately conserved by the representative Hester Malan Nature Reserve. The Tanqua Karoo would also rate high in priority if the degraded national park were not included. In the Nama-karoo the Noorsveld has no areas conserved, while Bushmanland and Gordonia have less than 0,1% of their area conserved. The Great Karoo and Steytlerville Karoo are the best conserved with 4,18% and 3,23%, respectively. However, in the Steytlerville Karoo the surrounding mountains are conserved while the Springbokvlakte, which have numerous rare and threatened taxa, are not protected.

Threatened taxa

The numbers of threatened taxa in the five major biotic groups are shown in Table 13.4 according to IUCN criteria. The succulent karoo has many more threatened plant taxa than the Nama-karoo. This may to some extent be an artefact of collecting intensity, since the Nama-karoo has been very poorly collected (Gibbs Russell et al 1984). Compared with the total diversity of the major biotic groups in the karoo, these totals represent 13,4% of succulent karoo taxa and only 2,8% of Nama-karoo taxa.

Table 13.5 shows the number of threatened taxa in relation to biogeographic area, while Figure 13.2 shows the relative concentrations of the threatened plant taxa (see previous comments). In the succulent karoo biome the Western Mountain Karoo has a high number of threatened taxa per unit area, with concentrations of these taxa in two areas: Nieuwoudtville-Calvinia and an area between Sutherland and Middelpos. The Western Mountain Karoo also has a low percentage area conserved in terms of both biogeographic area and Veld Type. The west coast Strandveld, Kamiesberg, Little Karoo, Worcester-Robertson Karoo and Gariep centre also have high numbers of threatened taxa per unit area (in decreasing order).

In the Nama-karoo biome, the Steytlerville, Great Karoo and Noorsveld areas have the greatest density of threatened taxa.

Nature and extent of threats

The nature and extent of threats in the succulent karoo and Nama-karoo biomes are not well documented and are poorly known. The main threats are agriculture, invasive alien plants, mining operations, succulent collectors and, to a limited degree, urbanization.

The major threat seems to be agriculture, both intensive and extensive. In both biomes irrigation schemes have been established along most perennial rivers allowing large-scale intensive cultivation. For example, the Olifants River scheme has converted some 9 114 ha of dry land into arable land (De Kock 1983). There is potential to increase the area currently irrigated, but this is not economically viable at present. There is also extensive dry-land cropping in the wetter parts of the succulent karoo (wheat and barley) and Nama-karoo (maize). This is a dangerous practice, since the unpredictability of rainfall leads to abandoned lands which take many years to become revegetated. This is particularly noticeable in southern Namaqualand (see Macdonald this volume).

The impact of crop herbicides and pesticides on the surrounding vegetation in the karoo is not documented. However, the effects of pesticides used to combat the brown locust *Locustana pardalina*, particularly benzine hexachloride, have received much attention recently (McKenzie and Longridge 1988). The major impact of these poisons is on the fauna. The long half-life and persistence of these poisons results in their concentration up the food chain and their toxic effects will therefore be evident well into the future.

Extensive agriculture in the form of pastoralism occurs throughout the karoo. The resultant problems of selective overgrazing, overstocking and trampling are therefore widespread. The effects of this are well documented (Vorster and Roux 1983; Hilton-Taylor and Moll 1986; Roux and Theron 1987). In general, mismanagement leads to irreversible vegetation change with loss of both cover and species diversity.

There are a number of invasive alien plant species that have become established in the karoo. *Opuntia aurantiaca, O rosea, O imbricata, O megacantha, Cereus peruviana* and *Alhagi camelorum* are all common in the eastern parts; *Prosopis* sp is dominant in riverine areas to the north; *Nicotiana glauca* and *Nerium oleander* are widespread in riverine areas; and *Acacia cyclops* is common in sandy areas to the south-west (Brown and Gubb 1986; Roux and Theron 1987). The overseeding of natural pastures with alien species such as *Atriplex nummularia,* in an attempt to raise carrying capacity, is cause for concern. Hobson (1988) has shown that *Atriplex* modifies the soil and surrounding environment. It is not known what effects this will have on natural vegetation. A conservative figure of 150 000 ha of land is estimated to be planted under *Atriplex* (Hobson 1988). With the deterioration of veld, many indigenous species have also become invasive, for example *Acacia mellifera* subsp *detinens, A karroo, Rhigozum trichotomum,* and *Chrysocoma ciliata*. These species have a competitive advantage and outcompete the favoured forage species (Roux and Theron 1987).

Mining operations, particularly open-cast mining, are a major threat to the succulent karoo and parts of the Nama-karoo. A large area of the northern Cape is restricted diamond mining area (essentially the entire coastline from Doringbaai northwards to Alexander Bay). This restriction is in itself a form of conservation. However, the effects, both direct and indirect, of the mining operations on vegetation and soil erosion are severe (Talkenberg 1982; Boucher and Le Roux in press). There are numerous large mineral deposits (gypsum, marble, monazite, kaolin, copper, limestone, ilmenite and titanium) in the Vanrhynsdorp, Namaqualand, Gariep, Karas and Gordonia areas. Associated with these deposits are unique habitats and numerous rare species. Exploitation of these resources will lead to loss of diversity. The mining procedures used, and more particularly the restoration/rehabilitation measures taken once mining is complete, are in need of attention.

A major threat to certain plant species in the succulent karoo and parts of the Nama-karoo are succulent collectors. Despite the fact that South Africa is a signatory to the CITES agreement which controls world trade in threatened species, large-scale illegal trade in wild collected plants continues. The prices fetched on overseas markets make this a very profitable business; some species of *Haworthia,* for example, sell for between 15 and 25 pounds sterling per plant. Until recently the penalties imposed on those apprehended were not sufficient to deter them from continuing, since the potential profits from sale of the plants were greater than the maximum fines imposed. There is also some illegal trade in species of lizards, for example *Cordylus* species. Control and legislation concerning trade in threatened taxa is inadequate at present.

Urbanization is a threat only in the Worcester-Robertson, Oudtshoorn and Vanrhynsdorp areas. Urbanization and development in the Worcester-Robertson area have reduced the extent of natural habitats, many of which are unique. A large number of threatened taxa are associated with these habitats. General infrastructure such as roads and railways also have some impact. The Sishen-Saldanha railway, for example, has reduced the population sizes and eliminated populations of several rare species in the Vanrhynsdorp area. De Kock (1983) highlights the effects on the environment of expanding urbanization in the Vanrhynsdorp area. He concludes that present regional planning in the Clanwilliam, Vanrhynsdorp and Vredendal magisterial districts, does not consider the vulnerability of the natural environment.

Conservation priorities

A low intensity of plant collecting (Gibbs Russell et al 1984), poor status of systematics for many taxa, and general lack of understanding of ecosystem processes and functioning characterize our present knowledge of the karoo. This makes the determination of conservation priorities for the karoo extremely difficult. However, using Veld Types, biogeographic areas and threatened taxa as guides, we shall outline some broad conservation priorities. A great deal more inventory work, systematics and collecting is necessary, particularly in the Nama-karoo biome, if we are to refine these priorities to the level achieved in the fynbos biome.

The degree of priority, based on the percentage of each Veld Type conserved, is indicated in Table 13.2, where no * implies low priority and *** implies high priority. These results should be used in conjunction with those from biogeographic areas and threatened taxa (Tables 13.3 and 13.5). Veld Types 24, 30, 35, 39, 40 and 41 have high priority (no area conserved).

In terms of biogeographic area conserved and density of threatened taxa, the Kamiesberg, Roggeveld, Tanqua, Vanrhynsdorp centre, Western Mountain Karoo, Gariep centre and Worcester-Robertson areas of the succulent karoo biome have high conservation priority.

The future of the Gariep centre seems assured in the form of two Richtersveld parks. The Tanqua has 2,23% conserved if the Tankwa Karoo National Park is included. However, additional areas of less degraded vegetation are required to ensure protection for the endemic rare species of the Tanqua. The Worcester-Robertson area is somewhat anomalous, since despite its very high number of threatened taxa per unit area, and its large number of conservation areas, many of these taxa are not in conservation areas. Also there are numerous unique habitats in the area, particularly the communities with an unusual mixture of fynbos, renosterveld and karoo elements. Many of these are severely threatened by development. The Western Mountain Karoo, the Vanrhysdorp centre, Kamiesberg and Roggeveld areas have the highest conservation priority.

In the Nama-karoo, the Noorsveld has the highest conservation priority (no area conserved, high number of threatened taxa, and serious degradation of vegetation from overgrazing and cutting of noors *Euphorbia coerulescens* for fodder). The Steytlerville area has high priority in terms of threatened taxa, while Bushmanland, Gordonia, Karas centre and the Central Upper Karoo (in decreasing order) have very little of their area conserved.

Within the biogeographic areas there are a number of specific threatened habitats which deserve protection. These include quartz patches south of Ladismith in the Little Karoo, dolerite outcrops in the Western Mountain Karoo, granite outcrops in the Kommaggas area of Namaqualand, the Nama geological system between the Spektakel and Annenous Passes in northern Namaqualand, and riverine areas throughout the region. The large pans in northern Bushmanland (eg Verneukpan) should receive some attention. The west coast Strandveld, which has a high number of threatened taxa, is threatened with destruction by mining activities. The proposed Groenrivier nature reserve would conserve a representative portion of west coast Strandveld, including many of the threatened taxa. However, it is essential that this reserve be extended inland to include not only the strand communities, but also the succulent karoo and fynbos patches.

At present plans are well advanced for the proclamation of a national park in the central Richtersveld (VT31, 33) and two provincial reserves, one in the eastern Richtersveld and the other at Groenrivier on the west coast (VT31, 34). Extensions to the Oorlogskloof Reserve (VT28), the newly acquired Anysberg area (VT26, 31, 43), the Akkerendam Reserve on the Hantamsberg (VT39, 43), and the Vrolijkheid Nature Reserve (VT26, 43) are also being investigated by the CDNEC. The new reserves and extensions will improve the conservation status of the succulent karoo biome.

National parks and nature reserves certainly perform an important function in the karoo in that they conserve genetic diversity and seedbanks which allow for recolonization of degraded areas. However, in view of the use of the karoo as rangeland, the major thrust of conservation should be aimed at correct landuse management.

CONCLUSIONS

Of the total area of the fynbos biome, 14% is conserved. Most of this conserved area is in the mountains, thus it is the lowlands of the south-western Cape that have the highest priority for conservation attention. The priorities within these vegetation types have been established and precise sites ranked according to a number of criteria. A detailed conservation guideplan is available. This is probably one of the few areas in South Africa with such a detailed plan, yet progress regarding the acquisition of the priority areas is extremely slow. The comment of Greyling and Huntley (1984), that the neglect of the floristically diverse fynbos lowlands reflects the traditional emphasis given to conserving areas with spectacular ungulate and carnivore species, is still true today.

The conservation status of the succulent karoo (1,2% conserved) and Nama-karoo (0,5% conserved) biomes is poor compared with that of the fynbos biome. Our knowledge of the biodiversity of these two biomes is inadequate, and our understanding of ecosystem processes and functioning is incomplete. Conservation priorities can be determined on a broad scale in terms of Veld Type and biogeographic area. However, few specific sites can be identified owing to our poor knowledge and the scattered state of the available data. It is clear, however, that the Western Mountain Karoo (particularly the area between Nieuwoudtville and Sutherland), the Vanrhynsdorp centre, Noorsveld, Steytlerville (Springbokvlakte), Bushmanland and Gordonia areas are in need of attention.

Cowling (1986) stated that the possibility of re-establishing pre-settlement animal communities and grazing patterns exists for only three reserves, which are larger than 15 000 ha. He argued that large biosphere reserves were required to improve the conservation status of the two biomes. Although this is certainly true, indications from a preliminary analysis of the threatened plants data base show that a system of smaller reserves is also necessary, if species conservation is one of the goals. A number of threatened taxa in the succulent karoo biome have small restricted populations.

The international significance of the poor conservation status of the fynbos and succulent karoo biomes is outlined by Huntley and Ellis (1984). Greyling and Huntley (1984) have therefore suggested that a co-ordinated national effort be focused on these regions. They argued further that funds utilized for conservation efforts should be directed to international rather than parochial priorities.

The overriding conservation effort in both the succulent karoo and Nama-karoo biomes should, however, be directed at improved utilization and management of the karoo shrublands. Only through wise utilization and correct management will veld deterioration, soil erosion and desertification be halted and the goal of conservation achieved.

ACKNOWLEDGEMENTS

We thank the Chief Director of the Chief Directorate Nature and Environmental Conservation, Cape Province, for permission for A Le Roux to co-author this paper. We also thank the CDNEC for permission to utilize and publish data from the CICEC data base, and for access to staff and facilities. We are indebted to D Laidler, who designed the CICEC data base, for his time and patience in helping with data retrieval and analysis. We also thank J W Lloyd, M J Simpson, R Pool, C J Burgers, A L de Villers and E H W Baard, all of the CDNEC, for inputting and checking of data and for preparing maps and tables. We are grateful to E J Moll and three anonymous referees for comments on the manuscript. C Hilton-Taylor gratefully acknowledges the financial support of the Committee for Nature Conservation Research of the CSIR National Programme for Ecosystem Research.

REFERENCES

ACOCKS J P H (1953). Veld Types of South Africa. *Memoirs of the Botanical Survey of South Africa* **28**, 192 pp.

ACOCKS J P H (1964). Karoo vegetation in relation to the development of deserts. In *Ecological Studies in Southern Africa.* (ed Davis D H S) Junk, The Hague. pp 100 – 112.

ACOCKS J P H (1975). Veld Types of South Africa. Second edn. *Memoirs of the Botanical Survey of South Africa* **40**, 128 pp.

BAARD E H W (1988). The status of some rare and endangered herpetological species of the south-western Cape Province, South Africa. Manuscript submitted to Biological Conservation.

BIGALKE R C (1979). Conservation in Fynbos ecology: a preliminary synthesis. (eds Day J, Siegfried W R, Louw G N and Jarman M L) *South African National Scientific Programmes Report* **40**, CSIR, Pretoria. 148 – 157.

BOND P and GOLDBLATT P (1984). Plants of the Cape Flora — a descriptive catalogue. *Journal of South African Botany Supplement* **13**.

BOUCHER C and LE ROUX A (in press). Dry coastal ecosystems: South African west coast strand vegetation. In *Ecosystems of the World 2. Dry Coastal Ecosystems.* Elsevier, Amsterdam.

BROWN C J and GUBB A A (1986). Invasive alien organisms in the Namib Desert, Upper Karoo and the arid and semi-arid savannas of western southern Africa. In *The Ecology and Management of Biological Invasions in Southern Africa.* (eds Macdonald I A W, Kruger F J and Ferrar A A) Oxford University Press, Cape Town. pp 93 – 108.

BURGERS C J, NEL G and POOL R (1987). Plan for Nature Conservation (Region A1). Recommendations for the conservation of lowland fynbos areas in the south western Cape. Unpublished report prepared for the NAKOR National Working Group for Region A1. Cape Department of Nature and Environmental Conservation, Cape Town.

CDNEC (1988). *Leopard Conservation in the Cape Province.* Pamphlet produced and published by the Chief Directorate Nature and Environmental Conservation. Cape Provincial Administration, Cape Town.

COWAN G I (1987). S A Plan for Nature Conservation. Unpublished Annual report (April 1986 – March 1987). Department of Environment Affairs, Pretoria. 14 pp.

COWLING R M (1986). A description of the Karoo Biome Project. *South African National Scientific Programmes Report* **122**, CSIR, Pretoria. 43 pp.

COWLING R M, CAMPBELL B M, MUSTART P, MCDONALD D J, JARMAN M L and MOLL E J (1988). Vegetation classification in a floristically complex area: the Agulhas Plain. *South African Journal of Botany* **54**, 290 – 300.

COWLING R M, PIERCE S M and MOLL E J (1986). Conservation and utilisation of South Coast Renosterveld, an endangered South African vegetation type. *Biological Conservation* **37**, 363 – 377.

COWLING R M and ROUX P W (1987). The karoo biome: a preliminary synthesis. Part 2 — vegetation and history. *South African National Scientific Programmes Report* **142**, CSIR, Pretoria. 133 pp.

COWLING R M, ROUX P W and PIETERSE A J H (1986). The karoo biome: a preliminary synthesis. Part 1 — physical environment. *South African National Scientific Programmes Report* **124**, CSIR, Pretoria. 115 pp.

DAY J, SIEGFRIED W R, LOUW G N and JARMAN M L (eds) (1979). Fynbos ecology: a preliminary synthesis. *South African National Scientific Programmes Report* **40**, CSIR, Pretoria. 165 pp.

DE KOCK G L (1983). Aspekte van lanskapverandering in die Landrosdistrikte van Clanwilliam, Vanrhynsdorp en Vredendal. Unpublished PhD thesis, University of Stellenbosch, Stellenbosch.

DEPARTMENT OF ENVIRONMENT AFFAIRS (1988). National Register of Protected Areas in South Africa. Unpublished internal report of the Chief Directorate: Environmental Conservation, Department of Environment Affairs, Pretoria.

EDWARDS D (1974). Survey to determine the adequacy of existing conserved areas in relation to vegetation types. A preliminary report. *Koedoe* **17**, 2 – 37.

FRANKEL O H and SOULE M E (1981). *Conservation and Evolution.* Cambridge University Press, Cambridge.

GIBBS RUSSELL G E (1987). Preliminary floristic analysis of the major biomes in southern Africa. *Bothalia* **17**, 213 – 227.

GIBBS RUSSELL G E, RETIEF G E and SMOOK L (1984). Intensity of plant collecting in southern Africa. *Bothalia* **15**, 131 – 138.

GOLDBLATT P (1978). Analysis of the flora of southern Africa: its characteristics, relationships and origins. *Annals of the Missouri Botanical Garden* **65**, 369 – 436.

GOOD R (1974). *The Geography of the Flowering Plants.* (Fourth edn). Longmans, London.

GREIG J C (1984). Conservation status of South African land tortoises; with special reference to the geometric tortoise *(Psammobates geometricus). Amphibia-Reptilia* **5**, 27 – 30.

GREYLING T and HUNTLEY B J (eds) (1984). Directory of southern African conservation areas. *South African National Scientific Programmes Report* **98**, CSIR, Pretoria. 311 pp.

HALL A V (1984). Proposals for Nature Conservation Areas in the Coastal Lowlands of the South-Western Cape Province. Report to the Minister of Environment Affairs, the Honorable Mr W E Wiley. Unpublished report, Department of Environment Affairs, Cape Town.

HALL A V (1987). Threatened plants in the fynbos and karoo biomes, South Africa. *Biological Conservation* **40**, 29 – 52.

HALL A V, DE WINTER B, FOURIE S P and ARNOLD T H (1984). Threatened plants in southern Africa. *Biological Conservation* **28**, 5 – 20.

HALL A V and VELDHUIS H A (1985). South African Red Data Book: Plants — fynbos and karoo biomes. *South African National Scientific Programmes Report* **117**, CSIR, Pretoria. 160 pp.

HILTON-TAYLOR C (1987). Phytogeography and origins of the karoo flora. In The karoo biome: a preliminary synthesis. Part 2 — vegetation and history. *South African National Scientific Programmes Report* **142**, CSIR, Pretoria. 70 – 95.

HILTON-TAYLOR C and MOLL E J (1986). The karoo — a neglected biome. *Veld and Flora* **72**, 33 – 36.

HOBSON C D (1988). Micro-environments in two Oldman Saltbush *Atriplex nummularia* plantations and adjacent karoo vegetation. Unpublished paper presented at the 4th Annual Research Meeting of the Karoo Biome Project.

HUNTLEY B J (1978). Ecosystem conservation in southern Africa. In *Biogeography and Ecology of Southern Africa.* (ed Werger M J A) Junk, The Hague. pp 1333 – 1384.

HUNTLEY B J (1980). Threatened terrestrial habitats today: status survey. Conservation of Threatened Natural Habitats. Unpublished HABCON Working Document 3. 15 pp.

HUNTLEY B J (1984). Characteristics of South African biomes. In *Ecological Effects of Fire in South African Ecosystems.* (eds Booysen P de V and Tainton N M) Springer, Berlin. pp 2 – 17.

HUNTLEY B J and ELLIS S (1984). Conservation status of terrestrial ecosystems in southern Africa. In *Proceedings of the Twenty-Second Working Session, Commission on National Parks and Protected Areas.* IUCN, Gland, Switzerland pp 13 – 22.

JARMAN M L (ed) (1986). Conservation priorities in lowland regions of the fynbos biome. *South African National Scientific Programmes Report* **87**, CSIR, Pretoria. 55 pp.

JARVIS J U M (1979). Zoogeography. In Fynbos ecology: a preliminary synthesis. (eds Day J, Siegfried W R, Louw G N and Jarman M L) South African National Scientific Programmes Report **40**, CSIR, Pretoria. 82 – 87.

KEAY R W J (1959). *Vegetation Map of Africa South of the Tropic of Cancer.* Oxford University Press, London.

KRUGER F J (1977). Ecological reserves in the Cape fynbos: toward a strategy for conservation. *South African Journal of Science* **73**, 81 – 85.

KRUGER F J (1982). Use and management of Mediterranean ecosystems in South Africa — current problems. In *Medecos: Dynamics and Management of Mediterranean-Type Ecosystems.* (eds Conrad C E and Oechel W C) USDA Forest Service General Technical Report PSW-58. Berkley, Ca. pp 42 – 48.

KRUGER F J and BIGALKE R A (1984). Fire in fynbos. In *Ecological Effects of Fire in South African Ecosystems.* (eds Booysen P de V and Tainton N M) Springer, Berlin. pp 67 – 114.

KRUGER F J and TAYLOR H C (1979). Plant species diversity in Cape fynbos: gamma and delta diversity. *Vegetatio* **41**, 85 – 93.

LE ROUX A (1988). Namaqualand Checklist. Unpublished document, Chief Directorate Nature and Environmental Conservation, Cape Province.

LOW A B and MCKENZIE B (1988). Conservation priority survey of the Cape Flats. *Veld and Flora* **74**, 24 – 25.

MACDONALD I A W and JARMAN M L (1984). Invasive alien organisms in the terrestrial ecosystems of the fynbos biome, South Africa. *South African National Scientific Programmes Report* **85**, CSIR, Pretoria. 66 pp.

MACLEAN G L (1985). *Roberts' Birds of Southern Africa.* John Voelker Bird Book Fund, Cape Town.

MCDOWELL C (1986). Bid to save *Protea odorata. Veld and Flora* **7**, 98 – 101.

MCDOWELL C (1988). The influence of agriculture on the decline of West Coast Renosterveld, south-western Cape, South Africa. Manuscript submitted to Biological Conservation.

MCKENZIE B M and LONGRIDGE M (eds) (1988). Proceedings of the Locust Symposium held at the Kimberley Museum, Kimberley. *South African Institute of Ecologists Bulletin,* Special Issue.

MOLL E J (ed) (1982). *Proceedings of a Symposium on Coastal Lowlands of the Western Cape, March 19 – 20, 1981.* University of the Western Cape, Bellville.

MOLL E J and BOSSI L (1984a). *Vegetation map of the fynbos biome.* Government Printer, Pretoria.

MOLL E J and BOSSI L (1984b). Assessment of the extent of the natural vegetation of the fynbos biome of South Africa. *South African Journal of Science* **80**, 355 – 358.

MOLL E J, CAMPBELL B M, COWLING R M, BOSSI L, JARMAN M L and BOUCHER C (1984). A description of vegetation categories in and adjacent to the fynbos biome. *South African National Scientific Programmes Report* **83**, CSIR, Pretoria. 29 pp.

MOLL E J and JARMAN M L (1984). Clarification of the term fynbos. *South African Journal of Science* **80**, 351 – 352.

OLIVER E G H, LINDER H P and ROURKE J P (1983). Geographical distribution of present-day Cape taxa and their phytogeographical significance. *Bothalia* **14**, 427 – 440.

ROUX P W and THERON G K (1987). Vegetation change in the karoo biome. In The karoo biome: a preliminary synthesis. Part 2 — vegetation and history. (eds Cowling R M and Roux P W) *South African National Scientific Programmes Report* **142**, CSIR, Pretoria. 50 – 69.

ROUX P W and VORSTER M (1983). Vegetation change in the karoo. *Proceedings of the Grassland Society of Southern Africa* **18**, 25 – 29.

RUTHERFORD M C and WESTFALL R H (1986). Biomes of southern Africa — an objective categorization. *Memoirs of the Botanical Survey of South Africa* **54**, 1 – 98.

SCHEEPERS J C (1983). The present status of vegetation conservation in South Africa. *Bothalia* **14**, 991 – 995.

SMITHERS R H N (1983). *The Mammals of the Southern African Subregion.* University of Pretoria, Pretoria.

TAKHTAJAN A (1969). *Flowering Plants: Origin and Disperal.* Oliver and Boyd, Edinburgh.

TALKENBERG W F M (1982). An Investigation of the Environmental Impact of Surface Diamond Mining Along the Arid West Coast of South Africa. Unpublished MSc thesis, University of Cape Town, Cape Town.

TANSLEY S A (1988a). The status of threatened Proteaceae in the Cape Flora, South Africa, and the implications for their conservation. *Biological Conservation* **43**, 227 – 239.

TANSLEY S A (1988b). An examination of the potential threats affecting rare Cape Proteaceae, South Africa. Manuscript submitted to the *South African Journal of Botany.*

TAYLOR H C (1978). Capensis. In *Biogeography and Ecology of Southern Africa.* (ed Werger M J A) Junk, The Hague. pp 171 – 230.

VAN JAARSVELD E (1987). The succulent riches of South Africa and Namibia. *Aloe* **24**, 45 – 92.

VON BROEMBSEN S L (1979). *Phytophthora cinnamomi* – a threat to the western Cape Flora? *Veld and Flora* **65**, 53 – 55.

VORSTER M and ROUX P W (1983). Veld of the karoo areas. *Proceedings of the Grassland Society of Southern Africa* **18**, 18 – 24.

WEIMARCK H (1941). Phytogeographical groups, centres and intervals within the Cape Flora. *Lunds Universitets Arsskrift.* NF Adv. 2 Bd. 37 Number 5, 1 – 143.

WERGER M J A (1978a). Biogeographical divisions of southern Africa. *Biogeography and Ecology of Southern Africa.* (ed Werger M J A) Junk, The Hague. pp 145 – 170.

WERGER M J A (1978b). The Karoo-Namib Region. *Biogeography and Ecology of Southern Africa.* (ed Werger M J A) Junk, The Hague. pp 231 – 299.

WHITE F (1983). *The Vegetation of Africa.* UNESCO, Paris.

WICHT C L (1945). *Report of the Committee on the Preservation of the Vegetation of the South Western Cape.* Special Publication of the Royal Society of South Africa.

CHAPTER 14

Conservation status of coastal and montane evergreen forest

C J Geldenhuys, D R MacDevette

INTRODUCTION

Indigenous forests occur along the eastern margin of southern Africa, from the Soutpansberg (22° 40' S) and the Tongaland coast (27° S) to the Cape Peninsula (34° S). It is the smallest biome in the area (Rutherford and Westfall 1986), covering less than 3 000 km^2 (Anon 1982; Huntley 1984). The large ratio of forest margin to forest area accentuates the importance of forest margins in forest survival. The forests occur in an area with an annual rainfall (summer, all-year or winter) ranging from 525 to 1 950 mm, and on a wide range of geological formations (Rutherford and Westfall 1986; Geldenhuys 1987).

   Mixed evergreen forest covered the major part of Africa until the eastern escarpment formed during the Cretaceous period. This separated the moist forested coastal belt from the arid interior which has no forest (Deacon et al 1983). Patterns of forest distribution were further complicated by climatic and landscape changes, even during the Holocene (Deacon et al 1983; Scholtz 1986). Acocks (1953) and White (1983) attributed the relic nature of the forests within the grassland and fynbos biomes to the destructive activities of man during the (relatively recent) past 100 to 300 years. However, Feely (1980, 1986) indicated that most of the present southern African grassland existed throughout the Holocene and was not induced by recent forest clearing. Fires associated with hot, desiccating winds have further fragmented the forests (McKenzie 1978; Scheepers 1978; Geldenhuys 1989b). The fragmentation has been aggravated by continuing landuse practices, such as clearing for agriculture, forestry and subsistence utilization, and veld burning practices for grazing and improved water runoff in catchments (Phillips 1963; Feely 1980, 1986; Cooper 1985).

The forest flora has strong affinities to the tropical forest flora. Their biogeographical relationships are revealed in the chorological subdivision of the forests (White 1983). The Indian Ocean Regional Mosaic (Moll and White 1978) along the Natal and eastern Cape coast includes Undifferentiated lowland forest, Sand forest, Dune forest, Swamp forest, and Fringing forest. The Afromontane Region (White 1978) includes the forests of the Transvaal Escarpment, the Natal Montane and Mist-belt, the eastern Cape Amatole and the southern Cape. Transitional forests between the two regions occur in the drier lowlands and river valleys such as Kaffrarian Subtropical Transitional Thicket in the eastern Cape (Cowling 1984; Everard 1987). Mountain chains and escarpments, river valleys, adjacent vegetation regions, and stands of alien trees provide corridors for plant species migration (Edwards 1967; Moll and White 1978; Geldenhuys et al 1986; Everard 1987). However, in several areas the climate in valleys between adjacent forest complexes, fires, mountain peaks and ridges, and landuse practices are barriers to the migration of forest biota.

The type and quality of management and possible impacts on the vegetation is determined largely by ownership (Cooper 1985; McKenzie 1988). The largest proportion of forests, and the largest forests, in South Africa are owned by public authorities. A relatively small portion (18,7% in Transvaal, 28,2% in Natal, and 23,3% in the southern Cape) are privately owned (Cooper 1985; Geldenhuys 1989a).

The purpose of this paper is to review information relating to the importance of the forests for protecting biodiversity of plants and animals, the patterns of biotic diversity in the indigenous evergreen forests, the impacts of landuse practices on these patterns, and the measures taken or required for the conservation of the forest biota.

PATTERNS OF BIOTIC DIVERSITY

Very few studies have analysed the species richness of indigenous forests in South Africa. Most of the sources list the number of species per regional forest complex, forest or unit area. Extensive use has been made in this paper of unpublished data to supplement the published data.

Plants and animals in forest complexes

Species richness of plants, mammals and birds remains relatively constant along the tropical-temperate gradient of southern Africa except for the south-western extreme (Table 14.1). However, the composition of the biota changes along the gradient owing to a turnover of species.

The animal data (Table 14.1) are less comprehensive and were compiled from distribution maps and descriptions of individual species (Smithers 1983; Maclean 1985; Sinclair 1985). The forest mammals represent 14,1% (40) and the forest birds 14,2% (106) of the total terrestrial component of southern Africa, excluding those of the shores and large bodies of water. However, Rautenbach (1978) listed 76 mammals associated with forests (26,6% of the total) from updated but unpublished distribution maps. He noted that the Forest Zone component was distinct from the others; that it contained 16,4% endemic species; that the majority of species occurred over several biotic zones; and that faunistically it was most closely related to the Southern Savanna Woodland subzone and the other zones through which it is scattered. A large proportion of the birds are also not confined to the forest habitat and both mammals and birds have wide distribution ranges.

Passmore and Carruthers (1979) recorded 12 frogs and Patterson (1987) recorded a few reptiles which are specifically associated with the forest habitat. Several insect genera are confined to the forest biome. Many are host-specific to certain plant species or genera (Geertsema 1964; Scholtz and Holm 1985; Swain and Prinsloo 1986). The lists of rare and endangered vertebrate species in South Africa indicate that 11 out of 93 reptiles and amphibians, 13 out of 102 birds, and 15 out of 92 mammals are characteristic of evergreen forests (Branch 1988; Brooke 1984; Smithers 1986).

TABLE 14.1   Numbers of plant, mammal and bird species in forest complexes of southern Africa

(a)	Number of plant species		
Forest complex	Woody	Herbaceous	Total
North-eastern Transvaal escarpment	154	170	324
Eastern Transvaal escarpment	193	182	375
Natal north coast	323	167	490
Umtamvuna river gorge	334	172	506
Transkei mountains	172	83	255
East London coast	156	85	241
Eastern Cape Amatole mountains	184	205	389
Southern Cape: Mossel Bay to Humansdorp	179	286	465
Southern Cape: Swellendam area	75	67	142
Cape Peninsula	47	43	90
Total	649	636	1 285

(b)	Number of animal species	
	Mammals	Birds
Transvaal	25	82
Natal mountains and midlands	24	71
Natal coast and lowlands	34	99
Eastern Cape coast and mountains	33	82
Southern Cape	29	56
Western Cape	15	31
Total	40	106

Very few studies record species richness of birds and mammals per unit area. Seydack (1984), using autotriggering cameras, recorded 13 mammal species in six 1-km^2 census blocks over a six-month period in Goudveld forest, southern Cape. Some of these animals are generally considered as non-forest mammals. Cody (1983) recorded a decline in bird species richness from 43 in Alexandria forest on the eastern Cape coast to 15 in the forest patches of the Cape Peninsula. Koen (1987) recorded 35 bird species on three study sites of about 6 ha each in the Knysna Forest: 28 for dry forest, 31 for moist forest, and 29 for wet forest. Moll (1974) attributed the richness — 54 bird species — of the Dwesa Forest in Transkei, the richest habitat in his study area, to the large variety of trees, shrubs and climbers. Frost (in Koen 1987) recorded 28 species for the coastal forest of Natal.

Comprehensive check-lists of plants are available for forests of the Cape Peninsula, southern Cape, eastern Cape mountains, Zululand coastal dunes, and eastern and north-eastern Transvaal escarpment (Taylor 1955; Van der Schijff and Schoonraad 1971; Venter 1972; Campbell and Moll 1977; McKenzie et al 1977; McKenzie 1978; Moll 1978, 1980a,b; Scheepers 1978; Weisser and Drews 1980; Nicholson 1982; Abbott 1985; Philipson 1987; Geldenhuys 1989c). Data are less comprehensive for the eastern Cape dunes (it excludes the Alexandria forest), the Transkei inland and coastal forests, and the Natal midlands (Von Breitenbach and Von Breitenbach 1983; Burns 1986; Cawe 1986; Johnson and Cawe 1987; Lubke and Strong 1988; Geldenhuys unpublished data).

A total of 649 woody and 636 herbaceous plants were included in the check-lists for forests (Table 14.1). This ratio of herbaceous plants in the total forest flora is much higher than the herbaceous ratio for individual forests. This suggests that either there is a higher turnover in herbaceous plants or that the herbaceous component is generally undersampled.

The forests have a relatively high species richness of 0,514 species km^{-2}. They cover only 0,08% of the land area and contain 5,35% (1 285) of the species. Gibbs Russell (1987) indicated a ratio of 0,0081 plant species km^{-2} overall for southern Africa with over 24 000 taxa. Only fynbos exceeds the forest value with 1,36 species km^{-2} (7 316 species). The third richest biome is grassland with 0,25 species km^{-2} (3 788 species).

Collecting intensity and size of the forest complexes varied considerably and makes direct comparisons unreliable. The Cape Peninsula, Swellendam and Umtamvuna forests total less than 1 200 ha each, whereas the southern Cape forests cover 60 500 ha. All other southern African forests are of intermediate size. Nevertheless, species richness remains relatively constant along the tropical-subtropical gradient except for the south-western extreme. Forests of the Natal north coast and the southern Natal/Pondoland sandstone complex (Umtamvuna) have the highest number of woody species, and the Swellendam and Cape Peninsula forests have the lowest number. The other forests all have similar numbers of woody species. The Umtamvuna and southern Cape forests have more species than might be suggested by the general southward attenuation of tree and shrub species, and the Umtamvuna forests have a large proportion of endemic plants (Van Wyk 1981). In the southern Cape the mountain and coastal forests are continuous in places (Geldenhuys 1989a) and contain elements of both the coastal and the montane zones of further north (Geldenhuys 1989c).

Quantitative studies of the tree and shrub layers of a number of Natal forests (Moll 1978, 1980a,b; D R MacDevette and co-workers unpublished data) indicate that the Zululand dune forests at Sodwana (115 species) are richer in species than Hawaan (63 species), Hlogweni (60), Krantzkloof (86) or Karkloof (74).

The forest flora is further characterized by the wide distribution of most taxa, as is shown by distribution maps for trees and shrubs (Von Breitenbach 1986) and for ferns (Schelpe and Anthony 1986). Furthermore, the southward attenuation of species (from a tropical-subtropical origin) is much larger than the northward attenuation from a temperate origin (Phillips 1931; Scheepers 1978; McKenzie 1978; Tinley 1985; Cawe 1986; MacDevette 1987; Geldenhuys 1989c). In the southern Cape forest complex 162 species drop out from Humansdorp to Mossel Bay, but only 23 species from Mossel Bay to Humansdorp (Geldenhuys 1989c). There are several centres where disjunct species and species with distribution limits are concentrated (Geldenhuys 1986, 1989c). Similarly, Tinley (1985) recorded a progressive reduction in the number of tree and shrub species southwards along the eastern continental margin. He lists several terminal areas with 131 species dropping out from Tongaland to Mtunzini in northern Natal, and 109 species dropping out along the eastern Cape coast. The centres where endemic species, disjunct species and species with distribution limits concentrate provide useful insights into the biogeography and ecology of the species and of the forests (Phillips 1927; Geldenhuys 1981, 1986; Cowling 1983; Tinley 1985; A E Van Wyk pers comm), and require special conservation management.

Alpha diversity

Alpha diversity refers to the number of species within a homogeneous community (Cowling et al this volume). Published quantitative studies of forests are based on plots of different sizes (10 x 10 m to 30 x 30 m), and different plant growth-form components of the forests (from trees ≥100 mm diameter at breast height (DBH) through all woody plants to all species). This complicates the determination of general patterns of species richness.

Studies of species-area curves in southern African forests to determine optimal plot size, or the standardized alpha diversity per 0,1 ha plot, were done opportunistically only, during studies on other vegetation types (Bond 1983; Cowling 1984; Cowling et al this volume). Van Daalen et al (1986) indicated that on the basis of species-area curves, the minimal plot area and the number of species per unit area increased from dune scrub, through bush clumps and forest margin, to forest in a dune-succession gradient at Mtunzini, Natal. In the first full-scale, but as yet incomplete, attempt MacDevette et al (1989a, b) determined species richness of higher plants in 0,1 ha plots for Natal forest (Tables 14.2 and 14.3) using the Whittaker method (Schmida 1984). The Natal forests represent a gradient from cool montane *Podocarpus* forests through mist-belt mixed *Podocarpus* forests to the hot and humid lowland coastal and dune forests.

TABLE 14.2   Comparison of species richness per 0,1 ha plot in forests and other vegetation types

Vegetation	Mean	Range
Australian tropical rain forest [1]	140	
Zimbali dune forest, Natal [2]	97	93 – 100
Zululand coastal forest [2]	87	79 – 98
Renosterveld, Cape [3]	83	57 – 103
Natal mist-belt *Podocarpus* forest [2]	79	62 – 91
North Carolina forests on fertile soil [4]	75	
Cape fynbos [3,5,6]	65	31 – 126
Subtropical transitional thicket [3]	61	37 – 98
Hawaan dune forest [2]	61	57 – 64
Natal montane *Podocarpus* forest [2]	59	46 – 68
SE Cape Afromontane forest [3]	53	
S Cape platform forest [5]	52	
Australian temperate forests and woodlands [7]	48	29 – 105
Australian temperate forests [8]	34	32 – 36

(1) Webb et al 1967; (2) MacDevette et al 1989b; (3) Cowling 1984; (4) Peet and Christenson 1980; (5) Bond 1983 (6) Kruger 1979; (7) Whittaker 1977; (8) Rice and Westoby 1983

TABLE 14.3   Species richness for different growth forms on 0,1 ha plots in major forest types of Natal (MacDevette et al 1989b)

| Growth form | Forest type | | |
	Montane	Mist-belt mixed	Coastal lowlands and dunes
Trees and shrubs	21 – 29	28 – 42	33 – 61
Lianes	4 – 6	7 – 11	10 – 14
Vines	5 – 7	7 – 9	9 – 19
Forbs	7 – 11	8 – 14	0 – 11
Epiphytes	2 – 3	2 – 6	0 – 1
Ferns	4 – 14	4 – 10	1 – 6

Global comparisons of species richness at a 0,1 ha scale can be made from the limited data available (Table 14.2). It is probable that the small sample size for some of the Natal forest types may account for their relatively high mean species richness. However, it is clear that the Natal forests have some of the highest values for mean species richness in South Africa. The completion of the work in the Natal forests (MacDevette et al 1989b) will provide a better perspective on those forests. Although beta diversity (species turnover) was not measured, we presume that for the forests it will be low, at least when compared with fynbos.

Comparison of species richness on 0,1 ha, 0,04 ha and 0,01 ha plots for southern Africa shows that montane forests in general have fewer species than lowland, coastal and dune forests (Tables 14.2, 14.3, 14.4 and 14.5). Furthermore, drier and/or warmer forests are richer than wetter and/or cooler forests. Of the coastal forests, the Natal coastal and dune forests (particularly at Sodwana) are appreciably richer than the southern Cape forests.

TABLE 14.4   Species richness of woody plants per 0,04 ha plot (mean and range) for forests in southern Africa

Forest	Plants ≥10 mm DBH	Plants ≥100 mm DBH
S Cape[1] (montane)	8,3 (5–14)	
(foothill)	16,1 (8–25)	
(platform)	20,5 (8–29)	
(river scarp)	26,2 (20–37)	
(river terrace)	22,3 (19–27)	
(coast scarp)	20,0 (17–23)	
(dune)	21,3 (13–30)	
Diepwalle[2]		9,2 (5–15)
Amatole[2]		10,9 (7–15)
Fort Grey[1]		12,5 (11–14)
Transkei [3] (disturbed moist)	12,7 (7–26)	
(undisturbed moist)	13,1 (7–20)	
(disturbed dry)	16,2 (8–28)	
(undisturbed dry)	18,5 (8–30)	
Sodwana dune forest[4]	38,9 (23–64)	
Mlalazi dune forest[4]	23,0 (14–30)	
Yengele dune forest[4]	29,7 (19–40)	
Mtentweni coastal forest[4]	29,5 (16–39)	
Dukuduku coastal forest[4]	29,0 (24–39)	
Ngome mist-belt forest[4]	24,2 (17–23)	
Wonderwoud (escarpment)[5]		7,6 (4–13)
Hanglip forest (climax)[6]		8,2 (2–12)
(regrowth)[6]		9,0 (3–17)

(1) C J Geldenhuys unpublished data; (2) J C van Daalen unpublished data; (3) Cawe 1986; (4) MacDevette et al 1989a; (5) Geldenhuys and Pieterse 1989; (6) Geldenhuys and Murray 1989.

TABLE 14.5   Species richness for all species per 0,01 ha plot for forests in southern Africa

Western Cape	
Peninsula[1] (foothill, valley)	18,3 (11–25)
Peninsula[2] (montane)	16,1 (12–23)
(foothill, valley)	19,1 (11–32)
Hangklip[3] (montane)	14,0 (6–20)
South-western Cape[4]	
Wet montane (deep soils)	20,7 (14–28)
(shallow soils)	14,3 (8–22)
Moist-dry foothill	23,5 (11–32)
Dry foothill	20,2 (10–27)
Boulder zones	14,1 (9–20)
W Cape area	13,4 (8–20)
Swellendam area	19,5 (9–30)
Jonkersberg (Mossel Bay)	24,5 (18–32)
South-eastern Cape[5]	
Elandsberg (montane, moist)	21,2 (19–24)
(lowland, drier)	32,8 (30–36)
Eastern Cape	
Thicket[6] (mesic kaffrarian)	49,1 (35–59)
(xeric kaffrarian)	37,3 (26–54)
(mesic succulent)	37,5 (28–55)
(xeric succulent)	28,2 (19–43)
East London[7] (riverine forest)	(20–32)
(riverine thicket)	(18–44)
(coastal forest)	(10–39)
(coastal scrub)	(13–28)
Natal[8]	
Ballito dune	(51–58)
Zululand coast	(53–73)
Hawaan coast	(31–38)
Midlands	(44–63)
Drakensberg	(35–47)

(1) McKenzie et al 1977; (2) Campbell and Moll 1977; (3) Boucher 1978 (10 x 20 m plots); (4) McKenzie 1978; (5) Cowling 1984; (6) Everard 1987; (7) Lubke and Strong 1988; (8) MacDevette et al 1989b

Lubke and Strong (1988), showed that in East London coastal vegetation, gradients of increasing species richness were associated with an increase in moisture and a decrease in salinity and wind velocity. Everard (1987) showed an increased species richness with increased rainfall in eastern Cape subtropical transitional thicket. The data of McKenzie (1978) show a sharp decline in species richness from Mossel Bay to the Cape Peninsula although all the sites lie at approximately 34° S. The decline probably relates to two major factors: less favourable climatic conditions due to an increasingly large proportion of winter rain (colder and drier) to the west; and increasingly small and fragmented forests to the west and therefore a sharp drop-out of species owing to island effects.

Patterns of species richness of different growth forms are related to climatic, edaphic and disturbance gradients (Tables 14.3 and 14.6). In the southern Cape, rainfall decreases and temperature and soil fertility increase from the mountains to the dunes. The mountain and foothill forests are frequently exposed to fire disturbance, whereas platform and scarp forests are relatively undisturbed (Geldenhuys 1989b). River terrace forests are sometimes disturbed by floods. Coastal scarp forests are exposed to cold salt-spray bearing winds, and dune forests are all relatively young. Highest richness of all species was associated with intermediate rainfall and temperature, low fire disturbance, and high habitat diversity (platform and river scarp forests). The montane forests consisted of single- or few- species stands of pioneer species such as *Virgilia divaricata* and/or *Cunonia capensis* (with *Ocotea bullata*) in the early stages of regrowth, which are enriched as stand development progresses. The richness of herbaceous species was greater in the moister areas, and the richness of woody species was greater in the drier areas. Total species richness for woody plants followed the same general pattern irrespective of the minimum size of the cohort included in the analysis. In both the Natal and southern Cape forests fern species richness decreased, and richness of vines, graminoids, geophytes and forbs increased, from the mountain to the coastal forests; epiphyte richness, however, was highest in the mist-belt mixed, platform and river scarp forests.

TABLE 14.6   The contribution to total richness in 0,04 ha plots of different-sized cohorts of woody species and different growth forms of herbaceous species in different landscapes of the southern Cape (Geldenhuys unpublished data)

				Landscape			
	Mountain	Foothill	Platform	River		Coast scarp	Dune
				Scarp	Terrace		
Growth form							
All woody	10,4	19,7	26,4	32,9	31,3	27,7	27,7
Woody >1 cm DBH	8,3	16,1	20,5	26,2	22,3	20,0	21,3
Woody >5 cm DBH	6,9	12,7	15,4	17,9	17,5	15,0	14,3
Woody >10 cm DBH	5,0	10,4	11,5	12,4	11,5	9,1	7,3
All herbaceous	9,9	12,1	16,5	17,3	15,5	14,3	12,3
Vines	0,5	1,7	2,3	3,5	3,3	3,6	3,3
Epiphytes	1,0	1,9	3,9	2,8	0,8	0,7	0,0
Ferns	4,9	3,0	3,4	2,6	2,0	2,0	1,0
Graminoids	1,9	2,8	3,3	3,1	4,0	2,6	2,7
Geophytes	0,6	1,2	1,3	2,2	2,0	2,6	2,7
Forbs	1,0	1,5	2,5	3,1	3,5	2,9	2,7

Disturbance regimes and diversity

Tree deaths, windfalls and lightning strikes regularly cause small gaps of between 0,005 and 0,1 ha in size (Geldenhuys and Maliepaard 1983; C J Geldenhuys unpublished data). Elephant *Loxodonta africana* preferentially destroyed particular tree species in the Knysna forests and caused many tree-fall gaps (Von Gadow 1973). Lightning- and man-induced fires, flooding and landslides on steep slopes during heavy rainstorms (Bosch and Hewlett 1980; Anon 1985; Pammenter et al 1985; C J Geldenhuys unpublished data) occasionally expose larger areas in the forests.

The general pattern suggests that undisturbed forest is somewhat richer than disturbed forest, and that mature forest is richer than regrowth or seral forest. However, this pattern does not always hold, and precisely what is meant by undisturbed forest is not always clear from the cited literature.

Few data are available on changes in species richness and species turnover following disturbance of the forest canopy. Pammenter et al (1985) compared a burnt coastal dune forest (8,4 ha) near Mtunzini, Natal, with adjacent unburnt forest, three years after the fire. The burnt area consisted of an almost monospecific stand of *Trema orientalis* (relative density of 73,5%). The next most important species, *Mimusops caffra* and *Euclea natalensis*, recovered vegetatively from trees present before the fire. Diversity was much higher in the unburnt community, but the number of species was not markedly higher (44 versus 37). In the Amatole forests, a monospecific closed-canopy mature stand of *Cunonia capensis* of 0,4 ha established in a fire gap of unknown history (C J Geldenhuys unpublished data). Some canopy trees of the surrounding forest are becoming established in the understorey, whereas the present ground flora is totally different from that of the mature forest.

No data are available on the effects of the utilization of forest products on the composition and diversity of the forests, but several papers indicate the selective nature of the exploitation (King 1938; Taylor 1961, 1962; Phillips 1963; Wells 1973; Scheepers 1978; Geldenhuys 1980; Cunningham 1985; McKenzie 1988).

Human population growth in tribal areas close to forests in Transkei, Ciskei and KwaZulu increases the need for forest products such as firewood, building material, food and traditional medicines. This has resulted in a decrease of species richness and even the disappearance of forests (Cunningham 1985; D R MacDevette pers obs).

A few studies have indicated the effect of disturbance of the forest margin on patterns of species richness across the forest boundary. In the fynbos islands in the Knysna Forest the boundary between fynbos and forest is very sharp with very few forest species regenerating in the relatively old fynbos (Van Daalen 1981). Unburnt forest margins in the Natal Drakensberg have definite margin-associated species and are richer in species than burnt margins (Everard 1986). Margins exposed to frequent fires lacked the typical margin species, contained more typical forest interior species, and were of intermediate richness between unburnt margin and forest interior communities. The margin species protect the forest interior from the penetration of fire and the drying effects of winds.

Several studies indicate the changes in species richness as well as species turnover during the recovery of forest after the exclusion of fire or other disturbances. Dune development and stabilization along a prograding coastline provide excellent insights into primary succession towards forest when disturbance decreases (Weisser et al 1982; Van Daalen et al 1986). Along the seral stages from dune scrub to mature forest, the species composition continually changes and species richness increases. Dune forest along the Zululand coast regrows rapidly following protection of the dune cordon from fire and grazing (Weisser 1978). This indicates the extent of human control over the species composition in the dune forests from the early Iron Age (approximately AD 300) to the present (Feely 1980; MacDevette and Gwala 1989). Highland sourveld grassland on the Thabamhlope Plateau progressed towards a *Leucosidea sericea* and a *Buddleja salviifolia* scrub forest as a result of the exclusion of fire and grazing for 30 to 40 years (Westfall et al 1983). Scrub communities at Orange Kloof on the Cape Peninsula, unburnt and undisturbed for more than 30 years, were physiognomically very similar to the fire-adapted fynbos but showed greater floristic affinity to the forest communities (McKenzie et al 1977). Several species were more or less exclusive to the scrub communities. Of the forest communities, the regrowth stands had a lower species richness than the mature stands.

Fast-growing alien trees planted in forest gaps and in plantations along forest margins protect forest margins against frequent fires, aid the rehabilitation of the forests and provide for timber, fibre and firewood needs (Geldenhuys et al 1986). They supplement the few true indigenous pioneer forest trees such as *Virgilia* spp (Phillips 1926), *Trema orientalis* (Scheepers et al 1968;

Scheepers 1978), and *Acacia karroo* (Weisser and Marques 1979) in ameliorating disturbed sites. They allow forest succession to proceed in a similar way to that of the indigenous pioneer trees. The initial pure stand of the indigenous or alien pioneer tree is gradually colonized and enriched by the shade-tolerant species of middle to late seral stages (Phillips 1926; Huntley 1965; Seagrief 1965; Scheepers et al 1968; Scheepers 1978; Weisser and Marques 1979; Knight et al 1987; C Jacobs unpublished data).

Small mammal species richness in different habitats in the Natal Drakensberg appeared to be related to succession (Rowe-Rowe and Meester 1982). It increased from the pioneer stage (grassland), reached an asymptote during the intermediate *Buddleja-Leucosidea* scrub stage, and then declined in the climax stage (forest).

CONSERVATION OF BIOTIC DIVERSITY IN RELATION TO LANDUSE PRACTICES

Recognition of the value of forests is a pre-requisite for the future conservation of forests. Forests offer many direct uses and indirect values (McKenzie 1988). The direct uses are furniture timber, building material, fuel wood, traditional medicines, food, material for home crafts, decorative material, hunting, recreation and burial sites. The indirect values include protection of water supply and soils in catchments, and the potential for the development of pharmaceuticals.

Forests in public ownership and many private forests are in an advanced stage of recovery from timber exploitation in the nineteenth and early twentieth centuries. Today timber and other minor but important forest products are utilized conservatively from small ecologically suitable areas of state forest, mainly in the southern Cape (Van Dijk 1987; Milton 1987a, b; Geldenhuys and Van der Merwe 1988).

In rural areas not controlled by government, forests have long been a source of material for subsistence (Le Roux 1981; Cooper 1985; Cunningham 1985; Feely 1986). In recent decades people in tribal areas have gathered food, medicinal plants and wood at increased levels and have caused significant disturbances within the forest (Moll 1972; Cunningham 1985; D R MacDevette pers obs; D Muir pers comm).

In the southern Cape a policy has been adopted to rehabilitate destroyed forest areas, to consolidate existing forest patches by reconverting abandoned plantation areas or to extend forest margins to manageable boundaries, and to eliminate alien vegetation (Geldenhuys et al 1986).

FORESTS IN PROTECTED AREAS

Forest distribution in southern Africa is characterized by several large forest complexes which are separated by zones of no forest or only small, fragmented and isolated forest patches. Unfavourable climatic and site conditions, and landuse practices not reconcilable with forest persistence, pose barriers to the gene flow between the larger, more viable forests. Forest conservation therefore has two major facets: maintenance of the components and critical processes within a forest ecosystem, and maintenance of gene flow among the different forests.

Protected forests in southern Africa range from forests in private and tribal ownership that are in good condition, to forests in conservancies and natural heritage sites, through to forests in nature reserves and wilderness areas proclaimed under the Forest Act. The forest types and forest complexes in southern Africa are generally well represented in protected areas (Table 14.7). However, many of the forest patches and the corridors that exist at present, which may be vital for the continued conservation of the forest species within the conserved areas, are not protected.

SYNTHESIS

Many plant and animal species are specifically adapted to the characteristic ecological processes of the forest environment and of different forest development stages. The diversity patterns discussed are determined by the climatic, edaphic, reproductive (gene flow) and disturbance processes of the forests.

TABLE 14.7   Forests under the control of public authorities

Forest type	Total area ha	Protected	
		Area ha	Percentage
South-western Cape[1,2]	no data	320	>80
Southern Cape[3]	60 561	44 005	72,7
Eastern Cape and Ciskei			
Montane and inland forests[4,5]	no data	28 665	>90
Coastal, dune and lowland forests[2]	no data	15 609	>80
Subtropical transitional thicket[6]	900 000	82 314	9,1
Mesic succulent thicket		(28 337)	
Xeric succulent thicket		(46 434)	
Mesic kaffrarian thicket		(1 663)	
Xeric kaffrarian thicket		(5 880)	
Transkei[7]	100 000	73 505	73,5
Natal and Kwazulu[8,9]	113 695	66 897	58,8
Montane *Podocarpus* forest	9 273	4 106	35,2
Mist-belt mixed *Podocarpus* forest	30 868	15 055	48,8
Coast scarp forest	21 266	14 882	70,0
Coast lowlands forest	10 839	8 087	74,6
Sand forest	5 986	2 325	38,8
Riverine forest	8 720	8 400	96,3
Swamp forest	4 843	4 106	84,8
Dune forest	18 900	9 936	52,6
Orange Free State[8]	300	0	0
Transvaal[8]	35 385		
Transvaal Drakensberg Escarpment	32 700	25 982	79,5
Soutpansberg	2 435	2 206	90,6
Coast scarp forest (Lebombo)	250	250	100,0

(1) McKenzie et al 1977; (2) Cape Province Department of Nature and Environmental Conservation unpublished data; (3) Geldenhuys 1989a; (4) Eastern Cape Forestry Region and Ciskei Department of Agriculture and Forestry unpublished data; (5) National Parks Board unpublished data; (6) G la Cock pers comm; (7) Von Breitenbach and Von Breitenbach 1983; (8) Cooper 1985; (9) Porter 1983; MacDevette 1987; Gordon and Bartholomew 1988; A Wills and M Ward pers comm

Disturbance is a key factor in the maintenance of viable populations of many of the early and mid-seral forest plant and animal species. Forests have been continuously cleared and exploited for timber, plant food and medicines over at least the last 1 400 years and they still persist in complex mosaics (Feely 1986). At present seral species are well represented in regularly disturbed areas. They could become rare in many areas following protection and the progression of succession. Existing anecdotal information suggests that forests, for example the Hawaan Forest on the Natal north coast, have lost species owing to the policy of protecting the forest area from man-made disturbance. Undisturbed forest is not always richer than disturbed forest: richness depends on the development stage. The very early seral stages may start off with monospecific stands, whereas advanced regrowth phases can have more species than the mature phase. Continual changes occur in the species present in the different development phases. Several species are adapted to develop optimally in particular seral stages and require those seral stages for persistence. It is a challenge to reserve managers to maintain a balance between areas of early, middle and late forest seral stages in order to maintain the existing species richness of the forests.

The natural fragmentation of the forests over geological periods may have had its effect on the loss of species. We may be at the stage at which the forests are in a new steady state or may even gain species. However, the disjunctions between local forest patches, and between forests of different landscape zones in a region, and of regional complexes may prevent or slow the natural gene flow among the forests. This process may also be responsible for the steep decline in species richness of forests in the west. Furthermore, it is probable that very few forests exceed the minimum critical size for the survival of rarer, specialized forest biota. Island biogeographic theory suggests that the size of and distance between the isolated forests will determine the future survival of the more sensitive taxa (MacArthur and Wilson 1967). However, several thicket formations bridge the gap between coastal and montane forests, often along river valleys. Tinley (1985) suggested that the thickets form refugia for forest elements during arid periods and are initial sites from which these elements expand during wet periods. They also provide dispersal links either linearly along rivers, valleys, scarps and coast dunes, or across the stepping stones of bush-clump archipelagos. These considerations require careful study.

Unlike most regions with large areas of tropical and temperate forests, the forests of southern Africa are relatively well protected. In most areas the era of forest destruction and mismanagement has passed. The large and important forests are in public ownership under management by various government authorities. There is general awareness of the value of the forests, and guidelines have been developed for their sustained use (McKenzie 1988). However, several ecologically important forests in private ownership, such as Grootbos in the south-western Cape, Nelson's Kop in the Orange Free State, Blouberg in northern Transvaal and several in Natal, do not have adequate protection and action is needed for their conservation (Cooper 1985). Furthermore, it should be considered that conservation management policies of forests are likely to be greatly influenced by the management of the vegetation adjacent to the forests and of forest margins.

ACKNOWLEDGEMENTS

This work formed part of the conservation forestry research programme of the Department of Environment Affairs, Forestry Branch. We acknowledge the assistance of our colleagues in our own department and in other nature conservation departments in the compilation of the data.

REFERENCES

ABBOTT A (1985). Flora of the Umtamvuna Nature Reserve. Unpublished interim list of plants. Clearwater, Port Edward.
ACOCKS J P H (1953). Veld types of South Africa. *Memoirs of the Botanical Survey of South Africa* **28** 192 pp.
ANONYMOUS (1982). *Forestry guide plan for South Africa.* Department of Environment Affairs, Pretoria. 198 pp.
ANONYMOUS (1985). *Annual report of the Department of Environment Affairs.* 1983-84. Government Printer, Pretoria.
BOND W J (1983). On alpha diversity and the richness of the Cape Flora: a study in southern Cape fynbos. In Mediterranean Type Ecosystems — the role of nutrients. (eds Kruger F J, Mitchell D T and Jarvis J U M) *Ecological Studies* **43**, Springer, Berlin pp 337 – 356.

BOSCH J M and HEWLETT J D (1980). Sediment control in South African forests and mountain catchments. *South African Forestry Journal* **115**, 50 – 55.

BOUCHER C (1978). The Cape Hangklip area. 2. The vegetation. *Bothalia* **12**, 455 – 497.

BRANCH W R (1988). South African Red Data Book : Reptiles and Amphibians. *South African National Scientific Programmes Report* **151**, CSIR, Pretoria. 240 pp.

BROOKE R K (1984). South African Red Data Book — Birds. *South African National Scientific Programmes Report* **97**, CSIR, Pretoria. 213 pp.

BURNS M E R (1986). A synecological study of the East London Coast dune forests. MSc thesis, Rhodes University, Grahamstown. 247 pp.

CAMPBELL B M and MOLL E J (1977). The forest communities of Table Mountain, South Africa. *Vegetatio* **34**, 105 – 115.

CAWE S G (1986). A quantitative and qualitative survey of the inland forests of Transkei. MSc thesis, University of Transkei, Umtata.

CODY M L (1983) Bird diversity and density in South African forests. *Oecologia* **59**, 201 – 215.

COOPER K H (1985). *The conservation status of indigenous forests in Transvaal, Natal and O F S, South Africa.* Wildlife Society of Southern Africa, Durban. 108 pp.

COWLING R M (1983). Phytochorology and vegetation history in the south-eastern Cape, South Africa. *Journal of Biogeography* **10**, 393 – 419.

COWLING R M (1984). A syntaxonomic and synecological study in the Humansdorp region of the Fynbos Biome. *Bothalia* **15**, 175 – 227.

CUNNINGHAM A B (1985). The resource value of indigenous plants to rural people in a low agricultural potential area. PhD thesis, University of Cape Town, Cape Town.

DEACON H J, HENDY Q B and LAMBRECHTS J J N (eds)(1983). Fynbos palaeoecology: a preliminary synthesis. *South African National Scientific Programmes Report* **74**, CSIR, Pretoria. 216 pp.

EDWARDS D (1987). A plant ecological survey of the Tugela River Basin. *Memoirs of the Botanical Survey of South Africa* **36**, 285 pp.

EVERARD D A (1986). The effects of fire on the *Podocarpus latifolius* forests of the Royal Natal National Park. *South African Journal of Botany* **52**, 60 – 66.

EVERARD D A (1987). A classification of the subtropical transitional thicket in the eastern Cape, based on syntaxonomic and structural attributes. *South African Journal of Botany* **53**, 329 – 340.

FEELY J M (1980). Did iron age man have a role in the history of Zululand's wilderness landscapes? *South African Journal of Science* **76**, 150 – 152.

FEELY J M (1986). The distribution of Iron Age farming settlements in Transkei: 470 to 1970. MA thesis, University of Natal, Pietermaritzburg. 224 pp.

GEERTSEMA H (1964). The Keurboom moth *Leto venus* Stoll Order: Lepidoptera (Hepialidae). *Forestry in South Africa* **5**, 55 – 59.

GELDENHUYS C J (1980). The effect of management for timber production on floristics and growing stock in the southern Cape forests. *South African Forestry Journal* **113**, 6 – 15, 25.

GELDENHUYS C J (1981). *Prunus africana* in the Bloukrans river gorge, southern Cape. *South African Forestry Journal* **118**, 61 – 66.

GELDENHUYS C J (1986). Nature's Valley: a refugium for rare southern Cape forest species. *Palaeoecology of Africa* **17**, 173 – 181.

GELDENHUYS C J (1987). Distribution and classification of the indigenous evergreen forest and deciduous woodland in South Africa. In *South African Forestry Handbook.* (eds Von Gadow K, Van der Zel D W, Van Laar A, Schönau A P G, Kassier H W, Warkotsch P W, Vermaas H F, Owen D L and Jordaan J V) Southern African Institute of Forestry, Pretoria. pp 443 – 453.

GELDENHUYS C J (1989a). Distribution, size and ownership of forests in the southern Cape. Paper in preparation, Saasveld Forestry Research Centre, George.

GELDENHUYS C J (1989b). Bergwind fires determine forest patch pattern in the southern Cape landscape, South Africa. Paper in preparation, Saasveld Forestry Research Centre, George.

GELDENHUYS C J (1989c) Composition and phytogeography of the Knysna Forest flora, with annotated checklist. Paper in preparation, Saasveld Forestry Research Centre, George.

GELDENHUYS C J and MALIEPAARD W (1983). The causes and sizes of canopy gaps in the southern Cape forests. *South African Forestry Journal* **124**, 50 – 55.

GELDENHUYS C J and MURRAY B (1989). Analysis of the tree composition and structure of the Hanglip forest, northern Transvaal. Paper in preparation, Saasveld Forestry Research Centre, George.

GELDENHUYS C J and PIETERSE F J (1989). Analysis of the tree composition and structure of Wonderwoud forest, northeastern Transvaal. Paper in preparation, Saasveld Forestry Research Centre, George.

GELDENHUYS C J and VAN DER MERWE C J (1988). Population structure and growth of the fern *Rumohra adiantiformis* in relation to fern harvesting in the southern Cape forests. *South African Journal of Botany* **54**, 351 – 362.

GELDENHUYS C J, LE ROUX P J and COOPER K H (1986). Alien invasions in indigenous evergreen forest. In *The ecology and management of biological invasions in southern Africa.* (eds Macdonald I A W, Kruger F J and Ferrar A A) Oxford University Press, Cape Town. pp 119 – 131.

GIBBS RUSSELL G E (1987). Preliminary floristic analysis of the major biomes in southern Africa. *Bothalia* **17**, 213 – 227.

GORDON I G and BARTHOLOMEW R (1988) A preliminary report on the conservation status of the indigenous forests of Natal. Unpublished memo. Environment Conservation Office, Natal Provincial Administration, Pietermaritzburg.

HUNTLEY B J (1965) A preliminary account of the Ngoye Forest Reserve, Zululand. *Journal of South African Botany* **31**, 177 – 205.

HUNTLEY B J (1984). Characteristics of South African Biomes. In *Ecological effects of fire in South African Ecosystems.* (eds Booysen P de V and Tainton N M) Springer, Berlin. pp 2 – 17.

JOHNSON C T and CAWE S (1987). Analysis of the tree taxa in Transkei. *South African Journal of Botany* **53**, 387 – 394.

KING N L (1938). Historical sketch of the development of forestry in South Africa. *Journal of South African Forestry Association* **1**, 4 – 16.

KNIGHT R S, GELDENHUYS C J, MASSON P H, JARMAN M L and CAMERON M J (1987). The role of aliens in forest edge dynamics — a workshop report. *Ecosystem Programmes Occasional Report* **22**. CSIR, Pretoria. 41 pp.

KOEN J H (1987). Animal-habitat relationships in the Knysna Forest, South Africa: discrimination between forest types of birds and invertebrates. *Oecologia (Berlin)* **72**, 414 – 422.

KRUGER F J (1979). South African heathlands. In *Heathlands and related shrublands of the world*. A. Descriptive studies. (ed Specht R L) Elsevier, Amsterdam. pp 19 – 80.

LE ROUX P J (1981). Supply of fuel-wood for rural populations in South Africa. *South African Forestry Journal* **117**, 22 – 27.

LUBKE R A and STRONG A (1988). The vegetation of the proposed coastal National Botanic Garden, East London. *South African Journal of Botany* **54**, 11 – 20.

MACARTHUR R H and WILSON E O (1967). *The theory of island biogeography*. Princeton University Press, Princeton, NJ.

MACDEVETTE D R (1987).The dune woodlands of Zululand. In *South African Forestry Handbook*. (eds Von Gadow K, Van der Zel D W, Van Laar A, Schönau A P G, Kassier H W, Warkotsch P W, Vermaas H F, Owen D L and Jordaan J V). Southern African Institute of Forestry, Pretoria. pp 465 – 469.

MACDEVETTE D R and GWALA B R (1989). The dune forests of Tongaland. Paper in preparation, Natal Parks Board, Pietermaritzburg.

MACDEVETTE D R, MACDEVETTE D K and GORDON I G (1989a). A gradient analysis of a Zululand dune forest in Sodwana State Forest, South Africa. Paper in preparation, Natal Parks Board, Pietermaritzburg

MACDEVETTE D R, MACDEVETTE D K and GORDON I G (1989b). Diversity relations in the indigenous forests of Natal, South Africa. Paper in preparation, Natal Parks Board, Pietermaritzburg.

MACLEAN G L (1985). *Roberts' Birds of Southern Africa*. John Voelcker Bird Book Fund, Cape Town. 848 pp.

MCKENZIE B (1978). A quantitative and qualitative study of the indigenous forests of the southwestern Cape. MSc thesis, University of Cape Town, Cape Town. 178 pp.

MCKENZIE B (ed)(1988). Guidelines for the sustained use of forests and forest products. *Ecosystem Programmes Occasional Report* **35**. CSIR, Pretoria. 69 pp.

MCKENZIE B, MOLL E J and CAMPBELL B M (1977). A phytosociological study of Orange Kloof, Table Mountain, South Africa. *Vegetatio* **34**, 41 – 53.

MILTON S J (1987a). Effects of harvesting on four species of forest ferns in South Africa. *Biological Conservation* **41**, 133 – 146.

MILTON S J (1987b). Growth of seven-weeks fern (*Rumohra adiantiformis*) in the southern Cape forests: implications for management. *South African Forestry Journal* **143**, 1 – 4.

MOLL E J (1972). The current status of mistbelt mixed *Podocarpus* forest in Natal. *Bothalia* **10**, 595 – 598.

MOLL E J (1974). *A preliminary report on the Dwesa Forest Reserve, Transkei*. Wildlife Society of Southern Africa and University of Cape Town Wildlife Society, Cape Town.

MOLL E J (1978). A quantitative floristic comparison of four Natal forests. *South African Forestry Journal* **104**, 25 – 34.

MOLL E J (1980a). Additional quantitative ecological studies in the Hawaan forest, Natal. *South African Forestry Journal* **113**, 16 – 25.

MOLL E J (1980b). A quantitative ecological study of the Hlogwene forest, Natal. *South African Forestry Journal* **114**, 19 – 24.

MOLL E J and WHITE F (1978). The Indian Ocean Coastal Belt. In *Biogeography and ecology of Southern Africa*. (ed Werger M J A) Junk, The Hague. pp 561 – 598.

NICHOLSON H B (1982). The forests of the Umtamvuna River Reserve. *Trees in South Africa* **34**, 2 – 10.

PAMMENTER N W, BERJAK M and MACDONALD I A W (1985). Regeneration of a Natal coastal dune forest after fire. *South African Journal of Botany* **51**, 453 – 459.

PASSMORE N I and CARRUTHERS V C (1979). *South African frogs*. Witwatersrand University Press, Johannesburg. 270 pp.

PATTERSON R (1987). *Reptiles of southern Africa*. Struik, Cape Town. 128 pp.

PEET R K and CHRISTENSON N L (1980). *Hardwood forest vegetation of the North Carolina Piedmont*. Veroff Geobot Inst ETH, Stiftung Rubel, Zurich 69, 14 – 39.

PHILIPSON P B (1987). A checklist of vascular plants of the Amatole Mountains, eastern Cape Province/Ciskei. *Bothalia* **17**, 237 – 256.

PHILLIPS J (1963). *The forests of George, Knysna and the Zitzikama — a brief history of their management 1778-1939*. Government Printer, Pretoria.

PHILLIPS J F V (1926). *Virgilia capensis* Lam. ('Keurboom'): A contribution to its ecology and sylviculture. *South African Journal of Science* **23**, 435 – 454.

PHILLIPS J F V (1927). *Faurea macnaughtonii* Phill ('Terblanz'): a note on its ecology and distribution. *Transactions of the Royal Society of South Africa* **14**, 317 – 336.

PHILLIPS J F V (1931). Forest succession and ecology in the Knysna Region. *Memoirs of the Botanical Survey of South Africa*. **14**, 327 pp.

PORTER R N (1983). *The woody plant communities of Itala Game Reserve*. Natal Parks Board, Pietermaritzburg.

RAUTENBACH I L (1978). A numerical re-appraisal of the southern African biotic zones. *Bulletin of Carnegie Museum of Natural History* **6**, 175 – 187.

RICE B and WESTOBY M (1983). Plant species richness at tenth-hectare scale in Australian vegetation compared to other continents. *Vegetatio* **52**, 129 – 140.

ROWE-ROWE D T and MEESTER J (1982). Habitat preferences and abundance ratios of small mammals in the Natal Drakensberg. *South African Journal of Zoology* **17**, 202 – 209.

RUTHERFORD M C and WESTFALL R H (1986). Biomes of southern Africa — an objective categorization. *Memoirs of the Botanical Survey of South Africa*. 98 pp.

SCHEEPERS J C (1978). The vegetation of Westfalia Estate on the north-eastern Transvaal escarpment. *Memoirs of the Botanical Survey of South Africa* **42**, 230 pp.

SCHEEPERS J C, VAN DER SCHIJFF H P and KEET J D M (1968). An ecological account of the *Trema* plantations of Westfalia Estate. *Tydskrif vir Natuurwetenskappe* **8**, 105 – 120.

SCHELPE E A C L E and ANTHONY N C (1986). *Flora of southern Africa : Pteridophyta*. Botanical Research Institute, Pretoria. 292 pp.

SCHMIDA A (1984). Whittaker's plant diversity sampling method. *Israel Journal of Botany* **33**, 41 – 46.

SCHOLTZ A (1986). Palynological and palaeobotanical studies in the southern Cape. MA thesis, University of Stellenbosch, Stellenbosch. 280 pp.

SCHOLTZ C H and HOLM E (eds)(1985). *Insects of southern Africa*. Butterworths, Durban. 502 pp.

SEAGRIEF S C (1965) Establishment of *Podocarpus latifolius* in Blackwood plantations at the Hogsback. *South African Journal of Science* **61**, 433 – 437.

SEYDACK A H W (1984). Application of a photorecording device in the census of larger rain-forest mammals. *South African Journal of Wildlife Research* **14**, 10 – 14.

SINCLAIR J C (1985). *Field Guide to the birds of southern Africa*. Struik, Cape Town, 361 pp.

SMITHERS R H N (1983). *The mammals of the southern African subregion*. University of Pretoria, Pretoria. 736 pp.

SMITHERS R H N (1986). The South African Red Data Book: Terrestrial Mammals. *South African National Scientific Programmes Report* **125**. CSIR, Pretoria. 216 pp.

SWAIN V M and PRINSLOO G L (1986). A list of phytophagous insects and mites on forest trees and shrubs in South Africa. *Entomology Memoir* **66**, Department of Agriculture and Water Supply, Pretoria. 91 pp.

TAYLOR H C (1955). Forest types and floral composition of Grootvadersbosch. *Journal of the South African Forestry Association* **26**, 33 – 46.

TAYLOR H C (1961). The Karkloof forest, a plea for it protection. *Forestry in South Africa* **1**, 123 – 134.

TAYLOR H C (1962). A report on the Nxamalala forest. *Forestry in South Africa* **2**, 29 – 51.

TINLEY K L (1985). Coastal dunes of South Africa. *South African National Scientific Programmes Report* **109**, CSIR, Pretoria. 297 pp.

VAN DAALEN J C (1981). The dynamics of the indigenous forest-fynbos ecotone in the southern Cape. *South African Forestry Journal* **119**, 14 – 23.

VAN DAALEN J C, GELDENHUYS C J, FROST P G H and MOLL E J (1986). A rapid survey of forest succession at Mlalazi Nature Reserve. *Ecosystem Programmes Occasional Report* **11**, CSIR, Pretoria. 33 pp.

VAN DER SCHIJFF H P and SCHOONRAAD E (1971). The flora of the Mariepskop Complex. *Bothalia* **10**, 461 – 500.

VAN DIJK D (1987). Management of indigenous evergreen high forests. In *South African Forestry Handbook*. (Eds Von Gadow K, Van der Zel D W, Van Laar A, Schönau A P G, Kassier H W, Warkotsch P W, Vermaas H F, Owen D L and Jordaan J V) Southern African Institute of Forestry, Pretoria. pp 454 – 464.

VAN WYK A E (1981). Vir 'n groter Umtamvuna-natuurreservaat. *Journal of Dendrology* **1**, 106 – 108.

VENTER H J T (1972). Die plantekologie van Richardsbaai, Natal. DSc thesis, University of Pretoria, Pretoria. 274 pp.

VON BREITENBACH F (1986). *National list of indigenous trees*. Dendrological Foundation, Pretoria. 372 pp.

VON BREITENBACH F and VON BREITENBACH J (1983). Notes on the natural forests of Transkei. *Journal of Dendrology* **3**, 17 – 53.

VON GADOW K (1973). Observations on the utilization of indigenous trees by the Knysna elephants. *Forestry in South Africa* **14**, 13 – 17.

WEBB L J, TRACEY J G, WILLIAMS W T and LANCE G N (1967). Studies in the numerical analysis of complex rain-forest communities. 1. A comparison of methods applicable to site/species data. *Journal of Ecology* **55**, 171 – 191.

WEISSER P J (1978). A vegetation study of the Zululand dune areas. Conservation priorities in the dune area between Richards Bay and the Mfolozi River based on a vegetation survey. *Natal Town and Regional Planning Report Volume* **38**, Pietermaritzburg.

WEISSER P J and DREWS B K (1980). List of vascular plants of the forested dunes of Maputaland. In *Studies on the ecology of Maputaland*. (eds Bruton M N and Cooper K H) Rhodes University and Natal Branch of Wildlife Society of southern Africa. pp 91 – 101.

WEISSER P J, GARLAND I and DREWS B K (1982). Dune advancement 1937-1977 at Mlalazi Nature Reserve, Mtunzini, Natal, South Africa and a preliminary vegetation-succession chronology. *Bothalia* **14**, 127 – 130.

WEISSER P J and MARQUES F (1979). Gross vegetation changes in the dune area between Richards Bay and the Mfolozi River, 1937-1974. *Bothalia* **12**, 711 – 721.

WELLS M J (1973). The effect of the wagon building industry on the Amatole forests. *Bothalia* **11**, 153 – 157.

WESTFALL R H, EVERSON C S and EVERSON T M (1983). The vegetation of the protected plots at Thabamhlope Research Station. *South African Journal of Botany* **2**, 15.

WHITE F (1978). The Afromontane Region. In *Biogeography and ecology of southern Africa*. (ed Werger M J A). Junk, The Hague. pp 463 – 513.

WHITE F (1983). The vegetation of Africa : A descriptive memoir to accompany the UNESCO/AETFAT/UNSO vegetation map of Africa. *Natural Resources Research* **20**, UNESCO, Paris. 356 pp.

WHITTAKER R H (1977). Evolution of species diversity in land communities. *Evolutionary Ecology* **10**, 1 – 67.

CHAPTER 15

The conservation status of South Africa's continental and oceanic islands

J Cooper, A Berruti

INTRODUCTION

South African marine islands can be categorized as either continental or oceanic. Continental (offshore) islands occur on the continental shelf of southern Africa, within 11 km of the mainland. Oceanic islands are the sub-Antarctic Prince Edward Islands of the southern Indian Ocean.

Two important components of island biota, especially as regards ecosystem functioning (eg Siegfried 1982), are seabirds and seals, which come ashore to breed, and also in the case of penguins and elephant seals, to moult. Seabirds and seals are marine predators. Therefore the conservation of the ecosystems of the islands is linked with the conservation of their adjoining marine ecosystems.

In this review we attempt to describe the past exploitation and present management of South Africa's continental and oceanic islands, and tabulate their biota above the high-water mark. 'Case histories' are used as examples of likely future changes in biotic diversity at the two types of islands.

DESCRIPTION

Continental islands

There are 27 South African continental islands (>1 ha) extending from Hollamsbird, off the coast of Namibia to Bird, Algoa Bay, eastern Cape (Figure 15.1; Table 15.1). All the islands (and most of the offshore rocks) off the coast of Namibia belong to South Africa. The largest is Robben (507 ha) and the farthest offshore is Hollamsbird (10,7 km). Two islands, Bird at Lambert's Bay, and Marcus, are connected to the mainland by causeways. In addition, there are at least 58 southern African (50 South African) offshore rocks or stacks (<1 ha) on which seals or seabirds are known to breed. The continental islands have a warm temperate climate, but are not well vegetated. They can be divided into three geographical groups based on annual rainfall: islands off Namibia (least rainfall); western Cape; and southern/eastern Cape (most rainfall) (Figure 15.1). Descriptions of the islands are given by Rand (1963a, b) and Brooke and Crowe (1982).

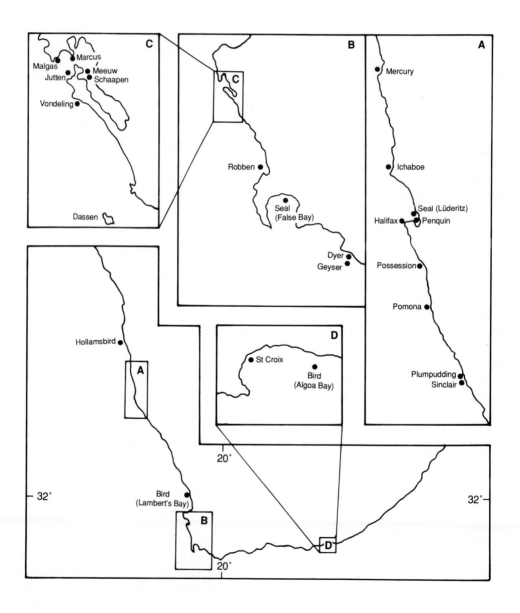

FIGURE 15.1 The South African continental islands

TABLE 15.1   The South African continental islands and their known free-living above-high-water mark taxa[1]

Name	Size (ha)	Distance offshore (km)	Indigenous taxa Plants	Indigenous taxa Animals	Alien taxa Plants	Alien taxa Animals	Total
Hollamsbird	1	10,7	0	3	not studied		
Mercury	3	1,0	0	12	0	4	16
Ichaboe	6,5	1,3		28	0	6	34
Seal, Luderitz	44	1,0	3	14	0	2	19
Penguin	36	0,9	1	12	0	2	15
Halifax	46	0,2	2	10	0	1	13
Possession	90	2,7	2	42	0	1	44
Pomona	3	0,2	0	6	not studied		
Plumpudding	1	0,6	0	7	not studied		
Sinclair	3	0,3		8	not studied		
Bird, Lambert's Bay	3	0,2		22	2	3	27
Malgas	9	0,9	3	30	6	5	44
Marcus	11	1,4	10	51	12	9	82
Schaapen	41	0,5	12	41	5	4	62
Meeuw	7	0,1	19	26	8	1	54
Jutten	46	1,3	7	48	12	9	76
Vondeling	21	1,0	2	44	3	3	52
Dassen	222	9,0	16	90	15	14	135
Robben	507	6,7	101	52	37	16	206
Seal, False Bay	2	5,7		14		3	17
Dyer	20	3,5	6	51	8	10	75
Geyser	3	4,8	0	0	1	9	8
St Croix	12	3,9	4	25	8	6	43
Brenton Rock	1	5,5	0	2	not studied		
Jahleel	2	1,3	2	5	not studied		
Seal, Algoa Bay	4	7,7	2	6	not studied		
Bird, Algoa Bay	19	8,4	6	12	4	6	28
All islands	-		134	208	53	35	430

1   Sources: Randall et al 1981; Cooper and Brooke 1986; Randall and Randall 1986–87 and Cooper et al in press

Oceanic islands

The Prince Edward Islands, comprising Marion (290 km^2), Prince Edward (44 km^2) and a few offshore stacks, are situated 22 km apart in the southern Indian Ocean at 47° S, 38° E, approximately 1 700 km from the African continent. They are mountainous with a peat-bog and mire vegetation at low levels, and bare rock, snow and ice at higher levels. The climate is sub-Antarctic. Detailed descriptions of the two islands may be found in Van Zinderen Bakker et al (1971), Gremmen (1981) and Smith (1987).

HISTORY OF EXPLOITATION

Continental islands

The continental islands and offshore rocks were first exploited in the seventeenth century by Europeans as sources of seal skins and food: seabirds and their eggs were eaten and a few larger islands (Schaapen, Dassen and Robben) were stocked with domestic animals (Frost et al 1976; Cooper and Brooke 1982, 1986; Cooper et al 1985). In the 1840s, the vast accumulations of guano were removed, resulting in much disturbance to seabirds and seals (Craig 1964; Shaughnessy 1984). Unsuccessful attempts at diamond recovery on several islands off Namibia (especially Possession) must have caused severe disturbances at the time. Thereafter, the removal of guano was controlled by concessions, and islands and most offshore rocks off southern Namibia were annexed to the Cape Colony in the 1860s (Frost et al 1976; Shaughnessy 1984). Direct government control was established over nearly all continental islands in the 1890s to protect the guano-producing seabirds and seal colonies (Shaughnessy 1984). As a result, not one island has been permanently lost as a potential breeding site for seals and birds. Only Robben Island supports a major human settlement. The commercial collection of jackass penguin *Spheniscus demersus* eggs continued until 1965 (Shelton et al 1984). By 1987, the removal of guano was restricted to six islands.

Indiscriminate sealing between the early seventeenth and the late nineteenth centuries eradicated about 23 island colonies of Cape or South African fur seals *Arctocephalus p pusillus* (Shaughnessy 1984). Cape fur seal populations recovered from very low levels at the turn of the century to an estimated 1,1 million individuals at 24 colonies in South Africa and Namibia in 1983, despite annual sealing (David 1987a). Fourteen colonies (nine off South Africa and five off Namibia) fall within South African territory, with an estimated 1983 population of 570 000 individuals. Although sealing still occurs, the demand for seal pelts has diminished (David 1987b).

Only one indigenous vertebrate species was lost from the continental islands as a result of exploitation, the dassie *Procavia capensis* at Dassen Island, although many colonies of seabirds were lost from individual islands (Shaughnessy 1984; Cooper et al 1985).

As a result of man's activities, alien plants and animals have become established at most islands (Cooper and Brooke 1982, 1986; Cooper et al 1985; Brooke and Prins 1986).

Oceanic islands

The Prince Edward Islands were heavily sealed (fur seal skins and elephant seal oil) in the nineteenth and early twentieth centuries (Cooper and Avery 1986; Cooper and Condy 1988; Richards in press). Seabirds and their eggs were taken as food by sealers. Sealing activity led to the introduction of aliens, most notably the house mouse *Mus musculus* (Cooper and Brooke 1986; Watkins and Cooper 1986).

Since 1947, when a meteorological base was established on Marion Island, there has been practically no exploitation (Cooper and Condy 1988). Commercial ranching, farming, mining and fishing have never taken place on either island.

PRESENT CONSERVATION STATUS

Continental islands

From the 1860s, the continental islands have been directly controlled by a succession of government organizations (Shaughnessy 1984). Seabird and seal populations were protected by legislation, and access to the colonies was controlled by permit and by the presence of a headman on most islands. Seabird and seal colonies are currently protected by the Sea Birds and Seals Protection Act (No 46 of 1973) and regulations as amended. Control of St Croix, Jahleel and Brenton Rock, Algoa Bay was transferred to the then Cape Department of Nature and Environmental Conservation in 1979. Four islands (Marcus, Malgas, Jutten and Schaapen) in Saldanha Bay were incorporated into the Langebaan National Park in 1986 and are subject to the provisions of the National Parks Act. Meeuw in Langebaan Lagoon is directly controlled by the Department of Defence. Robben has been used for various purposes (De Villiers 1971) and is currently controlled by the Department of Justice. The balance of the continental islands and some of the offshore rocks (see below) were ceded to the Cape Chief Directorate Nature and Environmental Conservation (CCDNEC) in 1987 and were gazetted as provincial nature reserves on 9 March 1988. Management plans are being developed for all islands under the control of the National Parks Board and CCDNEC. All but two continental islands are now administered by conservation organizations, while the CCDNEC is advising the Department of Justice on the management of Robben.

Of the 50 South African offshore rocks and stacks supporting breeding seabirds or seals, 29 (58%) are legally protected (Table 15.2), as are all seal colonies on offshore rocks. The 21 unprotected rocks and stacks, with only two exceptions, support cormorant colonies only (Cooper et al in press).

TABLE 15.2   Legal status of southern African offshore rocks and stacks supporting breeding seabirds and/or seals

Controlling authority	Numbers with breeding		
	Seabirds	Seals	Totals.
Namibia	3	5	8
National Parks Board	5	0	5
Cape Chief Directorate of Nature and Environmental Conservation	19	11	24[1]
Unalienated State Land	21	0	21[2]
South African totals	45	11	50[3]
Southern African totals	48	16	58[1]

1   Six CCDNEC-controlled offshore rocks support both breeding seabirds and seals, hence total does not tally
2   Includes four offshore rocks supporting breeding seabirds which were removed from the Seabirds and Seals Protection Act, No 46 of 1973 by Government Gazette Notice No R585 of 5 April 1974
3   Listed in Cooper et al in press

Oceanic islands

The Prince Edward Islands have not, as yet, been afforded legal nature reserve status (Cooper and Condy 1988). South African sovereignty was declared on 12 January 1948 (Prince Edward Islands Act, No 43 of 1948), and seabirds and seals are protected by the Sea Birds and Seals Protection Act, No 46 of 1973, which, inter alia, restricts landing to permit holders. The islands have a Territorial Sea of 12 nautical miles and an Exclusive Fishing Zone of 200 nautical miles (Heymann et al 1987). The islands fall outside the Antarctic Treaty area but are included within the area of the Convention on the Conservation of Antarctic Marine Living Resources (CCAMLR). According to Monteiro (1987) not all South African environmental legislation is formally applicable. A 'Code of Conduct' for the environmental protection of the Prince Edward Islands came into force in 1988 but has, as yet, no legal status, although its provisions are in the main adhered to (Heymann et al 1987; Cooper and Condy1988). Legal authority and responsibility for management of the Prince Edward Islands lies with the South African Department of Environment Affairs. Provisional management plans for the two islands are at present being drafted (P R Condy pers comm). Cooper and Condy (1988) review, in detail, the status of environmental conservation on the Prince Edward Islands. A recent proposal to construct an emergency landing strip for fixed-wing aircraft has been turned down (Heymann et al 1987).

INDIGENOUS AND ALIEN BIOTA

Continental islands

Incomplete surveys of the free-living above-high-water mark biota of South African continental islands (Brooke and Crowe 1982; Cooper and Brooke 1986; Randall and Randall 1986 – 87; Table 15.1) have identified 430 taxa, of which 20,5% are considered aliens (Table 15.3). Insects and birds form the larger part (42,8% combined) of identified taxa (Table 15.3). The only species known to be endemic to the islands is the Cape gannet *Sula capensis*. Numbers of nonvascular plants and microfaunal taxa are unknown, since practically no surveys have been undertaken. Known nonvascular plants are restricted to algae, and to lichens of which there are at least several taxa. The numbers of taxa present are significantly related to island size (Brooke and Crowe 1982; Cooper and Brooke 1986).

TABLE 15.3   Numbers of known indigenous and alien free-living above-high-water mark taxa on 20 South African continental islands, after Cooper and Brooke (1986)

Taxa	Indigenous	Alien	Total
Vascular plants	134	53	187
Gastropods	0	1	1
Earthworms	1	0	1
Centipedes	5	0	5
Insects	108	16	124
Crustaceans	3	1	4
Arachnids	27	1	28
Macroinvertebrates	144	19	163
Amphibians	1	0	1
Reptiles	8	1	9
Birds	53	7	60
Mammals	2	8	10
Vertebrates	64	16	80
Animals	208	35	243
Total	342	88	430

Oceanic islands

Marion Island supports 332 known free-living above-high-water mark taxa, including nonvascular plants which form 62,7% of the total (Table 15.4).

TABLE 15.4 Numbers[1] of known endemic, indigenous and alien free-living above-high-water mark taxa present on Marion Island

Taxa	Endemic	Indigenous	Alien	Total
Mosses	0	72	0	72
Liverworts	7	36	0	36
Lichens	2	*c* 100	0	*c* 100
Nonvascular plants	9	*c* 208	0	*c* 208
Vascular plants	1	22	10	32
Plants	10	*c* 230	10	*c* 240
Gastropods	0	2	0	2
Earthworms	0	2	1	3
Insects	6	21	10	31
Arachnids	3	13	2	15
Macroinvertebrates	9	38	13	51
Fish	0	0	1	1
Birds	0	25	0	25
Mammals	0	3	2	5
Vertebrates	0	28	3	31
Animals	9	66	16	82
Totals	19	*c* 296	26	*c* 322

1 Sources: Cooper and Brooke 1986; Crafford et al 1986; Gremmen 1981; Smith 1987; Van Zinderen Bakker et al 1971

Endemics, mainly nonvascular plants and macroinvertebrates, form 6,4% of the 296 indigenous taxa, and aliens, mainly vascular plants and macroinvertebrates, form 7,8% of the total identified taxa. There are no endemic birds or seals and only one endemic vascular plant, *Elaphoglossum randii*, at the Prince Edwards Islands. Endemicity in liverworts is high (19,4%; Table 15.4).

POPULATION SIZES AND TRENDS OF SEABIRDS AND SEALS

Continental islands

There are 14 breeding seabird species in southern Africa (Table 15.5). One species, the endemic Damara tern *Sterna balaenarum* breeds only on the mainland. Population estimates (Tables 15.5 and 15.6) show that, for most species, the islands and offshore rocks and stacks support most colonies and the majority of the marine population. Eight seabird taxa endemic to southern Africa (six species and two subspecies) breed on the continental islands. Four species, jackass penguin, great white pelican *Pelecanus onocrotalus,* roseate tern *Sterna dougallii* and Damara tern, are listed in the South African Red Data Book (Brooke 1984). The jackass penguin and Damara tern are also listed in the international Red Data Book (Collar and Stuart 1985).

TABLE 15.5   Numbers of coastal breeding localities of seabirds and seals in southern Africa, adapted from Cooper et al (in press) and references therein, and David (1987a, b)

Species	Southern African status	Number of coastal localities			% on islands
		Islands and rocks	Mainland	Total	
Jackass penguin *Spheniscus demersus*	endemic species	25	4	29	86,2
Great white pelican *Pelecanus onocrotalus*		1	2	3[1]	33,3
Cape gannet *Sula capensis*	endemic species	6	0	6	100,0
Cape cormorant *Phalacrocorax capensis*	endemic species	39	24	63	61,9
Whitebreasted cormorant *P carbo*		24	46	70[2]	34,3
Bank cormorant *P neglectus*	endemic species	44	3	47	93,6
Crowned cormorant *P coronatus*	endemic species	35	8	43	81,4
Kelp gull *Larus dominicanus vetula*	endemic subspecies	30	35	65	46,2
Greyheaded gull *L cirrocephalus*		2	9	11[2,3]	18,2
Hartlaub's gull *L hartlaubii*	endemic species	14	27	41[3]	34,1
Caspian tern *Sterna caspia*		8	12	10[2,3]	40,0
Swift tern *S bergii bergii*	endemic subspecies	15	6	21[3]	71,4
Roseate tern *S dougallii*		2	0	2[3]	100,0
Damara tern *S balaenarum*	endemic species	0	c 30	c 30	0,0
Totals for 14 species of seabirds		245	c 206	c 451	54,3
Cape fur seal *Arctocephalus pusillus pusillus*	endemic subspecies	18	6	24	75,0
Totals for all 15 species		263	c 212	c 475	55,4

1 Extinct colonies not included
2 Inland colonies exist in southern Africa, but exact numbers are unknown
3 Breeding does not occur at all localities every year

Trends in population size are known for some seabird species. Numbers of the great white pelican have decreased at Walvis Bay, but appear stable in the south-western Cape (Crawford et al 1981; J Cooper unpublished data). Populations of jackass penguins, Cape gannets, and Cape cormorants *Phalacrocorax capensis* off southern Namibia have decreased greatly. Off South Africa, jackass penguin and Cape cormorant populations have decreased west of Cape Point, whereas the Cape gannet population is increasing (Cooper et al 1982; Crawford et al 1983; Shelton et al 1984).

The crowned cormorant *P coronatus* population is stable (Crawford et al 1982). A recent decrease in the bank cormorant *P neglectus* breeding population is attributed, at least partially, to the invasion of an important colony by seals (Crawford et al in press; see below). Overall population trends in the populations of the whitebreasted cormorant *P carbo*, swift tern *Sterna bergii* and gulls *Larus* spp are less clear, although the gulls may be increasing in numbers (Brooke et al 1982; Crawford et al 1982; Ryan 1987; J Cooper et al unpublished data). The roseate tern population has decreased in size (Randall and Randall 1980).

TABLE 15.6   Total population estimates for breeding seabirds (excluding chicks) and seals (including pups) in coastal South Africa, including the Walvis Bay enclave, adapted from Cooper et al (in press) and references therein, and David (1987a, b)

Species	Total coastal population estimates			% on islands
	Islands and rocks	Mainland	Total	
Jackass penguin	152 642	100[1]	152 742[2]	99,9
Great white pelican	40	2 662	2 702	1,5
Cape gannet	185 751	0	185 751[2]	100,0
Cape cormorant	290 544	49 407	339 951	85,4
Whitebreasted cormorant	758	8 088	8 846	8,6
Bank cormorant	16 276	1 286	17 562	92,7
Crowned cormorant	4 617	3 427	8 044	57,4
Kelp gull	15 312	28 774	44 086	34,7
Greyheaded gull	20	2 496	2 516	0,8
Hartlaub's gull	4 010	29 616	33 626	11,9
Caspian tern	8	545	553	1,4
Swift tern	3 539	6 373	9 912	35,7
Roseate tern	150	2	152	98,6
Damara tern	0	206	206	0,0
Totals for 14 species of seabirds	673 667	132 982	806 649	83,9
Cape fur seal	253 844	317 696	571 540	55,6
Totals for all 15 species	927 511	450 678	1 378 189	67,3

1 Additional to Cooper et al in press
2 Does not include birds of prebreeding age, which do not usually come ashore

Southern African seal populations were greatly reduced by the beginning of the twentieth century (Shaughnessy 1984). Since then, and particularly since 1940, seal populations have increased greatly to an estimated 1,1 million individuals at 24 colonies despite annual sealing and the recolonization of only two former colonies in the twentieth century (David 1987a, b; Crawford et al in press). The population increase has occurred mainly at four large mainland colonies, which formed after 1940, whereas 13 of 18 island colonies have decreased in size (David 1987b). Only one mainland colony, Kleinzee, the largest, is in South African territory, the rest being in Namibia. Regular pup censuses have been carried out since 1971 and show a mean annual total population increase of about 3,7% (David 1987b). The islands support 75% of all colonies, but only 56% of the estimated South African seal population and 24% of the total southern African population (Tables 15.5 and 15.6). When breeding seabirds and Cape fur seals are treated together, the islands support approximately half the numbers of colonies and half the number of individuals (Tables 15.5 and 15.6)

Oceanic islands

Twenty-nine species of birds breed, or are thought to breed, at the Prince Edward Islands. Breeding has been definitely proven for only 25 species at Marion Island, of which one, the common diving petrel *Pelecanoides urinatrix*, is considered to have recently become extinct (Brooke 1984; Cooper and Condy 1988; J Cooper and C R Brown unpublished data; Table 15.7). Recent (1980s) population estimates vary from as little as 10 to 20 breeding pairs to '100s of 1 000s' (Table 15.7). Estimates are less precise for burrowing petrels and for all birds at Prince Edward Island.

Little is known of the population trends of seabirds at the Prince Edward Islands (Watkins 1987), but most species of burrowing petrels (four listed in the South African Red Data Book; Brooke 1984) have been reduced in numbers owing to the depredation of feral domestic cats *Felis catus* on Marion Island (Brooke 1984; Cooper and Condy 1988).

Of the three seal species, the two fur seals *Arctocephalus tropicalis* and *A gazella* are increasing in number, and the southern elephant seal *Mirounga leonina* is decreasing (Condy 1979; Kerley 1983; Skinner and Van Aarde 1983; Table 15.7). Increases in fur seal numbers (15% a year) are considered to reflect a recovery after the cessation of sealing (Kerley 1983). The decrease in elephant seal numbers (approximately 10% a year) seems to be related to a decrease in bull numbers, for reasons at present unclear, but possibly due to changes in food supply (Skinner and van Aarde 1983).

CONSERVING BIOTIC DIVERSITY

Continental islands

The future of the continental islands as relatively undisturbed breeding sites for seabirds and seals is secure and therefore direct human interference is unlikely to affect biotic diversity, especially if the collection of guano, and its attendant disturbance, is halted on all islands. Active management (eg the eradication of alien biota, especially cats) is likely to maintain or even increase the diversity of the indigenous biota of some islands (eg Dassen, Robben, Schaapen and Bird, Lambert's Bay; Cooper et al 1985; Berruti 1986). Pollution is at present not considered to be important (Cooper et al 1984), although a series of catastrophic oiling incidents could seriously affect jackass penguin populations (Morant et al 1981).

Increases in existing fur seal colonies, and recolonization of former colonies, may cause local decreases in seabird populations as a result of the physical invasion of areas occupied by breeding seabirds (Shaughnessy 1984; Crawford et al 1989; A B Berruti pers obs). An example is the recolonization of Mercury Island by seals between 1978 and 1981. By 1986, there were about 16 000 seals on the island, while numbers of jackass penguins decreased by 65% (17% of the population of the Namibian islands) and bank cormorants by 79% (15% of the world population)(Crawford et al 1989). Cape gannets were not badly affected, but might have been displaced earlier by seals at Hollamsbird Island (Shaughnessy 1984). Thus the gain of one abundant and widespread vertebrate species (Cape fur seal) to the island must be weighed against the likely loss of one or more less abundant seabird species.

The trends in population of jackass penguins, Cape gannets and Cape cormorants are dependent largely on the abundance of their prey, primarily pelagic shoaling fish of commercial importance (eg Crawford and Shelton 1981; Duffy et al 1987a, b). Populations of pelagic shoaling fish off Namibia have decreased greatly (Crawford et al 1987). Off South Africa, fish populations are apparently more stable (Crawford et al 1987) and are associated with a lesser decrease in seabird numbers (Crawford and Shelton 1981). The decrease of the jackass penguin population has caused particular concern (eg Collar and Stuart 1985) and has resulted in calls for a reduction in fishing quotas and the exclusion of fishing within 25 km of breeding colonies (Brooke 1984; Cooper et al 1984). However, the creation of fixed reserves as a measure to maintain seabird populations is questionable because the fish are migratory (Crawford et al 1987).

The 21 legally unprotected offshore rocks and stacks ('unalienated state land'; Table 15.1) should be added to the list of islands controlled by the CCDNEC, and proclaimed provincial nature reserves.

Oceanic islands

Compared with many oceanic islands, the Prince Edward Islands are relatively pristine. Only one species of bird, the common diving petrel, has been lost from Marion Island. Crafford and Scholtz (1987) suggest that introduced house mice might have been responsible for the extinction of a flightless moth *Pringleophaga kerguelensis* on Marion Island. Both species are present on mouse- and cat-free Prince Edward Island, and therefore no indigenous taxa have been lost from the island group.

Feral cats are at present being controlled by shooting at Marion Island and the population has been greatly reduced as a result, a total of 664 individuals having been shot in the first two summer seasons of the programme (Van Rensburg and Bester 1988a; M N Bester pers comm; Cooper and Condy 1988). However, unless the last female cat is eliminated, it seems most likely that the population will increase if control measures are halted. It is therefore essential that a periodic control programme continue indefinitely if eradication is not achieved through the current programme. To this end, additional control measures should be tested, and if found suitable, implemented (Veitch 1985; Cooper and Condy 1988; C R Veitch pers comm). Because burrowing petrels, the main prey of feral cats at Marion Island (Van Rensburg 1985), have delayed maturity and a one-egg clutch, the eradication of cats will not result in rapid increases in their population sizes. However, their breeding success should improve markedly (Van Rensburg and Bester 1988b) and, given enough time, the petrel populations of Marion Island should regain their former status.

It remains to be seen whether the increasing fur seal population will adversely affect bird populations and vegetation as they have done elsewhere (eg Bonner 1985; Crawford et al in press).

Population trends of seabirds, fur seals and elephant seals at the Prince Edward Islands are closely linked to the marine environment. The development of commercial fisheries within the foraging ranges of the breeding birds and seals of Prince Edward Islands could affect their conservation status and, perhaps, in the long run, their biotic diversity. However, no such fisheries are planned for the immediate future and, indeed, they may not be viable in view of the absence of a shelf region surrounding the islands (Cooper and Condy 1988).

Continuation of the present management policies at the Prince Edward Islands (Cooper and Condy 1988), the production and implementation of conservation-management plans, and the proclamation of the two islands as nature reserves, should result in the maintenance of biotic diversity for the forseeable future.

TABLE 15.7   Numbers of seabirds[1] (annual breeding pairs) and seals[2] (total population estimates) at the Prince Edward Islands

Species		Marion Island	Prince Edward Island
King penguin	Aptenodytes patagonicus	215 230	5 000
Gentoo penguin	Pygoscelis papua	888	655
Macaroni penguin	Eudyptes chrysolophus	405 084	17 000
Rockhopper penguin	E chrysocome	157 600	35 000
Wandering albatross	Diomedea exulans	1 533	1 277
Greyheaded albatross	D chrysostoma	4 881	1 500
Yellownosed albatross	D chlororhynchos	0	7 000
Sooty albatross	Phoebetria fusca	2 055	700
Lightmantled sooty albatross	P palpebrata	176	40
Northern giant petrel	Macronectes halli	314	180
Southern giant petrel	M giganteus	2 891	410
Fairy prion	Pachyptila turtur	100s	100s
Salvin's prion	P vittata salvini	100s of 1 000s	10s of 1 000s
Blue petrel	Halobaena caerulea	10s of 1 000s	100s of 1 000s
Greatwinged petrel	Pterodroma macroptera	10s of 1 000s	1 000s
Kerguelen petrel	P brevirostris	10s of 1 000s	br
Softplumaged petrel	P mollis	1 000s	1 000s
Grey petrel	Procellaria cinerea	1 000s	1 000s
Whitechinned petrel	P aequinoctialis	10s of 1 000s	1 000s
Blackbellied stormpetrel	Fregetta tropica	br	1 000s
Greybacked stormpetrel	Garrodia nereis	br	br
South Georgian diving petrel	Pelecanoides georgicus	100s	br
Common diving petrel	P urinatrix	0	br
Imperial cormorant	Phalacrocorax atriceps	589	120
Sub-Antarctic skua	Catharacta antarctica	720	60
Kelp gull	Larus dominicanus	200	30
Antarctic tern	Sterna vittata	<25	<25
Kerguelen tern	S virgata	10	20
Lesser sheathbill	Chionis minor	980	42
Southern elephant seal	Mirounga leonina	4 538[3]	386[4]
Sub-Antarctic fur seal	Arctocephalus tropicalis	19 857	14 761
Antarctic fur seal	A gazella	162	61+
Fur seal hybrids	A tropicalis/gazella	37	?

br – breeding suspected but not proven
1 After J Cooper and C R Brown unpublished data
2 After Condy 1978, 1979 and Kerley 1983
3 Last published estimate, for 1975; the population has subsequently decreased in size (Skinner and Van Aarde 1983) but no more recent estimates of the total population are available
4 Based on an incomplete 1977 census (Condy 1978)

CONCLUSIONS

Maintaining the conservation status and biotic diversity of the above-high-water mark fauna, especially seabirds and seals, of South Africa's continental and oceanic islands is linked to the marine environment. Overexploitation has seriously affected some fish populations in southern African waters, causing reductions in size in some seabird populations. Although the continental and oceanic islands were initially severely affected by sealing and other exploitive activities, they have been able to retain a high degree of their original biotic diversity. This situation appears likely to continue, especially if their conservation status is maintained and improved, management plans are drawn up and implemented, active and proposed alien eradication programmes are successful, and monitoring of seabird and seal populations continues. The Prince Edward Islands, and the continental offshore rocks and stacks which support breeding colonies of seabirds or seals, should be proclaimed nature reserves.

ACKNOWLEDGEMENTS

We thank the Sea Fisheries Research Institute, the South African Department of Environment Affairs, the South African National Committee for Oceanographic Research, and the South African Scientific Committee for Antarctic Research for financial and logistic support. Valuable comments were received from N J Adams, R K Brooke, P R Condy, J H M David, R M Randall, G A Robinson, P G Ryan, P A Shelton, W R Siegfried and B P Watkins.

REFERENCES

BERRUTI A (1986). The predatory impact of feral cats *Felis catus* and their control on Dassen Island. *South African Journal of Antarctic Research* **16**, 123 – 127.

BONNER W N (1985). Impact of fur seals on the terrestrial environment at South Georgia. In *Antarctic nutrient cycles and food webs.* (eds Siegfried W R, Condy P R and Laws R M) Springer, Berlin. pp 641 – 646.

BROOKE R K (1984). South African Red Data Book — Birds. *South African National Scientific Programmes Report* **97**, CSIR, Pretoria 213 pp.

BROOKE R K, COOPER J, SHELTON P A and CRAWFORD R J M (1982). Taxonomy, distribution, population size, breeding and conservation of the whitebreasted cormorant, *Phalacrocorax carbo*, on the southern African coast. *Gerfaut* **72**, 188 – 220.

BROOKE R K and CROWE T M (1982). Variation in species richness among offshore islands of the southwestern Cape. *South African Journal of Zoology* **17**, 49 – 58.

BROOKE R K and PRINS A J (1986). Review of alien species on South African offshore islands. *South African Journal of Antarctic Research* **16**, 102 – 109.

COLLAR N J and STUART N (1985). *Threatened birds of Africa and related islands.* The ICBP/IUCN Red Data Book, Part 1. (Third ed). International Council for Bird Preservation and International Union for Conservation of Nature and Natural Resources, Cambridge.

CONDY PR (1978). The distribution and abundance of southern elephant seals *Mirounga leonina* (Linn.) on the Prince Edward islands. *South African Journal of Antarctic Research* **8**, 42 – 48.

CONDY P R (1979). Annual cycle of the southern elephant seal *Mirounga leonina* (Linn.) at Marion Island. *South African Journal of Zoology* **14**, 95 – 102.

COOPER J and AVERY G (1986). Historical sites at the Prince Edward Islands. *South African National Scientific Programmes Report* **128**, CSIR, Pretoria, 82 pp.

COOPER J and BROOKE R K (1982). Past and present distribution of the feral European rabbit *Oryctolagus cuniculus* on southern African offshore islands. *South African Journal of Wildlife Research* **12**, 71 – 75.

COOPER J and BROOKE R K (1986). Alien plants and animals on South African continental and oceanic islands: species richness, ecological impacts and management. In *The ecology and management of biological invasions in southern Africa.* (eds Macdonald I A W, Kruger F J and Ferrar A A) Oxford University Press, Cape Town. pp 133 – 142.

COOPER J, BROOKE R K, SHELTON P A and CRAWFORD R J M (1982). Distribution, population size and conservation of the Cape cormorant *Phalacrocorax capensis*. *Fisheries Bulletin of South Africa* **16**, 121 – 143.

COOPER J and CONDY P R (1988). Environmental conservation at the sub-Antarctic Prince Edward Islands: a review and recommendations. *Environmental Conservation* **15**, 317 – 326.

COOPER J, HOCKEY P A R and BROOKE R K (1985). Introduced mammals on South and South West African islands: history effects on birds and control. In *Proceedings of the Symposium on Birds and Man, Johannesburg 1983.* (ed Bunning L J) Witwatersrand Bird Club, Johannesburg. pp 179 – 203.

COOPER J, HOCKEY P A R and RYAN P G (in press). Coastal birds of South Africa. An atlas and breeding site register, 1975 – 1984. *South African National Scientific Programmes Report.*

COOPER J, WILLIAMS A J and BRITTON P L (1984). Distribution, population sizes and conservation of breeding seabirds in the Afrotropical region. In Status and Conservation of the world's seabirds. (eds Croxall J P, Evans P G H and Schreiber R W) *International Council for Bird Preservation Technical Publication* **2**, 403 – 419.

COOPER J, WILLIAMS A J, CRAWFORD R J M and SUTER W (unpublished data). Distribution, population size and conservation of the swift tern in southern Africa.

CRAFFORD J E and SCHOLTZ C H (1987). Quantitative differences between the insect faunas of sub-Antarctic Marion and Prince Edward Islands: a result of human intervention? *Biological Conservation* **40**, 255 – 262.

CRAFFORD J E, SCHOLTZ C H and CHOWN S L (1986). The insects of sub-Antarctic Marion and Prince Edward Islands; with a bibliography of entomology of the Kerguelen Biogeographical Province. *South African Journal of Antarctic Research* **16**, 42 – 84.

CRAIG R (1964). The African guano trade. *The Mariner's Mirror* **50**, 25 – 55.

CRAWFORD R J M, COOPER J and SHELTON P A (1981). The breeding population of white pelicans *Pelecanus onocrotalus* at Bird Rock Platform in Walvis Bay, 1949-1978. *Fisheries Bulletin of South African* **15**, 67 – 70.

CRAWFORD R J M, COOPER J and SHELTON P A (1982). Distribution, population size, breeding and conservation of the kelp gull in southern Africa. *Ostrich* **53**, 164 – 177.

CRAWFORD R J M, DAVID J H M, WILLIAMS A J and DYER B M (1989). Competition for space: recolonising seals displace endangered, endemic seabirds off Namibia. *Biological Conservation.* **48**, 59 – 72.

CRAWFORD R J M, SHANNON L V and POLLOCK D E (1987). The Benguela ecosystem. 4. The major fish and invertebrate resources. In Oceanography and marine biology. (ed Barnes M) *An annual review* **25**, 353 – 505.

CRAWFORD R J M and SHELTON P A (1981). Population trends for some southern African seabirds related to fish availability. In *Proceedings of the Symposium on Birds of the Sea and Shore, 1979.* (ed Cooper J) African Seabird Group, Cape Town. pp 15 – 41.

CRAWFORD R J M, SHELTON P A, BROOKE R K and COOPER J (1982). Taxonomy, distribution, population size and conservation of the crowned cormorant, *Phalacrocorax coronatus*. *Gerfaut* **72**, 3 – 30.

CRAWFORD R J M, SHELTON P A, COOPER J and BROOKE R K (1983).Distribution, population size and conservation of the cape gannet *Morus capensis*. *South African Journal of Marine Science* **1**, 153 – 174.

DAVID J H M (1987a). Diet of the South African fur seal (1974-1985) and an assessment of competition with fisheries in southern Africa. *South African Journal of Marine Science* **5**, 693 – 713.

DAVID J H M (1987b). South African fur seal, *Arctocephalus pusillus pusillus*. In Status, biology, and ecology of fur seals. (eds Croxall J P and Gentry R L) *Proceedings of a international symposium and workshop, Cambridge, England, 23-27 April 1984*. NOAA Technical Report NMFS 51, 65 – 71.

DUFFY D C, SIEGFRIED W R and JACKSON S (1987a). Seabirds as consumers in the southern Benguela region. *South African Journal of Marine Science* **5**, 771 – 790.

DUFFY D C, WILSON R P, RICKLEFS R E, BRONI S C and VELDHUIS H (1987b). Penguins and purse seiners: competition or coexistence? *National Geographic Research* **3**, 480 – 488.

DE VILLIERS S A (1971). *Robben Island*. Struik, Cape Town.

FROST P G H, SIEGFRIED W R and COOPER J (1976). Conservation of the jackass penguin *(Spheniscus demersus* (L.)). *Biological Conservation* **9**, 79 – 99.

GREMMEN N J M (1981). The vegetation of the Subantarctic islands Marion and Prince Edward. *Geobotany* **3**, 1 – 149.

HEYMANN G, ERASMUS T, HUNTLEY B J, LIEBENBERG A C, RETIEF G DE F, CONDY P R and VAN DER WEST-HUYSEN O A (1987). Report to the Minister of Environment Affairs on an environmental impact assessment of a proposed landing facility on Marion Island — 1987. *South African National Scientific Programmes Report* **140**, CSIR, Pretoria. 209 pp.

KERLEY G I H (1983). Relative population sizes and trends, and hybridization of fur seals *Arctocephalus tropicalis* and *A gazella* at the Prince Edward Islands, Southern Ocean. *South African Journal of Zoology* **18**, 388 – 392.

MONTEIRO P M S (1987). Marion and Prince Edward Islands: the legal regime of the adjacent maritime zones. *Sea Changes. Institute of Marine Law Newsletter* **5**, 63 – 109.

MORANT P D, COOPER J and RANDALL R M (1981). The rehabilitation of oiled Jackass Penguins *Spheniscus demersus, 1970 – 1980*. In *Proceedings of the Symposium on Birds of the Sea and Shore, 1979.* (ed Cooper J) African Seabird Group, Cape Town. pp 267 – 301.

RAND R W (1963a). The biology of guano- producing seabirds. 4. Composition of colonies on the Cape islands, *Division of Sea Fisheries of South Africa Investigational Report* **43**, 32 pp.

RAND R W (1963b). The biology of guano-producing seabirds. 5. Composition of colonies on the South West African islands. *Division of Sea Fisheries of South Africa Investigational Report* **46**, 26 pp.

RANDALL B M and RANDALL R M (1986-87). *Mesembryanthemum aitonis* — the Sea Spinach of the Algoa Bay islands. *Veld & Flora* **72**, 116 – 118.

RANDALL R M and RANDALL B M (1980). Status and distribution of the roseate tern in South Africa. *Ostrich* **51**, 14 – 20.

RANDALL R M, RANDALL B M, BATCHELOR A L and ROSS G J B (1981). The status of seabirds associated with islands in Algoa Bay, South Africa, 1973 – 1981. *Cormorant* **9**, 85 – 104.

RICHARDS R (in press). The commercial exploitation of sea mammals at Iles Crozet, Marion and Prince Edward Islands before 1850. *Polar Monograph*.

RYAN P G (1987). The foraging behaviour and breeding seasonality of Hartlaub's gull *Larus hartlaubii*. *Cormorant* **15**, 23 – 32.

SHAUGHNESSY P D (1984). Historical population levels of seals and seabirds on islands off southern Africa, with special reference to Seal Island, False Bay. *Investigational Report of the Sea Fisheries Research Institute of South Africa* **127**, 61 pp.

SHELTON P A, CRAWFORD R J M, COOPER J and BROOKE R K (1984). Distribution, population size and conservation of the jackass penguin *Spheniscus demersus*. *South African Journal of Marine Science* **2**, 217 – 257.

SIEGFRIED W R (1982). The roles of birds in ecological processes affecting the functioning of the terrestrial ecosystem at sub-Antarctic Marion Island. *Comité National Français des Recherches Antarctiques* **51**, 493 – 499.

SKINNER J D and VAN AARDE R J (1983). Observations on the trend of the breeding population of southern elephant seals, *Mirounga leonina*, at Marion Island. *Journal of Applied Ecology* **20**, 707 – 712.

SMITH V R (1987). The environment and biota of Marion Island. *South African Journal of Science* **83**, 211 – 220.

VAN RENSBURG P J J (1985). The feeding ecology of a decreasing feral house cat, *Felis catus,* population at Marion Island. In *Antarctic nutrient cycles and food webs.* (eds Siegfried W R, Condy P R and Laws R M) Springer, Berlin. pp 620 – 624.

VAN RENSBURG P J J and BESTER M N (1988a). Experiments in feral cat population reduction by hunting on Marion Island. *South African Journal of Wildlife Research* **18,** 47 – 50.

VAN RENSBURG P J J and BESTER M N (1988b). The effect of cat *Felis catus* predation on three breeding Procellaridae species on Marion Island. *South African Journal of Zoology* **23,** 301 – 305.

VAN ZINDEREN BAKKER E M SR, WINTERBOTTOM J M and DYER R A (eds)(1971). Marion and Prince Edward Islands, Report on the South African Biological and Geological Expedition/1965-1966. Balkema, Cape Town.

VEITCH C R (1985). Methods of eradicating feral cats from offshore islands in New Zealand. *International Council for Bird Preservation Technical Publication* **3,** 125 – 141.

WATKINS B P (1987). Population sizes of king, rockhopper and macaroni penguins and wandering albatrosses at the Prince Edward Islands and Gough Island, 1951-1986. *South African Journal of Antarctic Research* **17,** 150 – 157.

WATKINS B P and COOPER J (1986). Introduction, present status and control of alien species at the Prince Edward islands, sub-Antarctic. *South African Journal of Antarctic Research* **16,** 86 – 94.

CHAPTER 16

Conservation status of southern African wetlands

C M Breen, G W Begg

INTRODUCTION

Although South Africa is a signatory of the RAMSAR convention, there has, until comparatively recently, been little interest in the conservation of wetlands. Consequently, in common with many other parts of the world, wetland degradation has proceeded apace and rather little is known of the nature of extant wetlands.

South African wetlands have not been formally classified or inventorised. In those that have been studied the emphasis has generally centred on waterfowl and very little is known of the other species. Consequently, there is a paucity of data on which to develop an analysis of diversity. In this paper we define the term wetland, comment on the value of wetlands, and present a broad overview of wetland diversity in South Africa.

We discuss the conservation status of wetlands and the implementation of conservation policy.

WETLAND DEFINITION

Wetlands occupy an intermediate position on the continuum between aquatic and terrestrial environments. Since the continuum is variable in space and time, it is not surprising that 'there is no single, correct, indisputable, ecologically sound definition for wetlands' (Cowardin et al 1979). The word wetland therefore is a generic term which is used to group those features of the landscape that are commonly referred to as marshes, swamps, bogs, floodplains, and in southern Africa also as dambos, vleis and pans, as a single type of ecosystem.

The single feature that separates wetlands from terrestrial systems is the persistence of water-saturated soil conditions. They are separated from aquatic systems by the depth of permanent free water which restricts development of emergent vegetation. Thus common characteristics such as high water table, hydric soil and hydrophilous vegetation have been incorporated into an internationally accepted definition of wetlands (Cowardin et al 1979):

Wetlands are transitional between terrestrial and aquatic systems where the water table is usually at or near the surface or the land is covered by shallow water . . .. Wetlands must have one or more of the following three attributes: (1) at least periodically, the land supports predominantly hydrophytes, (2) the substrate is predominantly undrained hydric soil, and (3) the substrate is nonsoil and is saturated with water or covered by shallow water at some time during the growing season of each year.

It is important, in the arid southern African context, to draw attention to the fact that not all three attributes have to coexist to justify designation as wetland. As will be shown later there are numerous shallow endorheic depressions in the arid parts of South Africa which support few if any hydrophytes (attribute 1), and may contain shallow water only once in five or more years (attribute 2). These systems are designated wetlands on the basis of hydric soils in which a component of rising capillary ground water carries dissolved solutes to the surface and results in 'salt desert' habitats (Etherington 1983).

THE VALUE OF WETLANDS

Although certain values of wetlands, for example fertile soils, have been recognized for millennia, many others, particularly those deriving from the manner in which wetlands function, have only recently been recognized and accepted. This is because of the relatively recent expression of problems, such as pollution, erosion and increased runoff, on scales that have potentially disastrous consequences for mankind. The value of wetlands to society can therefore be evaluated in the contexts of available resources and services accruing from the way in which the systems function (Table 16.1).

TABLE 16.1   A generalized depiction of resource and system values of wetlands and the predominant geographical scales over which society benefits. Adapted from Mitsch and Gosselink (1986).

Value	Geographical scale		
	Local	Regional	Global
Resources			
Water — storage, abstraction	———————— - - -		
Soil — crop production	———————— - - - - -		
Plants — timber, reed, pasture	———————— - - - -		
Peat — energy	——————————— - - - - - - -		
Animals — mammals, birds, fish, invertebrates	———————— - - - - - - -		
Endangered species	- - - - ———————————		
System			
Flood attenuation	——————————— - - -		
Aquifer recharge	——————————— - - -		
Water quality	——————————— - - -		
Social attributes (aesthetics, education etc)	- - ————————————		
Atmosphere quality	- - - - - ————————		

It is obvious that the value of a wetland will vary with the nature of the wetland and the needs of the population interacting with it. Thus, in undeveloped or underdeveloped areas (at a local scale) wetlands provide important and often life-supporting resources (water, food etc) and services (eg flood attenuation) may be less important or at least their value less well recognized. In more developed areas dependence on natural wetland resources decreases and the value of services increases because of the consequences of larger-scale (regional) environmental modification. Wetlands are also considered to be important factors in the regulation of global cycles including nitrogen, sulphur, methane and carbon dioxide as a result of the manner in which resources are utilized, for example peat, and of processes occurring in wetlands, for example denitrification (Mitsch and Gosselink 1986).

Wetland degradation is therefore of considerable concern because it impinges upon life-support systems at local, regional and global scales.

DIVERSITY

Wetlands of some description are likely to occur in the catchment of every river system in southern Africa, but their form and abundance vary considerably owing to regional differences in topography, climate, vegetation, soil, landuse and hydrological conditions.

Unlike the situation in the United States of America (Tiner 1984), Australia (Paijmans et al 1985) and Zimbabwe (Whitlow 1985) for example, few data exist on the distribution of wetlands in southern Africa. From the studies of the distribution of wetlands in South Africa that do exist (Scotney 1970; van der Eyk et al 1969; Begg 1988) it would appear that wetlands are best represented either in coastal environments (where the low river gradient, high rainfall and substantial runoff are factors conducive to wetland formation) or on footslopes in proximity to mountainous areas (where the geology and the high rainfall associated with highland areas frequently give rise to headwater wetlands near or at the source of rivers).

Noble and Hemens (1978) were the first to attempt a broad classification of the wetlands of South Africa and their classification provides a useful structure within which to examine the diversity of wetland systems. They recognized two broad categories, vleis and floodplains, and endorheic pans, each of which consisted of several subcategories (Table 16.2). Within the former category they identified river-source sponges, marshes and swamps and floodplains, depending on the nature of the interactions between hydrology and topography.

Dominant plant species, water depth and salinity were factors in the separation into subcategories. They also noted that because of the particular geomorphology, the distribution and variability of annual rainfall, and the generally high potential evaporation, there is a tendency towards the geographical location of certain types of wetlands. Thus bogs and sponges are generally confined to high-altitude regions, pans are common in the arid central basin (Orange Free State and northern Cape), and floodplains are evident wherever rivers attain grade (eg Natal midlands and coastal plains).

In 1983 Morant reviewed wetland classification and concluded that because of the potential diversity of wetlands in South Africa it would be appropriate to adopt, with relatively little modification, the system used by the United States Fish and Wildlife Services (Cowardin et al 1979). The system has, however, not been adopted and little further progress has been made in documenting wetland variability at a broad national scale.

River-source sponges and bogs of high altitudes (Jacot-Guillarmod 1962) are the least hydrologically dynamic wetland systems, and diversity is consequently brought about by other factors such as temperature and pH (Noble and Hemens 1978). It seems probable therefore that diversity of these systems at the local or regional scale is likely to be narrow. C Schwabe (pers comm)

TABLE 16.2   Classification of wetlands in South Africa (modified from Noble and Hemens 1978)

Vleis and Floodplains		
	River-source sponges (generally high altitude)	Bogs (sedges and mosses) Acid sponges (Restionaceae and Bruniaceae)
	Marshes and swamps (generally flat terrain)	Sedge marshes Restio marshes Reed marshes (vleis) Reed swamps Papyrus swamps Seasonal wetlands Swamp forests Salt marshes Mangrove swamps
	Floodplains (flat middle and lower reaches of rivers)	Karoo salt flats Floodplain vleis Storage floodplains
Endorheic pans		Salt pans Temporary pans Grass pans Sedge pans Reed pans Semi-permanent pans

recognizes three types: bogs, tarns in which the rare endemic *Aponogeton ranuncifolius* may be dominant, and sedge vleis in fairly flat terrain.

Pans, because of their importance as wildfowl habitats, were among the first wetlands to attract attention, and Geldenhuys (1976, 1981) and more recently Allan (1987a, b) have used vegetation physiognomy and species composition to separate different types of pans. Allan (1987a, b) recognizes four types on the highveld (reed pans, sedge pans, open pans and salt pans) and stressed the correlation between the type of pan and its avifauna, a synergistic effect which reinforces between-pan differences. A similar situation arises in some of the pans of the more arid Orange Free State and northern Cape (Geldenhuys 1981) where Seaman and Kok (1987) have reported a diversity of size and physico-chemical conditions which results in a complex pattern of inverte-brate diversity. Although pans are generally different (high diversity) they often share similar conditions for short periods (low diversity). The net result is that, whereas over the total inundation period pans may share taxa, pans with 'long' inundation periods have the greater diversity of species (alpha diversity).

Floodplains are the most hydrologically dynamic freshwater wetland systems because of their location in the middle and lower reaches of river catchments. Whenever a river attains grade and flow is dispersed horizontally, particularly during floods, sediment deposition, levee formation, channel switching and other fluvial processes can create a mosaic of varied land-forms and wetland habitats. On the Pongolo floodplain, for example, there is a mosaic of seasonally and permanently inundated habitats, and considerable standing water remains in pans after floods subside (Breen et al 1978; Furness and Breen 1980; Heeg and Breen 1982). Noble and Hemens (1978) referred to these as storage floodplains. Other floodplains are homogeneous, the diversity of wetland habitats is narrow, and little or no standing water remains after floods have subsided (eg Nylvlei,

Noble and Hemens 1978). Diversity apparently reflects the interactions between the hydrology and the hydraulics of the system. In some systems (eg Pongolo) high sedimentation and erosion rates force channel switching, whereas in the Okavango hydraulic resistance offered by the aquatic emergent vegetation is implicated (McCarthy et al 1986).

Elevational and substratum differences are considered to be the chief sources of diversity in wetlands (Gosselink et al 1981). Since these differences are most manifest in floodplains of high hydrological energy in which horizontal flow predominates, it is expected that the greatest species richness would occur in these systems. This accords with the observation of Harris et al (1983) that 'bird species diversity was related not only to the diversity of vegetation types, but also to the mosaic arrangement and distribution of those cover types within the marsh as measured by edge diversity'.

The data presented in Table 16.3, although of an incomplete and preliminary nature, show that diversity is higher in floodplains (Pongolo, Mkuze and Nyl) than it is in the less hydrodynamic endorheic pans of the Transvaal and Orange Free State. In the Mkuze Swamp system the mosaic of habitats has its origin in aeolian dune and fluvial depositional environments (van Heerden 1986; Stormanns and Breen 1987) and as such may on further investigation be shown to have the greatest diversity of any wetland system in South Africa.

TABLE 16.3   Preliminary analysis of the numbers of species associated with different types of wetlands in South Africa. Data adapted from various authors. 'Rare' and 'endangered' refer to status in South Africa.

	Wetland				
Species category	Transvaal pans[1]	Orange Free State pans[2]	Nylvlei [3]	Pongolo flood plain[4]	Mkuze system[5]
Plants					
Aquatic/semi-aquatic	13	10	no data	59	275
Birds					
Total	57	43	100	141	76
Rare/endangered	4	2	23	25	20
Fish					
Total	0	10	no data	40	35
Rare/endangered	0	10	no data	2	7
Reptiles/amphibians					
Total	no data	14	no data	48	102
Rare/endangered	no data	0	1	8	7

(1) Allan 1987a, b: (2) Geldenhuys 1976, 1981; (3) Tarboton 1987; (4) Heeg and Breen 1982; Musil et al 1973; D Johnson pers comm; (5) O Bourquin, D Johnson, N Robson and P Skelton pers comm; Stormanns and Breen 1987

The observed bird species diversity in the Nylvlei (Tarboton 1987) cannot be explained at present because of incomplete knowledge of vegetation structure and dynamics. It is of interest to note that in the Mvoti vlei, a wetland similar to but smaller than Nylvlei, canalization with concomitant reduction in habitat diversity resulted in a decline in the numbers and diversity of waterbirds (Scotcher 1987).

A number of wetlands are dominated by extensive stands of single species of genera such as *Phragmites*, *Typha* and *Cyperus*. Although these systems would be expected to show low species diversity, except perhaps along their margins, it has been observed that highest invertebrate species diversity and greatest equitability and evenness may occur in the centre of reedbeds (Mason and Bryant 1974). There are, however, no data to permit evaluation of this observation in South African wetlands.

It is evident from the literature on South African wetlands that there have been no attempts to measure either between-system or within-system diversity. Perhaps more significant is the absence of any attempt to develop an understanding of the mechanisms regulating diversity. It is, therefore, not possible at present to develop a strategy for the conservation of biotic diversity in wetlands that is based on knowledge and understanding of local systems. An intuitive approach offers the only real prospect for conservation under the present circumstances in which wetlands are being altered and degraded so rapidly.

CONSERVATION STATUS

A review of knowledge of and attitudes to wetlands shows that conservation organizations in South Africa are generally uninformed on the diversity of wetlands and have disparate approaches to the classification and conservation of wetlands (C H Stormanns pers comm). It is clear that without knowledge of the 'elements of diversity' (Jenkins 1978 in Noss 1987) it is not possible to construct a classificatory system, and without a technique for classification there is little hope for the formulation of a comprehensive strategy for the conservation of biotic diversity of the wetlands in South Africa. The most urgent need, therefore, is the development of a classification system that is 'both comprehensive and efficient in identifying the elements of concern' (Noss 1987). Furthermore, it is suggested that The Nature Conservancy approach suggested by Noss (1987) may be appropriate because the objective is inventory and protection rather than classification *per se*.

The need for implementation of a strategy for inventory that leads to conservation is urgent because of the large-scale destruction of wetlands in South Africa. Man-induced perturbations include drainage (for crop and timber production), infilling (for the construction of roads, housing schemes, industrial parks and the disposal of solid wastes), channelization (for flood control), dam construction (to provide for agricultural and domestic use of water), mining (for the extraction of clay, sand and peat), burning, overgrazing, chemical and nutrient pollution, and river regulation.

The extent of perturbation can be gauged from a recent study of the Mfolozi catchment (Begg 1988) in which it was estimated that 58% of the original wetland area (502 km^2) has been lost, and that only two per cent of this 10 000 km^2 catchment is presently occupied by wetland (Figure 16.1). Elsewhere, for example in parts of the Tugela basin (Scotney 1978) and Siyaya catchment (Begg 1986), it has been shown that over 90% of the original wetland resources have been destroyed.

The real implications of wetland alteration and degradation for biotic diversity in southern Africa therefore cannot be assessed, except in general terms. However, it can be stated that:

— wetlands have an important role to play in maintaining the diversity of plant and animal species;
— as in many other parts of the world, the rate of wetland loss that is suspected in southern Africa means that wetland-dependent species are threatened in this area;

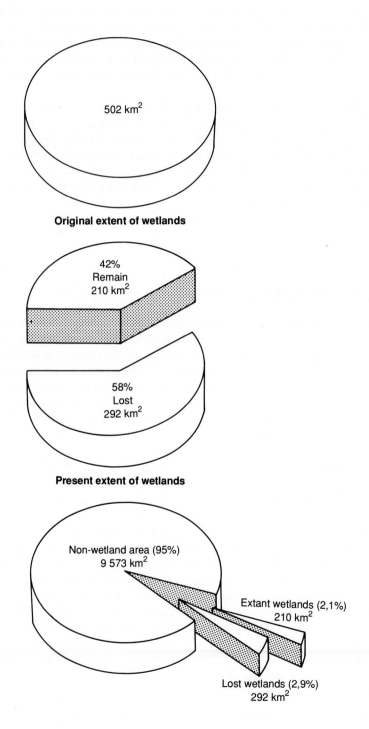

FIGURE 16.1 Original and remaining extent of wetlands in the Mfolozi catchment (Begg 1988)

— a high proportion of the species that are listed as threatened with extinction in South Africa (Table 16.3) are wetland dependent; and

— in the long term the implications of the reduced services provided by wetland environments may be far more serious for society than losses in biotic diversity.

IMPLEMENTATION OF CONSERVATION POLICY

The abundance of evidence for the declining status of wetlands in South Africa bears witness to the existence of four major problem areas:

— widespread ignorance of wetland functions;

— the ineffectiveness in South Africa of the pre-1983 legislation relating to wetland protection;

— the lack of an integrated wetland policy to link the multi-sectoral interests of the various government and non-government agencies in South Africa directly or indirectly involved in wetland regulation; and

— the fundamental lack of knowledge concerning the conservation requirements of wetlands, without which management strategies cannot be formulated, or implemented.

Wetland conservation measures in South Africa at present tend to rely heavily on the imposition of controls exercised by a single department, namely the Department of Agricultural Economics and Marketing, which administers the Conservation of Agricultural Resources Act of 1983. This Act, which became operative only on 26 May 1984, specifically provides for 'utilization and protection of vleis, marshes, water sponges and water courses', and in essence forbids any landuser throughout South Africa to drain or cultivate wetlands (or land within 10 m horizontally from the 1:10-year flood line of a watercourse) without written permission from the above-mentioned department.

Difficulties have arisen with regard to interpretation of the legislation and, because of the size of the country, from the lack of sufficient finance and manpower to implement the legislation. Consequently, it is not surprising to find that the legislation designed to protect wetlands that is currently available in South Africa has failed to arouse public consciousness. It is unlikely to do so in the future unless it is more rigorously and ruthlessly applied.

The fact that no effective strategy exists to ensure the conservation of South African wetlands implies that it is necessary to convince society (in general) and decision-makers (in particular) of the asset value of wetlands to human beings. If improved resource management is the legitimate objective of this book, it is proposed that emphasis should not be placed on the dependence of an undefined assemblage of plant and animal species upon wetlands, but instead on the bleak prospect of mankind's survival in a future without wetlands. Although apparently unaware of this at present, decision-makers will have to realize that the strain on future resources of this country (such as fresh water) means that in the face of exponential population growth, man's dependence upon wetlands is steadily increasing.

As stressed by Ehrenfeld (1986), 'the damage to diversity becomes yet more vivid if we look at the destruction of ecosystems and its projected consequences . . .'. In the case of a semi-arid region such as South Africa this means:

— lower agricultural productivity;

— poorer water quality;

— less reliable water supplies;

— increased incidence and severity of downstream flooding;

— threatened wildlife resources; and

— an inevitable increased incidence of species extinction.

Furthermore, the consequences of wetland destruction cannot be isolated because 'their destruction shifts economic and environmental costs to other citizens who have no voice in the decisions to alter them' (Maltby 1985). For ethical reasons, therefore, and especially in the South African context, which involves rural communities heavily reliant upon local sources of water to sustain themselves and their livestock throughout the year, good reason exists for much-improved, planned resource management.

Better balanced resource use in southern African catchments, with the attendant benefit of providing more favourable options for the maintenance of biodiversity, therefore holds the greatest promise for the future of mankind in the southern portion of the African continent.

REFERENCES

ALLAN D (1987a). Transvaal highveld pans. *African Wildlife* **42**, 233 – 235.

ALLAN D (1987b). The conservation status of the pans of the Lake Chrissie area. In The ecology and conservation of wetlands in South Africa (eds Walmsley R D and Botten M L) *Ecosystems Programmes Occasional Report* **28**, 91 – 100.

BEGG G W (1986). The Wetlands of Natal (Part 1). An overview of their extent, role and present status. *Natal Town Regional Planning Report* **68**. 144 pp.

BEGG G W (1988). The Wetlands of Natal (Part 2). The distribution,extent, and status of wetlands in the Mfolozi catchment. *Natal Town and Regional Planning Report* **71**. 262 pp.

BREEN C M, FURNESS H D, HEEG J and KOK H (1978). Bathymetric studies on the Pongolo River Floodplain. *Journal of the Limnological Society of southern Africa* **2**, 95 – 100.

COWARDIN L M, CARTER V, GOLET F C and LA ROE E T (1979). Classification of wetlands and deepwater habitats of the United States. US Fish and Wildlife Service (Department of the Interior). *Report FWS/OBS — 79/31.* US Government Printing Office. Washington, DC. 131 pp.

EHRENFELD D (1986). Thirty million cheers for diversity. *New Scientist* **64**, 38 – 43.

ETHERINGTON J R (1983). Wetland Ecology. *Studies in Biology No 154.* Edward Arnold, London.

FURNESS H D and BREEN C M (1980). The vegetation of seasonally flooded areas of the Pongolo River Floodplain. *Bothalia* **13**, 217 – 231.

GELDENHUYS J N (1976). Relative abundance of waterfowl in the Orange Free State. *Ostrich* **47**, 27 – 54.

GELDENHUYS J N (1981). Classification of the pans of the Orange Free State according to vegetation structure, with reference to avifaunal communities. *South African Journal of Wildlife Research* **12**, 55 – 62.

GOSSELINK J G, BAILEY S E, CONNER W H and TURNER R E (1981). Ecological factors in the determination of riparian wetland boundaries. In Wetlands of Bottomland Hardwood Forests. (eds Clark J R and Benfordo J) *Developments in Agricultural and Managed Forest Ecology* **11**. Elsevier, Amsterdam

HARRIS H J, MILLIGAN M S and FEWLESS G A (1983). Diversity: Quantification and Ecological Evaluation in Freshwater Marches. *Biological Conservation* **27**, 99 – 110.

HEGG J and BREEN C M (1982). Man and the Pongola floodplain. *South African National Scientific Programmes Report* **546**. CSIR, Pretoria. 117 pp.

JACOT-GUILLARMOD A (1962). The bogs and sponges of the Basutoland Mountains. *South African Journal of Science* **58**. 179 – 182.

MALTBY E (1986). Waterlogged wealth. Why waste the world's wet places? *Earthscan.* 1 – 200.

MASON C F and BRYANT R J (1974). The structure and diversity of the animal communities in broadland reedswamps. *Journal of Zoology, London* **172**, 289 – 302.

MCCARTHY T S, ELLERY W N, ROGERS K H, CAIRNCROSS B and ELLERY K (1986). The roles of sedimentation and plant growth in changing flow patterns in the Okavango Delta, Botswana. *South African Journal of Science* **83**, 579 – 584.

MITSCH W J and GOSSELINK J G (1986). *Wetlands.* Van Nostrand, Reinhold, New York, NY.

MORANT P D (1983). Wetland classification: towards an approach for southern Africa. *Journal of the Limnological Society of southern Africa* **9**, 76 – 84.

MUSIL C F, GRUNOW J O and BORNMANN C H (1973). Classification and ordination of aquatic macrophytes in the Pongola river pans. Natal. *Bothalia* **11**, 181 – 190.

NOBLE R G and HEMENS J (1978). Inland water ecosystems in South Africa — a review of research needs. *South African National Scientific Programmes Report* **34**. CSIR, Pretoria. 150 pp.

NOSS R F (1987). From Plant Communities to Landscapes in Conservation Inventories: A Look at The Nature Conservancy (U.S.A.). *Biological Conservation* **41**, 11 – 37.

PAIJMANS K, GALLOWAY R W, FAITH D P, FLEMING P M, HAANTJENS H A, HEYLIGERS P D, KALMA J D and LOFFLER E (1985). Aspects of Australian wetlands. CSIRO. *Australian Division of Water and Land Resources Technical Paper* **44**. 71 pp.

SCOTCHER J S B (1987). Nature Reserve Mvoti Vlei. In The ecology and conservation of wetlands in South Africa. (eds Walmsley R D and Botten M L) *Ecosytems Programmes Occasional Report* **28**. CSIR, Pretoria. 169 – 173.

SCOTNEY D M (1970) Soils and land-use planning in the Howick extension area. Unpublished PhD thesis. University of Natal, Pietermaritzburg.

SCOTNEY D M (1978). The present situation in Natal. Symposium on the relationship between agriculture and environmental conservation in Natal and KwaZulu. Durban. 19 – 20 October 1978, 16 – 34.

SEAMAN M T and KOK D J (1987). Ecological diversity in Orange Free State pans. In The ecology and conservation of wetlands in South Africa. (eds Walmsley R D and Botten M L) *Ecosystem Programmes Occasional Report* **28.** CSIR, Pretoria. 260 – 273.

STORMANNS C-H and BREEN C M (eds) (1986). *Proceedings of the Greater Mkuze Swamp Symposium and Workshop. 26 March 1986.* Natal Parks Board, Pietermaritzburg. Institute of Natural Resources. University of Natal. *Investigational report* **22.**

STORMANNS C-H and BREEN C M (1987). The use of two multivariate analytical techniques in the analysis of wetland vegetation for conservation evaluation. In The ecology and conservation of wetlands in South Africa. (eds Walmsley R D and Botten M L) *Ecosystem Programmes Occasional Report* **28.** CSIR, Pretoria. 274 – 282.

TARBOTON W R (1987). Nyl floodplain: its significance, phenology and conservation status. In The ecology and conservation of wetlands in South Africa. (eds Walmsley R D and Botten M L) *Ecosystem Programmes Occasional Report* **28.** CSIR, Pretoria. 101 – 114.

TINER R W (1984). *Wetlands of the United States: current status and recent trends.* US Fish and Wildlife Service (Department of the Interior) US Government Printing Office, Washington, DC. 59 pp.

VAN DER EYK J J, MACVICAR C N and DE VILLIERS J M (1969). Soils of the Tugela Basin — a study in sub-tropical Africa. *Natal Town and Regional Planning Report* **15.** 263 pp.

VAN HEERDEN I L (1986). The origin and development of the Mkuze Swamp system: a geological and geomorphological perception. In *Proceedings of the Greater Mkuze Swamp Symposium and Workshop. 26 March 1986.* (eds Stormanns C-H and Breen C M) Natal Parks Board, Pietermaritzburg. Institute of Natural Resources, University of Natal. *Investigational Report* **22.**

WHITLOW J R 1985. Dambos in Zimbabwe: a review. *Zeitschrift fur Geomorphologie.* **N F 52,** 115 – 146.

PART 5

Conservation status of river, coastal and marine ecosystems and their biota

CHAPTER 17

The conservation status of southern African rivers

J H O'Keeffe, B R Davies, J M King, P H Skelton

INTRODUCTION

Some rain that falls on land is lost into the air again, and the rest drains downhill, as surface flow, soil moisture or ground water, towards the lowest point of drainage. As the water drains downhill, pollutants are absorbed and its chemical quality changes in a way that is dependent on the geology, vegetation and landuse of the area. The lowest point of drainage in any drainage basin (catchment) is usually the river, and into it drains the chemically changed water, from its point of precipitation which can be metres to hundreds of kilometres away, depending on the size of the catchment. Rivers are therefore one of the most vulnerable of our ecosystems, since they are a reflection of all the activities in their catchments. For instance, agricultural activities can increase or decrease flow, increase nutrient and pesticide concentrations in the draining water, and increase sediment loads, while development of an urban area can increase storm flow, decrease dry-season flow, and increase concentrations in the water of pollutants such as sewage, oil and chemicals.

Adding to the vulnerability of rivers is their characteristic of unidirectional flow. Such flow ensures that any disturbance of a drainage system is reflected in all downstream parts of that system. A dam on a mountain stream can cause reverberations in ecosystem functioning in all parts of the river system downstream of the dam including the estuary (because of changes in flow and temperature regimes etc), and possibly in those parts of the system above the dam as well, if phenomena such as fish migrations are affected.

In most countries today it is rare to find a river that is undisturbed and unchanged in any part other than its high-altitude, fairly inaccessible source area, and even these areas may be influenced by factors such as acid precipitation and fire. In South Africa, by the time the necessary skills and manpower were available to study river ecosystems, most rivers had already been changed by a variety of landuse and water-supply practices, and it is probably already too late to accurately describe the natural communities, characteristics and ecosystem functioning of the lower parts of our river systems. Only in protected source areas can it be assumed that present conditions approximate those of the past, undisturbed situation.

With this in mind, we detail below why and how rivers have been the focus of so much development; how they are believed to function ecologically; the types of rivers, river zones and riverine biota in southern Africa; the major threats to the rivers and their biota; and an assessment of their conservation status. We conclude by discussing present trends and the potential for managing these systems wisely, in order to conserve ecosystem functioning and biotic diversity.

The water resources of southern Africa, and future demands

Southern Africa is generally resource rich, except for one vital component — water. Such water as there is is unevenly distributed (owing to rainfall and drainage patterns) and not necessarily near centres of urbanization, and the geological history and geomorphology of the subcontinent has resulted in a region generally devoid of natural lakes which might be used for water supply. In addition, rainfall is usually much lower than potential evaporation (Alexander 1985). Ground water reserves of fresh water are substantial but it is economically viable to extract only a small proportion of the resource (Department of Water Affairs 1986). These circumstances have led to river systems becoming the primary source of water for agricultural, industrial and domestic consumption, and power generation through the creation of numerous dammed in-stream lakes, canal systems, inter-basin transfers, farm dams and other abstraction schemes. Many river systems are now impounded at a number of points along their length, and most act as repositories for human wastes, suffer increased acidification and mineralization, have lost their indigenous riparian vegetation, or have had their flow regime altered beyond recognition from their natural state. Such circumstances hold grave implications for the integrity and continued functioning of river ecosystems, and indeed, even if disturbances were to halt today, a large proportion would probably not revert to their past condition because of the loss of riparian belts, massive catchment changes and possible extinction of undiscovered aquatic species.

The South African population is expanding at a rate of 2,7% per annum (World Bank 1980), indicating that it will double within 26 years unless the rate changes. Demands for water will therefore increase and any water reaching the sea may be considered by some to be 'wasted'. Efforts will obviously be increased to manipulate river flow for man's use, and in most cases these efforts will be opposed to the maintenance of the natural state, since such a state has not met human needs up to this time.

In such circumstances, the problem of conserving the biotic diversity of rivers actually reduces to one of conserving the quantity and quality of water in those rivers. Accordingly, in this paper we have concentrated not so much on species, communities and populations, but more on the conservation of habitats and ecosystems as emphasized by Siegfried and Davies (1982). The traditional view of nature conservation, that of the protection of species within fenced reserves, is inadequate for rivers, or indeed as a ruling ethic for conservation. Rivers serve to highlight this inadequacy because they are expressions of holistic catchment processes, and are longitudinal systems which cannot be fenced off. Rivers are going to be exploited for human uses, including water supply, effluent removal, recreation, and nature conservation. The aims of river conservation should therefore be:

— Careful planning for multiple uses. This could include zonation of rivers and/ or stretches of river for different priorities such as urban/industrial water supply, agriculture, forestry, recreation, nature conservation etc.

— The wise use of the renewable resources of rivers to prevent the breakdown of their essential ecological processes, such as water supply, nutrient cycling, etc.

THE PRINCIPLES OF RIVER ECOSYSTEM FUNCTIONING

This section briefly outlines current concepts and thinking on river ecosystem functioning (for more detailed overviews, see Ward et al 1984; Davies and Day 1986; Day, Davies and King 1986). It is necessary to explain these concepts in order to provide a rationale for our emphasis on processes/ecosystem conservation, and also to explain the level at which river ecologists find themselves in understanding their discipline. Behind all these concepts, however, lies the indisputable and inescapable truth, which we repeat here — rivers are the end expression of their catchments. The conservation of river systems is dependent entirely upon sound management of their catchments.

Contemporary views on river ecosystem functioning

PREDICTABLE VS UNPREDICTABLE SYSTEMS

There are at present two diametrically opposed views concerning the ecosystem level of the organization of rivers: the River Continuum Concept (RCC), originally postulated by Vannote et al (1980) for rivers in north temperate, forested regions, assumes gradual and predictable changes in the physical and chemical environment down rivers which lead to predictable biotic changes (see Figure 17.1). On the other hand Winterbourn et al (1983) believe that river ecosystems are driven by stochastic events, often by catastrophic climatic events, and as such are inhabited by opportunistic and generalist biotas. Although some aspects of these ideas are being tested for local rivers (eg by King et al 1989) the state of our understanding of southern African river ecosystems is such that we do not know which view is correct. We can, however, point out that within our subcontinent, climatic events are generally far less predictable than in north temperate zones (Alexander 1985), and that it is therefore possible that the Winterbourn et al (1983) model may be a more appropriate view of our systems. For example, in the coastal catchment of the Great Fish River in the eastern Cape, the month with the highest average rainfall is March, but between 1958 and 1976 the highest rainfall of the year occurred during nine out of the 12 months, including four times during the 'dry season' months of June, July and August (data for tertiary catchment Q93, Middleton et al 1981). This absence of seasonal predictability precludes any selection for species with lifestyles adapted to regular seasonal changes. If southern African rivers are governed more by environmental events, the biota may be resilient to most artificial perturbations, but may also be disadvantaged by river regulation, which removes much of the random hydrological variability.

FIGURE 17.1 A pictorial representation from Cummins (1979) of the River Continuum Concept (modified from Davies and Day 1986), showing the downstream gradient of physical factors and the corresponding biological adjustments. The diagram represents the river continuum as a single stem of increasing stream order, and orders have been roughly grouped into headwaters (orders 1–3), mid-sized rivers (4–6) and large rivers (7–12), considering the Mississippi River as order 12 at its mouth. The headwaters and large rivers are shown as heterotrophic (ratio of gross photosynthesis to community respiration P/R less than 1) because of restricted light in the headwaters and attenuation from depth and turbidity in the large rivers. The mid-sized rivers are depicted as autotrophic, with a P/R greater than 1, through a combination of reduced riparian shading and relatively shallow, clear water. The importance of terrestrial inputs of coarse particulate organic matter (CPOM) decreases and the transport of fine particulate organic matter (FPOM) increases down the river. The macro-invertebrate functional feeding groups shift from shredders and collectors in headwaters to collectors and grazers (scrapers) in mid-sized rivers, and to collectors in large rivers, which also develop plankton communities.

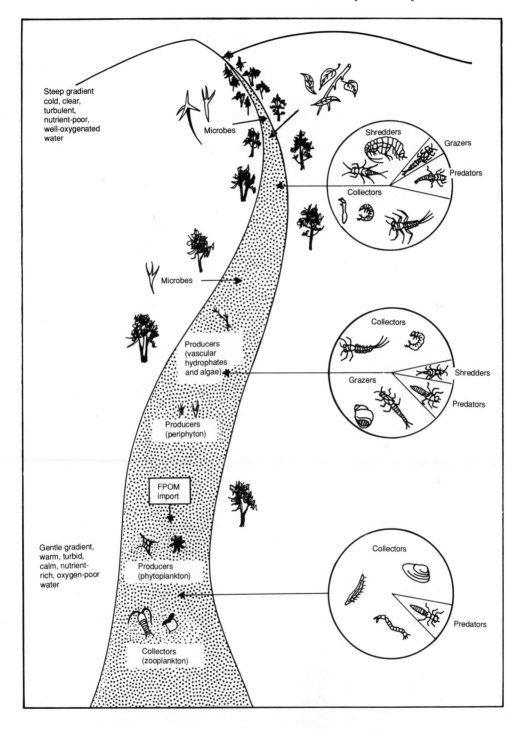

Steep gradient
cold, clear,
turbulent,
nutrient-poor,
well-oxygenated
water

Microbes

Shredders

Grazers

Predators

Collectors

Microbes

Producers
(vascular
hydrophates
and algae)

Collectors

Shredders

Grazers

Predators

Producers
(periphyton)

FPOM
import

Collectors

Gentle gradient,
warm, turbid,
calm, nutrient-
rich, oxygen-poor
water

Producers
(phytoplankton)

Predators

Collectors
(zooplankton)

The Serial Discontinuity Concept (SDC) of Ward and Stanford (1983), interprets impoundments as interruptions of the river continuum, causing shifts in abiotic and biotic variables such as temperature, with subsequent recovery at some unknown distance downstream from the perturbation. This idea of 'discontinuity' caused by impoundment may have profound implications for river conservation in southern Africa, particularly since almost every river ecosystem in the region has its flow regulated by one or more storage and power-production dams. The idea is currently being tested both in the south-eastern and western Cape by teams from Rhodes University (Buffalo River) and the University of Cape Town (Palmiet River). Initial results would seem to indicate that there are discontinuities and recovery processes operating in contrasting types of river system, but the intensity of these natural recovery processes appears to be weak and they seem to be easily overwhelmed by additional perturbations such as organic pollution; (Palmer and O'Keeffe in press(a); O'Keeffe et al, in press(b)).

MINIMUM FLOW REQUIREMENTS OF RIVERS (INSTREAM FLOW NEEDS)

So much water is extracted from southern African rivers that many that were once perennial now cease to flow for extended periods despite good rainfall. The Letaba and Luvuvhu rivers of the eastern Transvaal are classic examples of this situation. Both systems, essential life-support arteries of the Kruger National Park, have become a series of stagnating and isolated pools for periods of up to eight months in recent years, irrespective of whether the region is experiencing drought conditions, as a result of water extraction upstream of the Park. In such extreme (but unfortunately common) cases, the normal ecological functioning of a river may be severely modified or even destroyed.

Recent attempts to address such problems have involved examination of the so-called minimum flow requirements, or the in-stream flow needs (eg Gore 1978; Prewitt and Carlson 1980; Gore and Judy 1981) of rivers. A number of models have been developed, which assume that there is a point beyond which the abstraction of water from a river system will result in the progressive loss of habitat, disruption of ecological processes and disappearance of species. The need for an allocation of water for environmental needs in South African rivers was recognized by the Department of Water Affairs (Roberts 1981, 1983), and it was suggested that this allocation would represent approximately 11% of the country's MAR. Roberts pointed out that his suggestion was at best a first approximation, and that it needed verification through research. The North American work mentioned above has illustrated that different in-stream processes, species and communities, and indeed different river systems, have very different in-stream flow needs, and that the methods so far developed require a considerable research effort both to establish and to verify these needs. As most rivers will inevitably be maintained at functional rather than pristine levels, the objective for each river will have to be decided before any calculations are attempted. The dangers inherent in using any one method for the assessment of 'minimum flow requirements' are that a single figure may ultimately be applied indiscriminately and inappropriately, as a short-cut substitute for sound research.

In 1987, two workshops on minimum flow requirements of South African river systems produced two promising local approaches which are still being developed. The so-called 'Skukuza' approach developed for the stressed systems of the Kruger National Park (O'Keeffe et al in press(b)) requires a cumulative assessment of consumptive and non-consumptive water requirements for each ecosystem component (such as fish, riparian vegetation etc). Where time and resources are insufficient for this level of detail, the Flow Record Simulation approach (Ferrar and Chutter in press) simply uses a statistical treatment of hydrological records and simulations to arrive at an estimate of the total ecosystem requirements.

REGIONAL REVIEW OF RIVER TYPES, RIVER ZONES, AND RIVERINE BIOTA

River types

There are many different types of rivers in southern Africa, and all change in character between source and the sea. The need to classify types and sections of rivers in order to compare information between rivers, to identify rivers and sectors of rivers of conservation importance, and to allocate priority uses to different rivers, has been recognized for some time (eg Harrison 1959; Noble and Hemens 1978). Harrison (1959) defined 12 main hydrobiological regions within southern Africa based on differences in water chemistry and the aquatic biota (Figure 17.2). The river types can be summarized briefly as follows:

A   The Cape system — dark or clear acid rivers flowing over Table Mountain sandstone. There are many invertebrates and fish endemic to these systems. Several major studies have been undertaken, for example the Berg River (Harrison and Elsworth 1958; Fourie 1976); the Eerste River (King 1981); the Olifants River (Gaigher 1973; Skelton 1987; Coetzer 1982).

B   Recent limestone region near Bredasdorp — alkaline streams originating from springs (no studies).

C   Central arid regions — usually temporary, alkaline, and often saline rivers, with the exception of the Orange River. The Orange River has been studied by, for example, Cambray et al (1986) and Skelton (1986a).

D   The eastern Cape region — highly turbid, alkaline and often temporary rivers, with a fairly high degree of mineralization and depauperate faunas. The best-studied system is the Great Fish River (Scott et al 1972; Laurenson and Hocutt 1984; O'Keeffe and de Moor 1988).

E   The south-eastern coastal region — mostly short, precipitous, usually deeply incised rivers, in a high-rainfall region. The best-studied rivers are the Tugela River (Oliff 1963) and the Buffalo River (Palmer and O'Keeffe in press).

F   The Drakensberg Mountain region — torrential mountain streams which probably include many endemic species such as the Maluti minnow *Pseudobarbus quathlambae* (Skelton 1987), and the chironomid midge *Archaeochlus drakensbergensis* (Cranston et al 1987). Diatom ecology has been described by Schoeman (1971, 1972, 1973).

G   Highveld region — high rainfall in the east, low in the west. Most streams originate over Karoo sediments, but some originate on the Witwatersrand system and still others as springs in dolomite. Many river zones (see section below) are absent. Some systems have been intensively studied, in particular the Vaal River (Chutter 1963, 1967, 1972; Mulder 1971; de Moor 1982).

H   Eastern escarpment region — upper reaches of lowveld, subtropical rivers and rejuvenated reaches of highveld rivers. Fauna are thought to be similar to the foothill zones of region E, but with more tropical species (No studies).

J   Transvaal Mountain region — upper reaches of subtropical east- and north- flowing rivers. Invertebrates in the mountain streams of the Barberton area were studied by Hughes (1966). Endemic fish species include *Barbus treurensis* (Skelton 1987). Fish distribution and ecology in the region have also been described by Gaigher (1969) and Kleynhans (1979, 1980, 1983, 1986).

K   The Lowveld — subtropical rivers with a distinctive fauna. There are some recent and present studies by Pienaar (1978) and I Russell (pers comm) of the fish fauna, and by Moore and Chutter (1988) on the invertebrates. Extensive work on the Phongola River floodplain is summarized by Heeg and Breen (1982), and fish distributions are described by Pott (1969), Gaigher (1969) and Kleynhans (1986).

L   The Middleveld region — dry, low-gradient rivers, generally arising in regions G or J. The Jukskei/Crocodile was intensively studied by Allanson (1961) and Wilkinson (1979). Fish distributions are described by Kleynhans (1980, 1983, 1986).

M   The tropical-arid region — characterized by high temperatures, no permanent streams originate within the region. There are specialized endemics with tropical affinities, especially in the Okavango River, which has been studied by Fox (1976) and Skelton et al (1985). Fish distributions in the region are described by Kleynhans (1980, 1983, 1984, 1986).

Harrison's classification of hydrobiological regions has a low level of resolution. In a new classification exercise, O'Keeffe et al (unpublished) aim to use regional-level classifications (eg of climate and geology) to define the first level of a more sensitive hierarchical classification of rivers. The next two levels of the hierarchy, probably based on vegetation type and river zones, will be used to group similar stretches of river systems at a scale that will be useful to conservation and water managers.

River zones

Two main attempts have been made to classify the different environmental zones found along South African rivers. Harrison and Elsworth (1958) divided the Great Berg River into a series of zones from source to mouth, and later Harrison (1965) fitted these zones and data on other South African rivers to Illies' (1961) universal classification of river zones. Noble and Hemens (1978) presented a new approach, whereby all the country's rivers could be described, using five zones ranging from mountain stream to estuary. This system has been modified by O'Keeffe et al (unpublished), to give seven zones, some or all of which may occur in a river, and some of which may be repeated one or more times along a river. These zones are: mountain source, mountain stream, rocky foothill, sandy foothill, midland river, lowland river, and swamp; to these may be added the estuary.

Each of these zones has a distinct physical and hydrological character, and each upper zone, at least, has a distinct biota. The zones can be identified visually, although downstream changes from zone to zone tend to be gradual rather than abrupt. In general, the zones increase in length downstream, with the first two or three zones often being very short (for example in the 40-km Eerste River, the mountain stream zone is approximately 7 km long and the rocky foothill zone about 5 km long). The upper three zones (mountain source to rocky foothill) tend to support the rare, often endemic species, whereas more hardy cosmopolitan types occur in the lower zones. This is presumably because the environmental influences affecting cool, clean high-altitude streams are fewer and less complicated than those affecting mature rivers which are influenced by larger, more varied and more disturbed catchments. The higher zones, then, support species that are much less tolerant of, and more likely to be eradicated by, disturbance. These zones have frequently been invaded by cold-tolerant species, such as trout, introduced from the northern hemisphere (Bruton and Van As 1986).

Of the higher zones, the mountain source and mountain stream are often relatively undisturbed because of their protection in proclaimed mountain catchment areas. Generally, their biota is less threatened than that of any other zone, but as these zones are inhabited by many endemic and specialized species, the biota is likely to have low resilience to any changes. The most obvious exception to this is the high-altitude streams of the eastern escarpment, where commercial plantations may extend high into the mountains, and even to the source areas of the streams. This, however, does not presume that all such freshwater systems are not vulnerable. We do not know the biota of many freshwaters, or their sensitivities and resilience, and attention must be focused on them to ensure their maintenance.

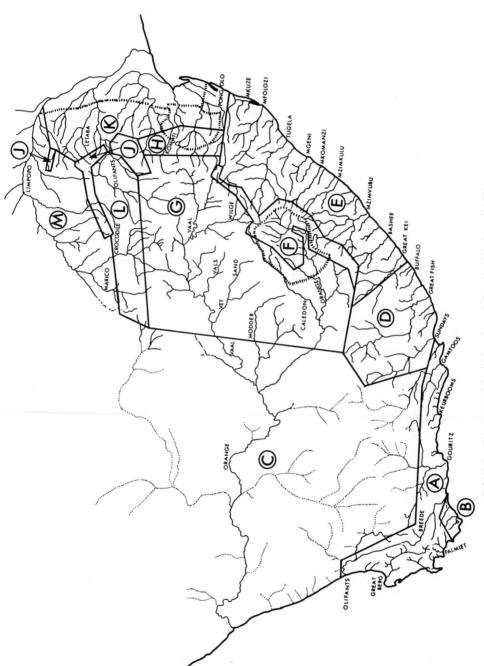

FIGURE 17.2 Hydrobiological regions defined by Harrison (1959)

The rocky foothill zone, on the other hand, is the most immediately vulnerable of our river zones. It supports a pollution-sensitive and often endemic fauna and yet represents what is usually the most suitable (geologically and hydrologically) reach for damming the river, and the highest reach beside which agricultural and urban development can take place.

This zone has been disturbed in river after river, causing the disappearance or at least the modification of distinctive aquatic communities. In the south-western Cape, at least, this must be regarded as an 'endangered zone'.

The lower zones cannot be ignored, either. Though not usually carrying a particularly rare or sensitive biota, they may be important migratory routes for fish, feeding and breeding grounds for waterfowl, and their riparian belts may be important sanctuaries for other forms of wildlife.

Biotic diversity

Oliff (1963) compared studies on invertebrate densities in the Tugela River with similar studies in Yorkshire (United Kingdom), Denmark, New Zealand, and Kenya, concluding that the productivity in the Tugela was not significantly different from that in other rivers. He also compared invertebrate density and diversity with those measured in the Berg River by Harrison and Elsworth (1958) and found similar numbers of animals, but great differences in species composition, which he attributed to zonal differences between the two rivers, and to the transitional nature of the Tugela, which falls between temperate (ie Berg River) and subtropical conditions.

Harrison (1978) identified two main groups of invertebrates in southern African streams:

— An old element of cold stenothermal species with Gondwanaland affinities, increasingly restricted to montane areas as a result of catchment perturbations.
— A pan-afro-tropical element of:

• widespread tolerant species, which can exploit almost all types of streams;

• warm stenothermal, tropical species, restricted to the Transvaal lowveld with a few species in the middleveld (1 000–1 300 m) and in the Natal coastal strip;

• highveld climate species, found in the highlands of Transvaal, Natal and the Orange Free State, but down to sea level in the Cape Province;

• montane species, typically restricted to permanent high-altitude cool streams, but extending to sea level in the Cape Province; and

• species of temporary mountain streams, showing a high degree of specialization; typical of Namibia and the drier parts of the Cape Province.

REGIONAL PATTERNS OF BIOTIC DIVERSITY

The indigenous fish species of southern Africa are the best known and surveyed component of the aquatic biota, owing mainly to the efforts of the provincial nature conservation agencies, the National Parks Board, and a few dedicated researchers. The distribution records are therefore sufficiently reliable for an examination of diversity patterns within the context of Harrison's (1959) hydrobiological regions (Figure 17.2). Using at least one example river for each hydrobiological region, Figure 17.3 maps numbers of indigenous species throughout southern Africa south of the Limpopo River.

It is immediately apparent from Figure 17.3 that species diversity declines from the tropical lowveld in the north-east, including the Mkuze River of northern Zululand, where over thirty species per river is the norm, to the relatively species-poor rivers of the central interior and the southern and south-western Cape, where ten species is the average complement for a river. This

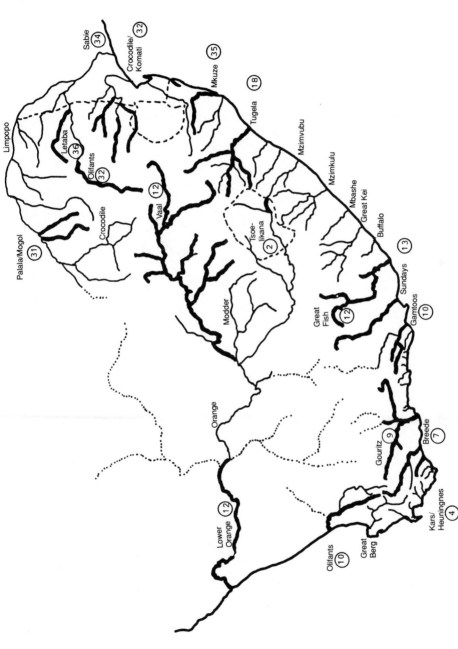

FIGURE 17.3 Species diversity of indigenous freshwater fish in selected rivers. Numbers refer to the number of species in each of the river systems marked by a thick line. Fish species diversity is shown for at least one river in each of Harrison's (1959) hydrobiological regions. Data from the following sources: Scott and Hamman (1984) for the Olifants (south-western Cape), Bree, Gouritz, Gamtoos, Kars/Heuningnes, Sundays and Great Fish rivers; Cambray (1984) lower Orange; Crass (1966) Tugela; Skelton et al (1988) Mkuze; Skelton (pers obs) Tsoelikana; Jubb (1967) Vaal; Gaigher (1969) Crocodile/Komati, Blyde/Treur, Sabie/Sand, Letaba, and upper Olifants (Limpopo) rivers; Kleynhans (1983) Palala/Mogol rivers

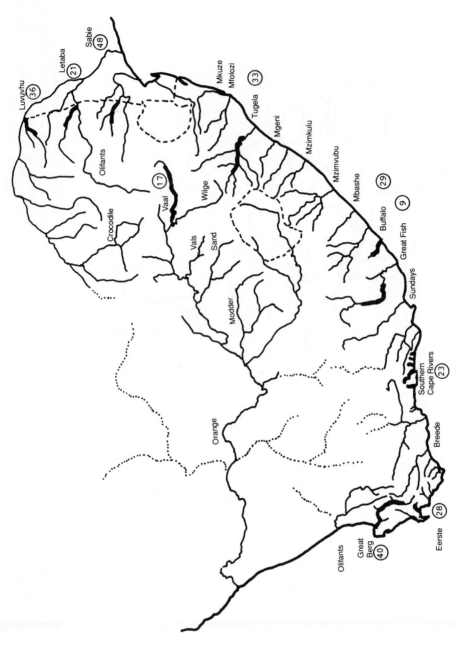

FIGURE 17.4 Species diversity for mayflies (Ephemeroptera) and caddis flies (Trichoptera) in selected rivers. To minimize the effects of different sampling regimes, the comparison is restricted to the stones-in-current biotope of the middle reaches of the rivers. Data from the following sources: Harrison and Elsworth (1958) for the Great Berg River; King (1983) Eerste; Harrison and Agnew (1962) southern Cape Rivers; O'Keeffe and de Moor (1988) Great Fish River; Palmer and O'Keeffe (unpublished) Buffalo; Oliff (1960) Tugela; Chutter (1967) Vaal; Moore and Chutter (1988) Sabie, Letaba and Luvuvhu rivers.

has previously been noted by Bowmaker et al (1978). This relatively low diversity is likely to be a result of the biogeographic isolation of the rivers of the southern Cape and the lower Orange River, an explanation that is supported by the high levels of endemicity in these systems. The Olifants River in the south-western Cape supports eight endemic fish species, and the lower Orange River one. None of the other rivers shown in Figure 17.3 supports more than one endemic fish species. A complicating factor is that the rivers of the southern and south-western Cape flow over leached, nutrient-poor Table Mountain sandstone, resulting in oligotrophic acid water, which may contribute to a low species diversity relative to the more productive tropical rivers. Species are also excluded at high altitudes, for example only two species are found in the Tsoelikana River, a high-altitude tributary of the upper Orange in the Drakensberg Mountains. Once again, geographic isolation may be a factor, and one of the two species, *Pseudobarbus quathlambae*, is endemic to these high mountain streams. In summary, Figure 17.3 shows high diversity in areas of low altitude and latitude, and low diversity in areas of high altitude and latitude. There is surprisingly little correlation between size of river system and number of fish species, although inclusion of the very large Zambezi River (156 fish species; Jackson (1986)) and the Okavango River (80 fish species; Skelton et al (1985)) in the analysis would indicate a positive size:diversity relationship.

The analysis of aquatic invertebrate diversity is more tentative because the taxonomy is incomplete. Surveys have been less frequent than for fish, and have often differed in method and intensity. We have therefore limited the analysis to two groups, the mayflies (Ephemeroptera) and caddis flies (Trichoptera), which have consistently been identified to species in the surveys used. Similarly, we have restricted the analysis to the stones-in-current biotope of the middle reaches of rivers surveyed so as to minimize the effects of different sampling effort in different surveys. The results are presented in Figure 17.4, and lead to conclusions different from those for fish distribution. Maximum species diversity was measured in the Sabie River (eastern Transvaal, 48 species) and the Great Berg River (south-western Cape, 40 species), with intermediate diversity in the Tugela (33 species) and Buffalo rivers (29 species). Lowest diversity was measured in the upper middle reaches of the Vaal River (17 species) and the middle reaches of the Great Fish River (9 species). Both these sections of river are characterized by long stretches of relatively homogeneous riverine habitat, consisting of long pools interspersed with occasional riffles. A tentative explanation for the patterns of diversity in Figure 17.4 may be that habitat diversity correlates with invertebrate diversity. As far as we are aware, no measurements of habitat diversity for southern African rivers have been attempted, but the middle reaches of the Sabie and Great Berg rivers seem from personal observation to provide more variable habitat than the Vaal and Great Fish rivers.

Diversity patterns are not simple to explain, but they are likely to be governed by different types of animals. Regional characteristics such as temperature and geographic isolation may be most important for animals such as fish, which are mobile within their ecosystems, but smaller-scale variables are likely to affect relatively immobile animals such as aquatic insect larvae, which have specific microhabitat requirements. The ecological consequences and important of natural biotic diversity are difficult to interpret. Biotic diversity *per se* should certainly not be seen as a measure of the importance of a system for conservation. For example, because the Great Fish River has 12 resident fish species, it is in no sense a more important conservation resource than the Olifants River (south-western Cape), which has only 10 species (Figure 17.3), 8 of which are endemic to that river system. The main conservation importance of biotic diversity is as an indicator of changes in the status of an ecosystem, and the next section investigates the evidence for such changes in southern African rivers over the past thirty years.

CHANGES IN BIOTIC DIVERSITY OVER TIME

Southern Africa has (or had) a rich diversity of river types and animal associations. To judge whether or not this diversity has been significantly reduced owing to man's impact is a difficult task. The invertebrate fauna of a number of rivers in South Africa were surveyed and then resurveyed after a sufficient interval to give an idea of degradation over time: Jukskei/Crocodile system (Allanson 1961; Wilkinson 1979); Berg River (Harrison and Elsworth 1958; Coetzer 1978); Buffalo River (Tugela System) (Oliff 1963; Fowles 1984); and Great Fish River (Allanson 1964 unpublished; O'Keeffe and de Moor 1988).

The first of these examples, the Jukskei/Crocodile, is of little use in a comparative study, since it was grossly polluted when Allanson (1961) reported on its invertebrate fauna, and was still grossly polluted during Wilkinson's (1979) resurvey. Wilkinson (1979) makes no comparisons of faunal diversity, but concludes that the types and extent of pollution had changed between the two surveys.

Coetzer (1978) concluded that the Berg River has become significantly more polluted since Harrison and Elsworth's (1958) survey. This conclusion was based on faunal changes as measured according to Chutter's (1972) Biotic Index. Annelids, ostracods, Chironomini, Orthocladiinae and caenid mayflies had all increased in abundance. Coetzer (1978) lists 60 invertebrate taxa from riffles, compared with 71 listed by Harrison and Elsworth (1958). Slightly different collecting methods, different levels of identification, and a less extensive sampling period preclude any conclusions from the small decrease in diversity.

Fowles (1984) was not able to show any reduction in invertebrate diversity in the Buffalo tributary of the Tugela River since Oliff's (1963) survey. He attempted to replicate Oliff's methods faithfully, and concluded that 'the results suggest only minor changes in water quality as indicated by the benthic fauna. At some stations improved water quality is suggested by a decrease in the Biotic Index value over the past 30 years' (Fowles 1984).

O'Keeffe and de Moor (1988) also replicated part of Allanson's 1964 survey of the invertebrates of the Great Fish River, with the intention of assessing the effects of the inter-basin transfer from the Orange River. The number of invertebrate taxa from riffles prior to the water-transfer scheme (41) was similar to the number after the scheme (47), but only 33% of the taxa were common to both surveys. In particular, the dominant species of Simuliidae, Chironomidae, and Trichoptera had changed, probably as a result of the transformation of the river's flow from seasonal to perennial.

The history of fish faunas of southern African rivers is better known than that of the invertebrates, owing to the observations of anglers, many of whom (such as R A Jubb and A C Harrison) were also expert naturalists. Although often unquantified, there are undoubted examples of local extinctions of fish species from many southern African river systems. Table 17.1 lists some of the better known examples. These alone amount to a probable loss of 25 populations in 14 river systems, but it is very likely that they form only a small proportion of the actual number of cases of local extinction. The impact of alien aquatic animals on indigenous fish populations is reviewed by Bruton and Van As (1986) and de Moor and Bruton (1988).

MAJOR THREATS TO SOUTHERN AFRICAN RIVERS AND THE CONSEQUENCES FOR THEIR BIOTA

River regulation: impoundment and inter-basin transfer of water

The flow of almost every river system in southern Africa has been regulated by one or more in-stream storage dams for agriculture, industry or domestic needs, or by structures associated with inter-basin (ie inter-catchment) transfers of water, from catchments with a surplus to those with

TABLE 17.1   Examples of local reduction or extinction of indigenous fish species in southern African rivers

1. Eerste River, south-western Cape:
Species now (apparently) extinct in this system
Pseudobarbus burgi
Sandelia capensis
(Skelton 1986b; 1987)

2. Berg River, south-western Cape:
Species eliminated from former range after introduction of smallmouth bass
Pseudobarbus burgi
Sandelia capensis
Barbus andrewi
(Harrison 1952; Skelton 1987)

3. Olifants River, western Cape:
Species reduced over much of their range
Pseudobarbus phlegethon
Barbus calidus
B capensis
B anoplus
B serra
Labeo seeberi
Austroglanis gilli
(Barnard 1943; Harrison 1963; Van Rensburg 1966; Jubb 1965, 1967; Gaigher 1973; Gaigher et al 1980; Skelton 1987)

4. Bree River, south-western Cape:
Species reduced over much of their range include
Pseudobarbus burchelli
Barbus andrewi
probably *Sandelia capensis*
(Skelton 1987; Gaigher et al 1980; Cambray and Stuart 1985)

5. Gamtoos River (also Kromme, Van Stadens and Swartkops rivers), southern Cape:
Species reduced over much of their range include
Pseudobarbus afer
Sandelia capensis
(Jubb 1959; Skelton pers obs)

6. Kowie River (and other eastern Cape rivers such as Kariega and Bushmans for *Myxus capensis*):
Species reduced over much of their range
Myxus capensis
Sandelia bainsii
(Bok 1979, 1983, 1984; Skelton 1987)

7. Buffalo and Keiskamma rivers, eastern Cape:
Species reduced over much of their range
Barbus trevelyaniani
Sandelia bainsii
(Gaigher 1975; Mayekiso 1986; Skelton 1987)

8. Umkomazana River, Natal:
Species locally extinct owing primarily to catchment changes
Pseudobarbus quathlambae
(Skelton 1987)

9. Orange River (Lesotho tributaries):
Species reduced after introduction of brown trout
Barbus aeneus
(Forson 1954)

10. Mooi River (Tugela), Natal:
Species reduced after introduction of trout
Amphilius natalensis
(Hey 1927; Crass 1964)

11. Blyde/Treur (Limpopo system), Transvaal:
Species reduced after introduction of bass and trout
Barbus treurensis
(Kleynhans 1987; Skelton 1987)

12. Apies River (Limpopo system), Transvaal:
Species disappeared from river probably owing to habitat alterations and pollution
Aplocheilichthys katangae
(Kleynhans 1986)

too little (Petitjean and Davies 1988a). To place this in perspective, there are 519 dams in South Africa with a capacity greater than 50 000 m³, which together impound half of the country's mean annual runoff (ie stream flow) (Department of Water Affairs 1986).

River regulation can have profound impacts on downstream river reaches in terms of flow (eg Foulger and Petts 1984), temperature (eg Ward and Stanford 1979), sediment regime (eg Simons 1979), channel morphology (eg Peiry 1987), and water chemistry (eg Krenkel et al 1979), as well as biotic community structure and functioning (eg Ward and Stanford 1979). Impacts vary enormously from one impoundment to another, depending on the way water is released downstream from the dam wall. Dams that release surface water downstream, for instance, will tend to increase stream temperatures, and hence to alter growth and the emergence of stream insects (thus affecting food chains — eg Ward 1985), whereas bottom-release systems in deep impoundments will tend to depress stream temperatures in receiving reaches, with opposite results.

South Africa is one of the world leaders in the technology of inter-basin transfers of water, but to date there is only a synthesis of available local and global information (Petitjean and Davies 1988a, b), and scattered items on the ecological consequences of these developments in the literature (Cambray and Jubb 1977; Laurenson and Hocutt 1984; O'Keeffe and de Moor 1988 — all on the Orange-Great Fish transfer scheme).

The main general effects of inter-basin transfers are to break down the natural barriers between catchments so that physical and biotic components are translocated and gene pools mixed or destroyed. The natural state of the system can never subsequently be assessed, and the effects on ecosystem processes are unknown.

Pollution

The assessment of the effects of organic pollution on South African rivers was an early focus of riverine research. There are a number of classic studies such as those of Harrison and Elsworth (1958) on the Great Berg River, Oliff (1960) on the Tugela River, Allanson (1961) on the Jukskei/Crocodile system and Chutter on the Vaal (1969a,b, 1971a,b), culminating in Chutter's Biotic Index (1972). The gross organic pollution problems associated with large cities tend to be fairly localized, but have dramatic effects on species diversity and ecosystem functioning. Probably more important to assess are the effects of moderate increases in organic loads, mostly from non-point sources, which are ubiquitous.

Other forms of pollution — acid rain, industrial heavy metals, mine dump effluents, herbicides and insecticides — are less well researched, although Toerien et al (1980) have described the effects of acidification from coal-mine dumps in the Olifants River. The potentially devastating synergistic effects of heavy metal, herbicide and insecticide pollution have not been researched in South Africa.

Catchment changes

Landuse changes on a large scale are characteristic of developing Africa. The development of large-scale agriculture, forestry and human settlement schemes has resulted in geographical changes on a broad scale throughout southern Africa. Overgrazing and high human populations have resulted in markedly accelerated rates of erosion in areas such as Natal-KwaZulu (Murgatroyd 1979), and Martin (1987) estimates that modern erosion rates in the Natal Valley are 12 to 22 times the geological average. Erosion may cause heavy siltation in streams and rivers with serious consequences for all forms of biota (Crass 1969; Chutter 1973). Not only are benthic organisms affected, but breeding, feeding and refuge areas for fishes are eliminated by sediments. The afforestation of upper catchments by introduced species such as *Pinus radiata* may reduce runoff by half (Wicht 1971), while clear-felling causes periodic catastrophic changes in runoff, sediment loads and organic input.

Alien biota

There are three categories of introduced biota which pose a threat to our river systems: animal — vertebrate (primarily fish), and invertebrate; large aquatic plants; and large terrestrial plant species which invade riparian zones.

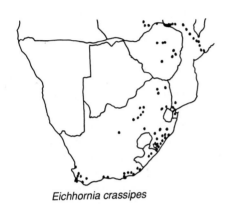

Eichhornia crassipes

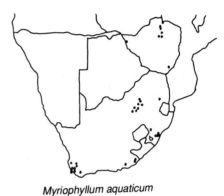

Myriophyllum aquaticum

Salvinia molesta

Azolla filiculoides

FIGURE 17.5 The distribution of the main aquatic problem plants in southern Africa. (Modified from Jacot Guillarmod 1979 and Ashton and Walmsley 1984.)

ANIMALS

De Moor and Bruton (1988) list 58 established species of introduced and translocated aquatic animals in South Africa, of which 41 are fish, 16 are invertebrates and one is a reptile (*Trachemys scripta*, the American terrapin). Potential problems associated with invasive animals may include habitat alteration, water-quality changes, the introduction of parasites and diseases, hybridization, and the elimination of indigenous species by predation or competition for resources. Introduced species such as carp *Cyprinus carpio*, bass *Micropterus dolomieui* and *M salmoides*, and trout *Oncorhynchus mykiss* and *Salmo trutta* may be a danger to as large a proportion as 60% of the threatened endemic freshwater fish in South Africa (Bruton and Van As 1986).

The literature on the impacts caused by introduced invertebrates is relatively small, but there are a number of species already at large in the region: Mollusca — *Helisoma duryi*, a planorbid (Appleton 1977); *Lymnaea columella*, (Appleton 1974); *Physa acuta*, (Appleton 1983); and Crustacea — *Charax tenuimanus*, (de Moor and Bruton 1988). The danger these species pose lies in their high reproductive potential and concomitant competition for resources, as well as the destructive predation by some species.

PLANTS

The devastating effects of invasive aquatic plants are well documented. The known distributions of the four major pest species are presented in Figure 17.5. Their adaptability to a wide range of conditions and their remarkably rapid vegetative growth allow them to colonize easily and then spread over a body of water extremely rapidly (eg Mitchell 1978). Their effects include habitat alteration, water-quality changes including deoxygenation below floating mats, waterway obstruction, provision of new habitat for other pest organisms (such as the vectors of schistosomiasis), competition for nutrients, and exclusion of light from the underlying water.

There is little research on the invasion of riparian zones by terrestrial alien plants, but it has been suggested that rivers lined by invasive acacias (*Acacia mearnsii* and *A longifolia*) may become relatively sterile, possibly as a result of the high tannin content of the leaf detritus of the *Acacia* (Skelton 1987). Other possible adverse effects include bank instability due to the replacement of deeper-rooted indigenous plants by relatively shallow-rooting invasives such as *Sesbania punicea, Acacia mearnsii, Populus* spp and *Salix babylonica*.

THE ASSESSMENT OF THE CONSERVATION STATUS OF RIVERS

A number of methods have been suggested for the assessment of the conservation status of rivers in many parts of the world. Savage and Rabe (1979) used geomorphological criteria, aquatic plants and invertebrates to classify streams in Idaho, USA. Blyth (1983) used physical and botanical features in Australian stream habitats, and Macmillan (1983), also in Australia, used a series of 'filters', including natural catchment vegetation, number of impoundments, mining activities, logging, road crossings, grazing, and the presence of exotic species, to screen out streams in terms of their conservation status. These systems produce broadly based indices of river status, but are inflexible in their response to special conditions, and in their information requirements.

Recent attempts have been made to develop a river conservation assessment scheme for South African rivers. The Nature Conservation Division of the Transvaal Provincial Administration is using an assessment system based on river zones in a provincial survey of river conservation status. Biotic and abiotic criteria are used to allocate zones to one of six classes, from pristine to totally modified (Kleynhans and Engelbrecht 1988). An experimental method has been developed using expert-system techniques which allow a variable input and response for specific rivers or stretches of rivers. The resulting computer program, known as the River Conservation System (RCS) (O'Keeffe et al 1987), has been tested on a number of rivers, providing useful results.

The RCS aims to assess the conservation status of a river or stretch of river, defined as 'a measure of the relative importance of the river for conservation, and the extent to which it has been disturbed from its natural state' (O'Keeffe et al 1987). The program poses up to 58 questions, covering all aspects of a river that may be relevant to its conservation. These questions are weighted according to their relative importance, and to their positive or negative value for conservation. Table 17.2 lists a selection of typical attributes on which questions are based. The questions and weightings were developed from a two-stage questionnaire sent to all river ecologists and conservationists in South Africa. A number of rules in the program may change weightings in response to specific conditions. For example, expectations of the diversity of the fish fauna will change with the size of river. The program calculates a size index based on mean annual runoff, stream order, stream length and catchment size, and reduces or increases the weighting for questions relating to the number of fish species in the river accordingly. Since few river systems in southern Africa have been researched in depth, there are often gaps and uncertainties in the available information. The RCS therefore asks for a maximum and minimum answer to each question, and also has a facility for dropping irrelevant questions from a particular assessment.

TABLE 17.2   Some of the 58 attributes of rivers used by the River Conservation System. Default weightings indicate the relative importance of each attribute in the assessment program. Positive and negative numbers indicate advantageous or disadvantageous attributes. These weightings may be changed within the program in response to special conditions in particular rivers.

Attribute	Default weighting
% of flow that is sewage effluent	-12
% unregulated river	+14
Number of mainstream dams	-8
Extent of rubbish	-6
Importance as a migration route	+5
% natural vegetation in the catchment	+18
Habitat diversity in the catchment	+11
Level of erosion in the catchment	-9
Mean population density in the catchment	-7
Number of endemic fish species	+17
Number of indigenous macrophytes	+12
Number of introduced fish species	-8
Importance of angling	-4

The RCS provides a global minimum score of 0 and a maximum of 100 for the river or stretch of river under consideration, and scores for the catchment, the river itself and the biota. The differences between the maxima and minima are a reflection of the level of information available about the river. The program also provides scores for each attribute, from which the most important positive or negative conservation aspects of the river can be identified. The final weighting, and the difference between the maximum and minimum answers for each question, are used to calculate a research-priority index indicating where further information is most needed.

There are still a number of flaws in the RCS, but it is nevertheless the most sophisticated tool for the assessment of conservation status in South Africa. Problems with the system include the following:

— There is no objective way of verifying the results of the program, except by comparing them with the opinion of an informed expert. The program is intended to emulate the judgement of such an expert.

— The attempt to assess both conservation importance and levels of degradation by a single method has not been successful, since they are not necessarily correlated variables.
— It appears from trial runs that very small streams are not assessed very realistically by the system. This may also be true for very large rivers, or for any extreme conditions in rivers.
— The workings of the program are far from simple, and require considerable understanding to achieve sensible results. The program should therefore not be used by anyone unfamiliar either with its workings or with the principles of river ecology.

The main successes of the RCS have been that it provides a consistent but flexible way of assessing rivers, and that it simplifies the diffuse intuitions and value judgements of conservationists in a way that is easily comprehensible to the target audience of engineers and water managers.

CONCLUSIONS

Fresh water is perhaps the major limiting resource for development in much of southern Africa. As a result, our rivers have been exploited and will continue to be exploited with increasing intensity. To attempt to maintain most river systems in a pristine condition is therefore unrealistic. Some parts of a few rivers may be maintained unmodified, but the major aim for river conservation should be to provide guidelines for the wise multiple use of a limited renewable resource. It follows that there are a number of questions for which quantitative answers must be provided by river ecologists, and these are:

— How much water can be abstracted from a river before its form and functions are seriously modified?
— How much effluent can be disposed of in a river before its self-purifying capacity is overwhelmed?
— How do catchment landuse changes affect the hydrology and water quality of rivers?
— What are the effects of impoundments and inter-basin transfers on the biota and ecological processes of rivers?
— What are the effects of introduced plants and animals on the indigenous biota of our rivers?

All but the last of these questions deal with ecosystem processes and habitat modification rather than biotic interactions. It has become obvious, during the writing of this review, that 'conserving biotic diversity' in rivers is a logical end point (rather than an aim) of a holistic conservation policy whose real aim must be to maintain habitat integrity and the efficient functioning of ecological processes. The meagre research cited shows little evidence of a decrease in biotic diversity, even in highly modified rivers. It does, however, show strong evidence of the replacement of indigenous and endemic species by translocated and alien species. This is a problem for nature conservationists who value the preservation of natural communities, but does it affect the efficiency of ecosystem processes such as nutrient cycling? Common sense indicates that a naturally adapted community would use resources more efficiently than a loose collection of introduced opportunists, but at present there is no research to confirm this view.

In practical terms, a holistic conservation plan for rivers must be supported at four levels: classification; monitoring; management; and research.

Classification has been discussed earlier. A map of subjective conservation status of southern African rivers has been produced by the Foundation for Research Development of the CSIR (O'Keeffe 1988). This map has five colour categories which represent states from pristine to badly degraded, and unknown status. It provides a starting point, but until we can classify our rivers and river zones in detail, management for different priorities will at best be haphazard.

Continuous long-term monitoring of conditions in rivers should aim to provide reliable data from which to assess changes in the conservation status of rivers, to develop guidelines for river management, and to design conservation research. Sensible management of rivers requires direction by a single authority for each catchment. Southern African rivers are at present subject to several layers and many different units of central and local government authority, and until a single co-ordinating authority manages a catchment, there will be little hope of preventing the overexploitation of river resources.

The preceding review has emphasized the depressing truth that knowledge of our rivers has progressed only locally since the original co-ordinated research of the 1950s and 1960s (Harrison and Elsworth 1958; Oliff 1960; Allanson 1961; Chutter 1967). This is primarily a result of the lack of resources (trained manpower, money and time) available in river research. Some of the questions listed earlier in this section are now being investigated, but many gaps remain.

Our concept of river conservation emphasizes the maintenance of essential ecological processes (such as water flow, nutrient cycling, material transport and vegetation maintenance). Perhaps the best illustration of the value of this concept of conservation comes from the examination of a river in which these processes have broken down. The Black River, flowing through Cape Town, is canalized for much of its length and receives the effluent from the Athlone sewage works. For most of the year its flow is 90% sewage effluent, and its biota is confined to oligochaete worms, hardy species of dipteran larvae, and microbes. It contains concentrations of faecal coliforms of up to 400×10^6 per 100 ml (City Engineer's Department 1985) compared with the recommended upper limit of the EEC for European rivers, which is 1 000 cells per 100 ml! The river supplies no water, its capacity for self-purification has been overwhelmed, it has no recreation or nature conservation potential, it is unsightly, it has an unpleasant odour, and it is a threat to public health. The only function of the river is to transport sewage to the sea, which could be done by a pipeline. The Black River exemplifies the conversion of a valuable natural resource into a public liability.

Most southern African rivers have not yet reached the degraded state of the Black River, but without a proper appreciation of their limits for exploitation, many will become less and less useful as their ecological functioning is broken down

ACKNOWLEDGEMENTS

The authors are very grateful to Dr Neels Kleynhans of the Transvaal Provincial Administration for his thorough review of the manuscript, and for many additional ideas and information. We would also like to thank the other, anonymous, referees for their constructive comments.

REFERENCES

ALEXANDER W J R (1985). Hydrology of low latitude Southern Hemisphere landmasses. In *Perspectives in Southern Hemisphere limnology*. (eds Davies B R and Walmsley R D) Junk, Dordrecht, Chapter 6.

ALLANSON B R (1961). Investigations into the ecology of polluted inland waters in the Transvaal. Part 1. *Hydrobiologia* **19(1)**, 1 – 76.

ALLANSON B R (1964). Lists of invertebrate samples from the Great Fish River. Unpublished lists. Institute for Freshwater Studies. Rhodes University, Grahamstown.

APPLETON C C (1974). A check-list of the flora and fauna of the Gladdespruit, Nelspruit district, eastern Transvaal. *Newsletter of the Limnological Society of Southern Africa* **22**, 49 – 58.

APPLETON C C (1977). The influence of abiotic factors on the distribution of biomphalaria pfeifferi (Krauss, 1848) (Mollusca: Planorbidae) and its life cycle in south-eastern Africa: A review of the role of temperature. *Journal of South African Biological Society* **18**, 43 – 55.

APPLETON C C (1983). Introduced aquatic and terrestrial molluscs in Natal. *Report to the Natal Parks Board.* 4 pp.

ASHTON P J and WALMSLEY R D (1984). The taxonomy and distribution of *Azolla* species in southern Africa. *Botanical Journal of the Linnean Society* **89**, 239 – 247.

BARNARD K H (1943). Revision of the indigenous freshwater fishes of the S.W. Cape region. *Annals of the South African Museum* **36(2)**, 101 – 262.

BLYTH J D (1983). Rapid stream survey to assess conservation value and habitats available for invertebrates. In *Proceedings of a workshop on survey methods for nature conservation. Adelaide, September 1983.* (eds Myers K, Margules C and Mustoe I) 343 – 375.

BOK A H (1979). The distribution and ecology of two mullet species in the eastern Cape, South Africa. *Journal of the Limnological Society of Southern Africa* **5**, 97 – 102.

BOK A H (1983). The demography, breeding biology and management of two mullet species (Pisces: Mugilidae) in the eastern Cape, South Africa. PhD thesis, Rhodes University, Grahamstown. 268 pp.

BOK A H (1984) Freshwater mullet in the eastern Cape — a strong case for fish ladders. *The Naturalist* **28**, 31 – 35.

BOWMAKER A P, JACKSON P B N and JUBB R A (1978). Freshwater Fishes. In *Biogeography and Ecology of southern Africa.* (eds Werger M J A and Van Bruggen A C) Junk, The Hague. pp 1181 – 1230.

BRUTON M N and VAN AS J G (1986). Faunal invasions of aquatic ecosystems in southern Africa, with suggestions for their management. In *The Ecology of Biological Invasions in South Africa.* (eds Kruger R J, Macdonald I A W and Ferrar A A) Oxford University Press. Cape Town. pp 47 – 62.

CAMBRAY J A (1984). Fish populations in the middle and lower Orange River, with special reference to the effects of stream regulation. *Journal of the Limnological Society of Southern Africa* **10(2)**, 37 – 49.

CAMBRAY J A, DAVIES B R and ASHTON P J (1986). The Orange River system. In *The Ecology of River Systems* (eds Davies B R and Walker K F) *Monographiae Biologicae* **60**, 89 – 122.

CAMBRAY J A and JUBB R A (1977). Dispersal of fishes via the Orange-Fish Tunnel, South Africa. *Journal of the Limnological Society of southern Africa* **3(1)**, 33 – 35.

CAMBRAY J A and STUART C T (1985). Aspects of the biology of a rare redfin minnow. *Barbus burchelli.* (Pisces, Cyprinidae), from South Africa. *South African Journal of Zoology* **20**, 155 – 165.

CHUTTER F M (1963). Hydrobiological studies on the Vaal River in the Vereeniging area. Part 1. *Hydrobiologia* **21 (1,2)**, 1 – 65.

CHUTTER F M (1967). Hydrobiological studies on the Vaal River. Unpublished PhD thesis, Rhodes University, Grahamstown.

CHUTTER F M (1969a). The distribution of some stream invertebrates in relation to current speed. *Internationale Revue der Gesamten Hydrobiologie* **54(3)**, 413 – 422.

CHUTTER F M (1969b). The effects of silt and sand on the invertebrate fauna of streams and rivers. *Hydrobiologia* **34(1)**, 57 – 76.

CHUTTER F M (1971a). Hydrobiological studies in the catchment of Vaal Dam, S. Africa Part 2. The effects of stream contamination on the fauna of stones-in-current and marginal vegetation biotopes. *Internationale Revue der Gesamten Hydrobiologie* **56(2)**, 227 – 240.

CHUTTER F M (1971b). Hydrobiological studies in the catchment of Vaal Dam, S. Africa Part 3. Notes on the Cladocera and Copepoda of stones-in-current, marginal vegetation and stony backwater biotopes. *Internationale Revue der Gesamten Hydrobiologie* **56(3)**, 497 – 508.

CHUTTER F M (1972). An empirical biotic index of the quality of water in South African streams and rivers. *Water Research* **6**, 19 – 30.

CHUTTER F M (1973). An ecological account of the past and future of South African rivers. *Newsletter of the Limnological Society of Southern Africa* **21**, 22 – 34.

CITY ENGINEER'S DEPARTMENT (1985). *Greening the city: storm water drainage report.* Town Planning Branch, City Engineer's Department, Cape Town City Council, 11 p.

COETZER A H (1978). The invertebrate fauna and biotic index value of water quality of the Great Berg River, Western Cape. *Journal of the Limnological Society of Southern Africa* **4(1)**, 1 – 7.

COETZER A H (1989). Hydrobiological report on the Olifants River (western Cape). Internal Report. Cape Provincial Administration, Nature Conservation Department.

CRANSTON P S, EDWARD D H D and COLLESS D H (1987). Archaeochlus Brundin: a midge out of time (Diptera: Chironomidae). *Systematic Entomology* **12**, 313 – 334.

CRASS R S (1964). *Freshwater fishes of Natal.* Shuter and Shooter, Pietermaritzburg.

CRASS R S (1966). Features of freshwater fish distribution in Natal and a discussion of controlling factors. *Newsletter of the Limnological Society of Southern Africa* **7**, 31 – 35.

CRASS R S (1969). The effects of land use on freshwater fish in South Africa, with particular reference to Natal. *Hydrobiologia* **34**, 38 – 56.

CUMMINS K W (1979). The natural stream ecosystem. In *The Ecology of Regulated Streams.* (eds Ward J V and Stanford J A) Plenum Press, New York, NY. pp 7 – 24.

DAVIES B R and DAY J A (1986). *The Biology and Conservation of South Africa's Vanishing Waters.* Published through the Centre for Extra-mural Studies, University of Cape Town, Cape Town. 186 pp.

DAY J A, DAVIES B R and KING J M (1986). Riverine ecosystems. In The conservation of South African rivers. (ed O'Keeffe J H) *South African National Scientific Programmes Report* **131**, CSIR, Pretoria. Chapter 1.

DE MOOR F C (1982). A community of *Simulium* species in the Vaal River near Warrenton. PhD thesis, University of Witwatersrand, Johannesburg.

DE MOOR I J and BRUTON M N (1988). Atlas of alien and translocated indigenous aquatic animals in southern Africa. *South African National Scientific Programmes Report.* **144**. CSIR, Pretoria. 310 pp.

DEPARTMENT OF WATER AFFAIRS (1986). *Management of the Water resources of the Republic of South Africa.* Published by the Department of Water Affairs. Pretoria.

FERRAR A A(1989). Ecological flow requirements for South African Rivers. *South African National Scientific Programmes Report* **162**. CSIR, Pretoria. 118 pp.

FORSON R (1954). Seven brown trout founded a fisherman's paradise in the Basutoland wilds. *Piscator* **32**, 116 – 119.

FOULGER T R and PETTS G E (1984). Water quality implications of artificial flow fluctuations in regulated rivers. *Science of the Total Environment* **37**, 177 – 185.

FOURIE J M (1976) Mineralisation of western Cape Rivers: an investigation into the deteriorating water quality related to drainage from cultivated lands along selected catchments, with special reference to the Great Berg River. Unpublished PhD thesis.

FOWLES B K (1984). Chemical and Biological resurvey of the rivers of the Tugela basin. Unpublished report for NIWR, CSIR, Pretoria.

FOX P J (1976). Preliminary observations on fish communities of the Okavango Delta. In *Proceedings of the Symposium on the Okavango Delta and its Future Utilisation*. Published by the Botswana Society, Gaberone, pp 125 – 130.

GAIGHER C M (1973). Voorlopige verslag: Die Status van die inheemse vis in die Olifants Rivier. Internal report, Cape Department of Nature and Environmental Conservation. 18 pp.

GAIGHER I G (1969). Aspekte met betrekking tot die ekologie, geografie en taksonomie van varswatervisse in die Limpopo- en Incomatiriviersisteem. PhD thesis, Rand Afrikaans University, Johannesburg.

GAIGHER I G (1975). The ecology of a minnow, *Barbus trevelyani* (Pisces), in the Tyume River, eastern Cape. *Annals of the Cape Provincial Museums (Natural History)* **11(1)**, 1 – 19.

GAIGHER I G, HAMMAN K C D and THORNE S C (1980). The distribution, conservation status and factors affecting the survival of indigenous freshwater fishes in the Cape Province. *Koedoe* **23**, 57 – 88.

GORE J A (1978). A technique for predicting in-stream flow requirements of benthic macroinvertebrates. *Freshwater Biology* **8**, 141 – 151.

GORE J A and JUDY R D (1981). Predictive models of benthic macroinvertebrate density for use in instream flow studies and regulated flow management. *Canadian Journal of Fisheries and Aquatic Science* **38**, 1363 – 1370.

HARRISON A C (1952). Sandelia species of the Cape of Good Hope. *Piscator* **23**, 82 – 91.

HARRISON A D (1959). General statement on South African hydrobiological regions. *National Institute for Water Research Report* **1**, Project 6.8H.

HARRISON A D (1965). Geographical distribution of riverine invertebrates in Southern Africa. *Archiv für Hydrobiologie* **61**, 387 – 394.

HARRISON A D (1963). The Cape Province. In *Freshwater Fish and fishing in Africa*. Nelson, Johannesburg. 54 pp.

HARRISON A D (1978). Freshwater Invertebrates (except molluscs). In *Biogeography and Ecology of Southern Africa*. (ed Werger M J A) Junk, The Hague, Chapter 34.

HARRISON A D and AGNEW J D (1962). The distribution of invertebrates endemic to acid streams in the western and southern Cape Province. *Annals of the Cape Provincial Museums* **2**, 273 – 291.

HARRISON A D and ELSWORTH J F (1958). Hydrobiological studies on the Great Berg River, Western Cape Province Part I. A general description, chemical studies on the water and main features of the fauna and flora. *Transactions of the Royal Society of South Africa* **34** 125 – 226.

HEEG J and BREEN C M (1982). Man and the Pongolo floodplain. *South African National Scientific Programmes Report* **56**, 117 pp.

HEY S A (1927). Fisheries survey 1925 – 1926. *Inland Waters Report* **1**, Department Mines and Industries, Union of South Africa.

HUGHES D A (1966). Mountain streams of the Barberton area, Eastern Transvaal. Part 1. A Survey of the Fauna. *Hydrobiologia* **27**, 401 – 438.

ILLIES J (1961). Versuch einer allgemein biozonatischen Gleiderung der Fliessgewasser. *Internationale Revue der Gesamten Hydrobiologie* **46**, 205 – 213.

JACKSON P B N (1986). Fish of the Zambezi System. In *The Ecology of River systems*. (eds Davies B R and Walker K F) *Monographie Biologike* **60**, pp 269 – 288.

JACOT GUILLARMOD A (1979). Water weeds in southern Africa. *Aquatic Botany* **6**, 377 – 391.

JUBB R A (1959). The plump redfin (rooivlerke) *Barbus* asper Boulenger 1911. *Piscator* **45**, 34 – 35.

JUBB R A (1965). Freshwater fishes of the Cape Province. *Annals of the Cape Provincial Museum* **4**, 1 – 72.

JUBB R A (1967). *Freshwater Fishes of Southern Africa*. Balkema, Cape Town. 247 pp.

KING J M (1981). The distribution of invertebrate communities in a small South African river. *Hydrobiologia* **83**, 43 – 65.

KING J M (1983). Abundance, biomass and diversity of benthic macro-invertebrates in a western Cape river, South Africa. *Transactions of the Royal Society of South Africa* **45**, 11 – 34.

KING J M, DAY J A, HURLY P R, HENSHALL-HOWARD M P and DAVIES B R (1989). Macroinvertebrate communities and environments in a Southern African mountain stream. *Canadian Journal of Fisheries and Aquatic Sciences* **45(12)**, 2168 – 2181.

KLEYNHANS C J (1979). The distribution and status of *Kneria auriculata* (Pellegrin) (Pisces: Kneriidae) in the Transvaal. *Journal of the Limnological Society of Southern Africa* **5**, 27 – 29.

KLEYNHANS C J (1980). A checklist of the fish species of the Crocodile and Matlabas Rivers (Limpopo System) of the Transvaal. *Journal of the Limnological Society of Southern Africa* **6**, 33 – 34.

KLEYNHANS C J (1983). A checklist of the fish species of the Mogol and Palala Rivers (Limpopo System) of the Transvaal. *Journal of the Limnological Society of Southern Africa* **9**, 29 – 32.

KLEYNHANS C J (1984). Die verspreiding en status van sekere seldsame vissoorte van die Transvaal en die ekologie van sommige spesies. DSc thesis, University of Pretoria, Pretoria.

KLEYNHANS C J (1986). The distribution, status and conservation of some fish species of the Transvaal. *South African Journal of Wildlife Research* **16**, 135 – 144.

KLEYNHANS C J (1987). A preliminary study of aspects of the ecology of a rare minnow *Barbus treurensis* Groenewald, 1958 (Pisces: Cyprinidae) from the eastern Transvaal, South Africa. *Journal of the Limnological Society of Southern Africa* **18**, 7 – 13.

KLEYNHANS C J and ENGELBRECHT J S (1988). A preliminary assessment of the conservation status and fish populations of the Olifants River (Limpopo System), Transvaal. Unpublished report of the Transvaal Nature Conservation Division. 31 pp.

KRENKEL P A, LEE G F and JONES R A (1979). Effects of TVA impoundments on downstream water quality and biota. In *The Ecology of Regulated Streams*. (eds Ward J V and Stanford J A) Plenum Press, New York, NY, pp 289 – 306.

LAURENSON L J B and HOCUTT C H (1984). The introduction of the rock catfish, *Gephyroglanis sclateri*, into the Great Fish River via the Orange/Fish Tunnel, South Africa. *The Naturalist* **28(1)**, 12 – 15.

MACMILLAN L A (1983). A method for identifying small streams of high conservation status. In *Proceedings of a Workshop on Survey Methods for Nature Conservation, Adelaide, September 1983*. (eds Myers K, Margules C and Mustoe I) pp 139 – 342.

MARTIN A K (1987). Comparison of sedimentation rates in the Natal Valley, South-West Indian Ocean, with modern sediment yields in east coast rivers of southern Africa. *South African Journal of Science* **83**, 716 – 724.

MAYEKISO M L du T (1986). Some aspects of the ecology of the eastern Cape Rockey *Sandelia bainsii* (Pisces, Anabantidae) in the Tyume River, eastern Cape, South Africa. Unpublished MSc thesis, Rhodes University, Grahamstown.

MIDDLETON B J, LORENTZ S A, PITMAN W V and MIDGLEY D C (1981). *Surface water resources of South Africa*. Volume V. Drainage regions MNPQRST The eastern Cape. Part 2 (Appendices). Report 12/81. Hydrological Research Unit, University of the Witwatersrand, Johannesburg.

MITCHELL D A (1978). Freshwater plants. In *Biogeography and Ecology of Southern Africa*. (ed Werger M J A) Junk, The Hague. pp 1113 – 1138.

MOORE C A and CHUTTER F M (1988). A survey of the conservation status and benthic biota of the major rivers of the Kruger National Park. *Contract Report of the National Institute for Water Research*. CSIR, Pretoria.

MULDER P F S (1971). 'n Ekologiese studie van die hengelvisfauna in die Vaalriviersisteem met spesiale verwysing na *Barbus kimberleyensis* Gilchrist en Thompson. PhD thesis, Rand Afrikaans University, Johannesburg.

MURGATROYD A L (1979). Geologically normal and accelerated rates of erosion in Natal. *South African Journal of Science* **75**, 395 – 396.

NOBLE R G and HEMENS J (1978). Inland water ecosystems in South Africa — A review of research needs. *South African National Scientific Programmes Report* **34**, CSIR, Pretoria. 150 pp.

O'KEEFFE J H (1988). (ed and compiler). *The Conservation Status of South African rivers*. 1:2,500,000 map printed by Surveys and Mapping, Cape Town.

O'KEEFFE J H, BYREN B A , DAVIES B R and PALMER R W (in press a). The effects of impoundment on the physico-chemistry of two contrasting southern African river systems. *Regulated Rivers: Research and Management*.

O'KEEFFE J H, DANILEWITZ D B and BRADSHAW J A (1987). An expert system approach to the assessment of the conservation status of rivers. *Biological Conservation* **40**, 69 – 84.

O'KEEFFE J H and DE MOOR F C (1988). Changes in the Physico-Chemistry and benthic invertebrates of the Great Fish River, South Africa following an interbasin transfer of water. *Regulated Rivers: Research and Management* **2**, 39 – 55.

O'KEEFFE J H, NEILSON I, SKELTON P H, KLOPPERS J, NEL F, VON DEM BUSCHE G, RUSSELL I, and GROBBELAAR G (in press). The Luvuvhu River — critical flow needs. *Proceedings of a workshop on "Flow Requirements of Kruger National Park Rivers and impact of proposed water resources development"*. Department of Water Affairs, Pretoria.

OLIFF W D (1960). Hydrobiological studies on the Tugela River System. Part 1. The main Tugela System. *Hydrobiologia* **14**, 281 – 392.

OLIFF W D (1963). Hydrobiological studies on the Tugela River system. Part III. The Buffalo River. *Hydrobiologia*. **21(3-4)**, 355 – 379.

PALMER R W and O'KEEFFE J H (in press b). Downstream effects of impounds on the water chemistry of the Buffalo River, South Africa. *Hydrobiologia*.

PEIRY J-L (1987). Channel degradation in the Middle Arve River, France. *Regulated Rivers: Research and Management* **1**, 183 – 188.

PETITJEAN M O G and DAVIES B R (1988a). Ecological impacts of inter-basin water transfers: some case studies, research requirements, and assessment procedures in southern Africa. A review of the ecological and environmental impact of inter-basin water transfer schemes in southern Africa. Synthesis (Part 1) and International Bibliography (Part 2). *Occasional report* **38**. FRD, Pretoria. 106 pp.

PETITJEAN M O G and DAVIES B R (1988b). A bibliography of inter-basin water transfer literature, with special reference to southern Africa and other arid regions of the world. *Report to Foundation for Research Development*. CSIR, Pretoria. 73 pp.

PIENAAR U DE V (1978). *The freshwater fishes of the Kruger National Park*. National Parks Board, Sigma Press, Pretoria.

POTT R McC (1969). The fish life on the Pongolo River and the effect of the erection of a dam on the fish populations. MSc thesis, University of the Witwatersrand, Johannesburg.

PREWITT C G and CARLSON C A (1980). Evaluation of four instream flow methodologies used on the Yampa and White Rivers, Colorado. Bureau of land management, *Biological Sciences Series* **2**, Denver Colo. 65 pp.

ROBERTS C P R (1981). Environmental considerations of water projects. *Department of Water Affairs Technical Report* **114**.

ROBERTS C P R (1983). Environmental constraints on water resources development. *The S.A. Institution of Civil Engineers 7th Quinquennial Convention*.

SAVAGE N L and RABE R W (1979). Stream types in Idaho: an approach to classification of streams in natural areas. *Biological Conservation* **15**, 301 – 315.

SCHOEMAN F R (1971). A hydrobiological study of the Diatomaceae of Lesotho. DSc thesis. University of Pretoria, Pretoria.

SCHOEMAN F R (1972). A hydrobiological study of the Diatomaceae of Lesotho with special reference to the water quality. *Civil Engineer in South Africa* **14(2)**.

SCHOEMAN F R (1973). *A systematical and Ecological study of the diatom Flora of Lesotho with special reference to the water quality*. V and R Printers, Pretoria.

SCOTT K M F, ALLANSON B R, and CHUTTER F M (1972). Orange River Project, working group for ORP hydrobiology of the Fish and Sundays Rivers. CSIR NIWR *Research Report* **306**, 461 pp.

SCOTT A H and HAMMAK C D (1984). Freshwater fishes of the Cape. *Cape Conservation Series* **5**.

SIEGFRIED W R and DAVIES B R (ed 1982). Conservation of ecosystems: theory and practice. *South African National Scientific Programmes Report* **61**, CSIR, Pretoria. 97 pp.

SIMONS D B (1979). Effects of stream regulation on channel morphology. In *The Ecology of Regulated Streams*. (eds Ward J V and Stanford J A) Plenum Press, New York, NY, pp 95 – 112.

SKELTON P H (1986a). The fishes of the Orange River. In *The Ecology of River Systems*. (eds Davies B R and Walker K F) *Monographiae Biologicae* **60**, 143 – 162.

SKELTON P H (1986b). The impact of trout and other introduced predatory fishes on indigenous fishes in South Africa. In Trout in South Africa. (eds Skelton P H and Davies M T T) *Ichthos Special Edition* **1**, 3 – 5.

SKELTON P H (1987). South African Red Data Book — Fishes. *South African National Scientific Programmes Report* **137**, 199 pp.

SKELTON P H, BRUTON M N, MERRON G S and VAN DER WAAL B C (1985).The fishes of Okavango drainage system in Angola, S W Africa and Botswana: taxonomy and distribution. *Ichthyological Bulletin* **50**, 23 pp.

SKELTON P H, WHITFIELD A K and JAMES N P E (1988). A survey of the fishes of the Mkuze Swamps, Maputaland, South Africa. J L B Smith Institute of Ichthyology. *Investigational Report* **27**, 31 pp.

TOERIEN D F, PIETERSE A J H, BARNARD J H, VAN DER SPUY P and ROOS J C (1980). Die invloed van bestaande en beoogde ontwikkeling op die wateromgewing van die bo-Olifantsopvanggebied, Transvaal, *Report for the Institute for Environmental Sciences.* University of the Orange Free State, Bloemfontein.

VANNOTE R L, MINSHALL G W , CUMMINS K W, SEDELL J R and CUSHING C E (1980). The River Continuum Concept. *Canadian Journal of Fisheries and Aquatic Sciences* **37**, 130 – 137.

VAN RENSBURG K J (1966). Die vis van die Olifantsrivier (weskus) met spesiale verwysing na die geelvis (*Barbus capensis*) en saagvin (*Barbus serra*). Cape Provincial Administration, Department of Nature Conservation. *Research Report* **10**, 14 pp.

WARD J V (1985). Thermal characteristics of running waters. *Hydrobiolgia* **125**, 31 – 46.

WARD J V, DAVIES B R, BREEN C M, CAMBRAY J A, CHUTTER F M, DAY J A, DE MOOR F C, HEEG J, O'KEEFFE J H and WALKER K F (1984). Limnological criteria for management of water quality in the southern hemisphere. Chapter 2. Stream Regulation. *South African National Scientific Programmes Report* **93**. CSIR, Pretoria. 32 – 63.

WARD J V and STANFORD J A (1979). Limological considerations in reservoir operation: Optimization strategies for protection of aquatic biota in the receiving stream. In *Proceedings of the Mitigation Symposium, U.S. Dept. Agriculture, Ft. Collins, U.S.A.* pp 496 – 501.

WARD J V and STANFORD J A (1983). The serial discontinuity concept of lotic ecosystems. In *Dynamics of Lotic Ecosystems.* (eds Fontaine T D and Bartell S M) Ann Arbor Science Publishers, pp 29 – 42.

WICHT C L (1971). The influence of vegetation in South African Mountain catchments on water supplies. *South African Journal of Science* **67(3)**, 201 – 209.

WILKINSON R C (1979). The indicator value of the stones-in-current fauna of the Jukskei-Crocodile River system. Transvaal, South Africa. MSc thesis, University of the Witwatersrand, Johannesburg.

WINTERBOURN M J, ROUNICK, J R and COWIE B (1983). Are New Zealand stream ecosystems really different? *New Zealand Journal of Marine and Freshwater Research* **15**, 321 – 328.

WORLD BANK (1980). *World Development Report.* Oxford University Press, New York, NY.

CHAPTER 18

The conservation status of southern African estuaries

A E F Heydorn

THE VARIABLE CHARACTER OF THE SOUTHERN AFRICAN COASTLINE

The nature and hence also the biotic diversity of estuarine systems is directly related to the characteristics of the coastal environment in which they occur. This is particularly evident on the southern part of the African continent in which sharply contrasting coastal conditions occur. The variable characteristics of the South African coastline were described by Heydorn and Flemming (1985) as follows (see also Figure 18.1).

The South African coastline, between the Orange River in the west and Ponta do Ouro in the east, covers a distance of about 3 000 km. This includes the coasts of the two independent states, Transkei, between latitudes 31° 04'S and 32° 04'S (250 km), and Ciskei, between 32° 40'S and 33° 30'S (150 km). To the west of Cape Agulhas the coastline borders on the South Atlantic, and to the east on the Indian Ocean. The east coast waters are characterized by the warm waters of the southward-flowing Agulhas Current, those of the west coast by sporadic upwelling of cold, nutrient-rich waters typical of the Benguela Current regime. Along the south-west and south coasts extensive mixing of water masses occurs.

These oceanographic conditions determine the coastal climate of southern Africa, with summer rainfall along the east coast, bimodal rainfall along the south coast, winter rainfall along the south-west coast, and semi-arid conditions along the west coast. The variability of these environmental conditions is reflected in the composition of animal and plant communities both on land and in the sea. For example, a wide variety of Indian Ocean corals harbouring a diverse Indo-pacific fish fauna, occur in the subtropical waters of the northern east coast, whereas dense kelp beds provide a habitat for commercially important rock lobster and abalone stocks in the cool upwelling regime of the west and south-west coast. In general the east coast waters are characterized by great biotic diversity, while the main focus of commercial fisheries is centred in the more productive waters of the south-west and south coasts, where fewer species occur in greater profusion.

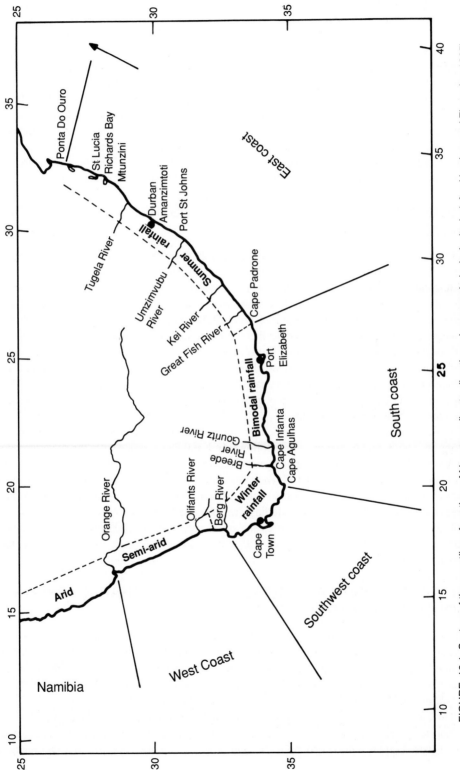

FIGURE 18.1   Sectors of the coastline of southern Africa according to climatic and geomorphological criteria (after Heydorn and Flemming 1985)

Besides climatic conditions, the geomorphology of coastal regions can also have a profound influence on the biotic diversity of estuarine systems. Because of the relatively steep seaward tilting of Natal and the northern Transkei and heavy summer rainfall, the rivers there flow at a greater velocity over a shorter distance, have greater cutting action, and tend to carry a bigger silt load while their estuaries are more dependent on the filtering action of their floodplains. Rivers in the Tsitsikamma/Outeniqua/Langeberg area also tend to be short and steep. However, they generally drain hard Table Mountain group sandstone/quartzite formations with little unbound soil cover and therefore carry a relatively small silt load. There are also profound differences between the vegetation surrounding the estuaries in subtropical Natal, the more temperate south coast, and the arid west coast. Similarly, substantial differences in faunal composition occur. This is illustrated even more vividly by the 'estuaries' of Namaqualand and Namibia, which are dry river courses or riverbed/delta areas that carry water only at times of exceptional rainfall.

ESTUARINE CONFIGURATION AND DISTRIBUTION IN SOUTHERN AFRICA

Under normal conditions an estuary is defined as that portion of a river system entering the sea where there is, within the boundaries of the land, a gradual transition in physical, chemical and biological characteristics from fresh water to sea water. The influence of sea water determines the nature of the vegetation, for example in mangrove swamps or halophytic plant communities. The term river mouth is used in the context of a river flowing into the sea in such a manner that the sea water does not mix with fresh water in the mouth region (eg Tugela and Orange Rivers). During a flood an estuary can be temporarily transformed into a river mouth. Coastal lakes are connected with the sea via one or more narrow channels (eg the Kosi Lake System, St Lucia and the Wilderness Lakes). Some coastal lakes have lost their connection with the sea (eg Lake Sibayi and Groenvlei). The term coastal lagoon is generally used in the context of an estuary characterized by a predominantly closed mouth.

If estuaries are considered to include river mouths, coastal lakes and lagoons, their distribution along the southern African coast is as follows: in Natal 73 estuaries enter the sea over a distance of 570 km; Transkei has 56 estuaries over a distance of 430 km; in the Cape Province between the Kei River and the Orange River 167 rivers enter the sea over a distance of about 2 000 km; and in Namibia, 20 river courses (mostly dry) enter the sea over a distance of about 1 000 km. The total area of estuaries in South Africa (including Transkei and Namibia) has been estimated at between 500 and 600 km^2. The extent of estuaries in Natal is approximately 400 km^2. It is clear that the east coast of the country is better endowed with estuaries than the south and, particularly, the west coast.

BIOTIC DIVERSITY AS A FUNCTION OF THE CHARACTERISTICS OF SOUTH AFRICAN ESTUARIES

The biotic diversity of southern African estuaries is obviously closely related to the characteristics of the coastal sector with which they are associated, and it has already been pointed out that, in general, east coast waters are characterized by a greater biotic diversity than those of the west coast. Synthesis reports on Natal and Cape estuaries have been published by Begg (1978, 1984a), Heydorn and Grindley (1981 – 85) and Heydorn and Morant (1986 – ongoing).

Differences in the biotic diversity of estuaries in their natural state are to be expected, since each estuarine system fulfils a specific ecological role in the coastal sector in which it is situated. This role is based on the natural driving forces and processes taking place within each system and, if biotic diversity is to be maintained at a high level, maintenance of these natural and functional processes is of the utmost importance.

What then are these forces and processes?

— Physical forces of tides, waves, wind and currents are responsible for water and sediment movement along the coast, including accretion and erosion of beaches and dunes, estuary mouths and in some cases floodplains and coastal lowlands. River flow fulfils a similar function from the land side.

— Mixing processes, especially in estuaries and the tidal sections of rivers where:

• fresh water from land and sea water mix;

• sediments of fluvial and marine origin mix, border upon or overlie each other, depending on the dynamic processes operative in the estuary or river system concerned; and

• nutrients of terrestrial, estuarine or marine origin interact.

— Chemical processes both in the sea and in estuaries which are dependent on the availability of nutrients, light penetration in the water, and fluxes both in the water column and between the water column and bottom sediments. Obviously the chemical processes relate closely to the mixing processes already mentioned.

— Biological processes which usually represent a response to the above-mentioned processes and which lead to plant and animal successions and the formation of communities and foodwebs which may be benthic, pelagic, aquatic or terrestrial. The biological processes are complex and include aspects such as primary production, grazing, predation, secondary production, decomposition, detritus formation, energy flow and nutrient cycling (Knox 1986). Figure 18.2 (after Berjak et al 1977), is a schematic representation of an estuarine foodweb, with man as a component. Note that man is the only predator exploiting every trophic level.

Breakdown of the processes summarized above is obviously deleterious to the ecological functioning of the estuary concerned, and one of the consequences will be a reduction in biotic diversity. If, for example, this involves a breakdown in plant communities that provide specialized habitats for microfauna upon which crustaceans and fish feed, higher trophic levels such as birds or mammals may be affected. If the fauna are components of marine recruitment stocks, there may be disruption of marine foodwebs as well (De Freitas 1984; Wallace et al 1984). Similar chain reactions may occur if light penetration in estuarine waters is reduced or if chemical characteristics such as salinity regimes are altered.

Although there will be little argument about the gravity of such ecological disruption, there is a conceptual problem in recognizing that it is actually taking place. This is usually not the case when a natural or man-induced catastrophe occurs, but the recognition of disruption is extremely difficult when slow and insidious changes take place, for example as a result of a progressive reduction of freshwater inflow into an estuary due to increasing water abstraction in the catchment or to progressive destruction or infilling of peripheral estuarine wetlands. This is a field that deserves far more attention from researchers. The community structure studies in Natal estuaries, utilizing multivariate data analysis techniques by Begg (1984b) and the habitat evaluation procedures (HEP) developed by Schamberger and Kumpf (1980), point in the right direction. Given adequate data, HEP provide a methodology to supply quantitative evaluation of baseline habitat conditions from which various changes through time can be predicted and tested. The development of a 'degradation index' for Natal estuaries takes this kind of work further. (A E L Ramm pers comm) and is therefore highly relevant.

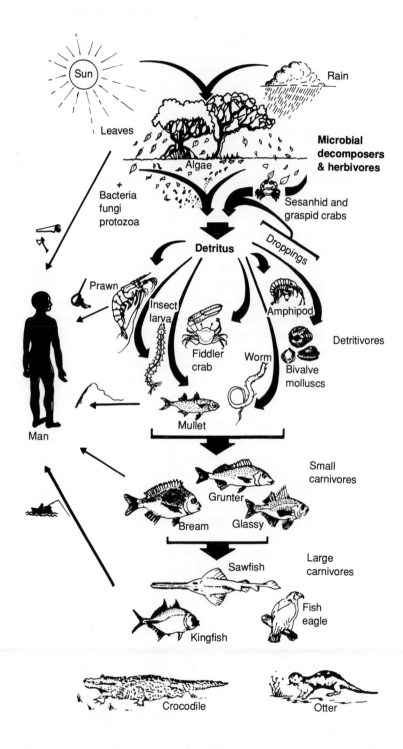

FIGURE 18.2   Diagrammatic depiction of trophic levels within an estuarine ecosystem (modified after Berjak et al 1977)

HUMAN ACTIVITY AND THE STATE OF SOUTHERN AFRICAN ESTUARIES

As estuaries exist in a state of dynamic equilibrium and are places of constant interaction between land and sea, they will obviously be affected not only by human activity in their immediate environs but also by man-induced changes in catchments and in the sea. Figure 18.3 illustrates this point. The following activities frequently have far-reaching effects on estuarine environments.

— excessive water abstraction from catchments, which deprives estuaries of the freshwater input they require, and in particular of the regular high-flow conditions required for maintenance of the important scour mechanisms of the mouth;

— agricultural practices that lead either to accelerated soil erosion and thus silt deposition in estuaries (eg through overgrazing, the ploughing of steep slopes, or destruction of the vegetation that binds river banks), or to encroachment on floodplains (eg by the planting of sugar cane, as in many places in Natal);

— residential or industrial developments in the immediate environment of estuaries, which either reduce the size and natural functioning of floodplains or require premature opening of sand-bars to avoid flooding as a result of natural rises in water level (eg at Wilderness and Swartvlei);

— road and rail construction where swamps and floodplains are infilled or bisected by embankments not capable of coping with the hydraulic forces operating during floods, either because of inadequate design or because of short-term savings in construction costs (eg at the Seekoei Estuary, Aston Bay and many Natal estuaries);

— the fixing of formerly dynamic mouths by retaining walls or groynes (eg as at the Berg and Kowie estuaries);

— various forms of pollution from the land or the sea, including the use of estuaries as dumping grounds.

It is not surprising to note that in an assessment of the state of the estuaries of the Cape and Natal (Heydorn 1986) it was found that, in general, the ecological condition of estuaries is closely related to the degree of degradation of their catchments. Thus, healthy estuaries were generally associated with catchments in good condition, and vice versa.

It was shown that in the Cape Province only 24% of the estuaries were still in good condition in 1985 – 86. In Natal the figure was 28%. With the current trend of rapidly increasing pressure on coastal resources and especially estuaries, these figures are disconcerting.

SUMMARY AND CONCLUSIONS

The main points that emerge are:

— Biotic diversity in estuarine environments is:

• associated with the natural conditions of the part of the coast where they occur; and

• linked to the ecological condition of the estuary concerned.

— The ecological condition of an estuary is not dependent only upon natural events or human activity in its immediate environs but can generally be associated with the condition of the catchment.

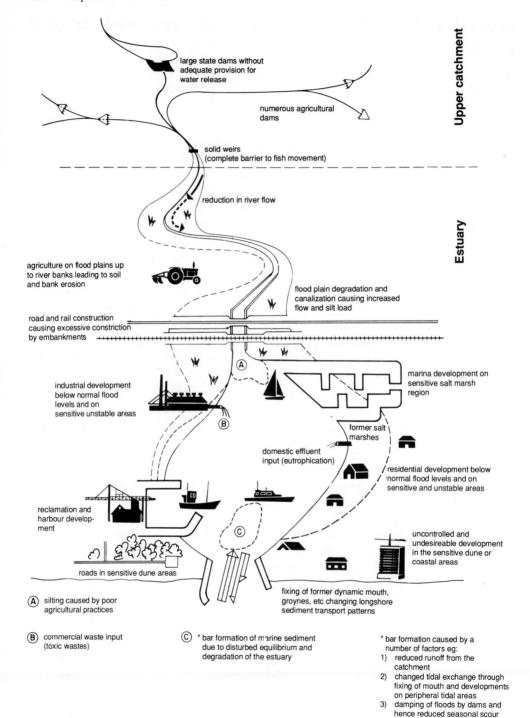

FIGURE 18.3 Human activities that can have a negative influence upon estuarine ecosystems (T J E Heinecken pers comm)

— Changes in the ecological condition of an estuary are obvious when a natural or man-induced episodic event has taken place, but can be difficult to detect when degradation is insidious and cumulative.

— Community structure studies (Begg 1984b) and habitat evaluation procedures (Schamberger and Kumpf 1980) indicate the direction of research aimed at the identification of progressive changes in estuarine conditions, and such research is, in fact, at present being conducted in Natal (Ramm pers comm).

— Sound management procedures for estuarine environments should be based on:

• maintenance of the natural processes operative in estuaries to the greatest possible extent;

• the determination of changes in community structure and habitat availability within estuaries; and

• improving catchment management.

— The reduction of biotic diversity in estuaries can lead to the breakdown of estuarine foodwebs. As marine organisms such as panaeid prawns (De Freitas 1984) and certain marine fish species (Wallace et al 1984) utilize estuaries during parts of their life cycle; this may also influence marine foodwebs.

The conservation of biotic diversity in estuaries therefore cannot be approached in terms of estuarine environments alone, but must be seen in the context of the associated catchments and marine areas.

REFERENCES

BEGG G (1978). The estuaries of Natal. *Natal Town and Regional Planning Report* **41,** Pietermaritzburg, 657 pp.

BEGG G (1984a). The estuaries of Natal, Part 2.*Natal Town and Regional Planning Main Series Report* **55,** Pietermaritzburg, 631 pp.

BEGG G (1984b). The comparative ecology of Natal's smaller estuaries. *Natal Town and Regional Planning Report* **62,** Pietermaritzburg, 182 pp.

BERJAK P, CAMPBELL G K, HUCKETT B I and PAMMENTER N W (1977). *In the mangroves of Southern Africa.* Wildlife Society of Southern Africa (Natal Branch) Durban, 72 pp.

DE FREITAS A J (1984). The Penaeoidea of Southeast Africa. I: The study area and key to the southeast African species. *Oceanographic Research Institute Report* **56,** Durban, 31 pp.

HEYDORN A E F (ed) (1986). An assessment of the state of the estuaries of the Cape and Natal 1985/86. *South African National Scientific Programmes Report* **130.** CSIR, Pretoria. 39 pp.

HEYDORN A E F and FLEMMING B W (1985). South Africa. In *The world's coastline.* (eds Bird E C F and Schwartz M L) Van Nostrand Reinhold, New York, NY. pp 653 – 667.

HEYDORN A E F and GRINDLEY J R (eds)(1981 – 85). Estuaries of the Cape: Part II: Synopses of available information on individual systems. *CSIR Research Reports,* Stellenbosch.

HEYDORN A E F and MORANT P D (eds) (1986 – ongoing). Estuaries of the Cape: Part II: Synopses of available information on individual systems. *CSIR Research Reports,* Stellenbosch.

KNOX G A (1986). *Estuarine ecosystems: A systems approach.* CRC Press, 230 pp.

SCHAMBERGER M L and KUMPF H E (1980). Wetlands and wildlife values: a practical field approach to quantifying habitat values. In Estuarine perspectives (ed Kennedy VS) *Proceedings of Fifth Biennial International Estuarine Research Conference, Jekyll Island, Georgia.* Academic Press, New York, NY. pp 37 – 46.

WALLACE J H, KOK H M, BECKLEY L E, BENNETT B, BLABER S J M and WHITFIELD A K (1984). South African estuaries and their importance to fishes. *South African Journal of Science* **80,** 203 – 207.

CHAPTER 19

Conserving biotic diversity on southern Africa's coastline

P A R Hockey, C D Buxton

DEFINITION, SCOPE AND DIVISION

The coastal zone is a broad band of land and water where ecological processes, such as biological production, consumption and the exchange of materials, occur at very high rates (Ray 1984). The boundaries of the coastal zone as defined in this framework are not fixed and will vary in relation to geomorphological and hydrological conditions at any particular site. In this regard it is important to recognize the interdependence between marine, coastal and terrestrial environments and the need to manage all three in conjunction. However, although the holistic ecosystem approach to management is desirable, ecologists should strive to separate coastal ecosystems into components that can be used by administrators (Heydorn 1983).

This study considers the intertidal and shallow subtidal biota of the open coastline between the Orange River and Kosi Bay. Estuaries and coastal lagoons are excluded.

Biological communities of the open coast vary with regard to several factors including geology and relief, water temperature, exposure, and nutrient status (eg Menge 1976; McLachlan et al 1981; McQuaid and Branch 1984, 1985; Bosman et al 1987; Denny 1987; Bosman and Hockey 1988a). The biota themselves exhibit tolerance limits to all of these extrinsic factors, and intrinsic variation in life-history characteristics such as reproductive output and dispersal mode, as well as competitive and predator-prey interactions, further complicate patterns of distribution and abundance (Branch 1985; Bosman and Hockey 1988b). Thus, although the biome has physical continuity, it is, like other biomes, biologically patchy.

The broadest level at which this biological patchiness can be defined is that of marine biogeographical provinces, the primary distinguishing influence being the temperature regime of the sea. A biogeographical zonation system for southern Africa was first established by Stephenson and Stephenson (1972), and the field is reviewed comprehensively by Brown and Jarman (1978) (see also McLachlan et al 1981; Branch and Griffiths 1988). The distribution tracts and geographical affinities of several groups, including algae, polychaetes, amphipods, hydrozoans and fish, have been invoked to identify the limits to biogeographical provinces.

Three provinces are recognized: a west-coast cool temperate province; a south-coast warm temperate province; and an east-coast warm subtropical province. The precise boundaries between these provinces are indeterminate and zones of overlap exist. The west- and south-coast provinces overlap between Cape Point and Cape Agulhas, and the south- and east-coast provinces overlap along the extreme eastern Cape and Transkei coasts. For the purpose of this study the respective boundaries are considered to lie midway between Cape Point and Cape Agulhas (at Hermanus), and at Port St Johns. A caveat is required here in that subtidal distributions may not mirror intertidal patterns. For example, Day (1967) found no similarity in subtidal and intertidal distributions of polychaetes, and Smale and Buxton (1989) concluded that subtidal fishes generally have wider distributions than intertidal species.

DISTRIBUTION, ENDEMISM AND DIVERSITY

Although some intertidal species are ubiquitous and occur along the entire southern African coast, others are restricted to one or two of the biogeographical provinces. Some species occur only in a portion of one of the provinces. Thus, from a conservation viewpoint, the three biogeographical provinces are best considered as discrete biomes.

Algal diversity is high around the coast, with the west coast supporting the highest proportion of endemic species (43% of 199 species) (Brown and Jarman 1978; Bolton 1987). The distribution patterns of many of the larger invertebrate species which have either commercial or subsistence value are well known, but many other species are known only from one locality or, in some cases, one specimen. These latter species, and species with apparently disjunct distributions, probably reflect in most instances a lack of collecting activity.

Different invertebrate groups show different patterns of diversity around the coast. Isopods are most diverse on the west and south-west coasts, bivalves exhibit the opposite pattern, and the diversity of patellid limpets is fairly constant, although few species are cosmopolitan in their distribution. There are slight peaks in patellid limpet diversity in overlap regions between biogeographical provinces (Figure 19.1). Several groups exhibit a high degree of endemism: 36% of marine polychaetes recorded in southern Africa are endemic to the subregion (Day 1967); 46% of amphipods are endemic (Griffiths 1974); and, as many as 87% of isopods may be endemic (Kensley 1978). The high diversity of some groups, notably gastropods and decapod crustaceans, on the east coast reflects incursion of tropical forms, but the majority of South Africa's coast is temperate, and it is in these regions that the highest degree of invertebrate endemism occurs.

Fishes that inhabit rock pools also show clear trends in both diversity and endemism in the movement from cold temperate waters of the southern Cape to the warmer subtropical waters of northern Natal. Data are unavailable for most of the west coast, but on the western Cape Peninsula only 13 species are found (Bennett and Griffiths 1984), all of which are endemic. In False Bay, 16 species have been recorded (Bennett and Griffiths 1984), and just east of Cape Agulhas the number increases to 21 (Bennett 1987). Again, all of the fish recorded in these areas are endemics. On the eastern Cape coast Beckley (1985) recorded 35 species at Port Elizabeth, and Christensen and Winterbottom (1981) recorded 38 species at Port Alfred. The proportion of endemics in these areas was 77% and 39%, respectively. In the Coffee Bay area on the Transkei coast, 105 species have been recorded, only 15 of which are endemic (T Hecht unpublished data). Farther east on the Maputuland coast 125 species have been recorded, of which only four are endemic to southern Africa (data extracted from J L B Smith collections made in 1976 and 1979).

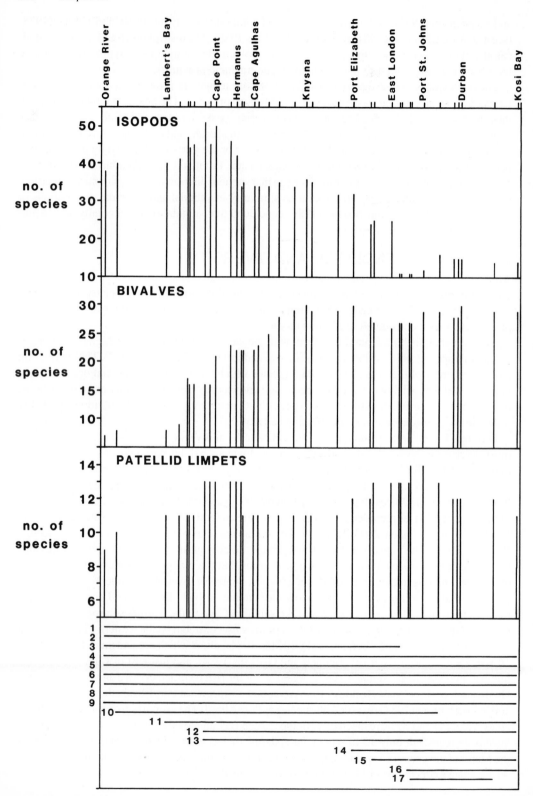

Rock-pool fishes may be classified as resident or transient (Gibson 1982). Residents are those species that occur in the pools as adults and juveniles, and breed there; transients use the pools either seasonally or at a particular stage of their lives, for example, as juveniles. Bennett (1987) showed that all of the cold temperate fishes recorded in rock-pools were residents, and that as the warmer subtropical regions were approached, so the proportion of resident species decreased. Data from Transkei and Natal support this observation: in Maputaland, only about 10 to 15% of the rockpool fishes are residents. In addition, Bennett (1987) shows a correlation between the number of resident species and endemism, and suggests that resident species are specifically adapted to the rock-pool environment and have more restricted geographical ranges than transient species. Although there is considerable overlap between the fishes found in rock-pools and those found subtidally (cf Beckley 1985; Smale and Buxton 1989), the high degree of endemism and the greater dependence of cool temperate species on rock pools indicates that these areas may be of critical importance to these resident fishes. Recent debate in the literature has also focused on the utilization of rock-pools and shallow subtidal areas by transient species (Beckley 1985; Bennett 1987; Clarke 1988; Smale and Buxton 1989). These studies show that a number of species, some of which are important to recreational and commercial fisheries, may rely heavily on shallow inshore waters and rock-pools as nurseries. Also important in this role are the surf zones of sandy beaches and shallow inshore waters with soft substrata (Lasiak 1981; Buxton et al 1984).

COASTAL UTILIZATION

Effective conservation demands perpetuation of functional ecosystems. A logical extension of this is that conservation needs are most immediate in areas under the greatest threat. The three principal threats to intertidal communities are development and recreation, exploitation, and pollution. Pollution risks are generally associated with development (industrial, agricultural, urban and recreational). Pressures on the coast for development and recreation are most intense along the east and south-east coasts, particularly the Natal south coast (Jackson and Lipschitz 1984). The west coast is at present relatively undeveloped and extensive stretches of shore owned by mining concerns are uninhabited, although landbased and marine mining operations may render these areas far from pristine. The greatest pressure from exploitation, both subsistence and commercial, occurs on the coasts of the eastern Cape, Ciskei, Transkei and, to a lesser extent, Natal (eg Hockey and Bosman 1986; Hockey et al 1988; de Freitas and Martin 1988).

Subsistence exploitation intensity has been quantified in Transkei: in the southern and central regions, an estimated 5,57 tonnes (max >14t) of shellfish is removed annually per kilometre of rocky coast (Hockey et al 1988). This activity has had a marked effect on shore communities, increasing the representation of sessile, inedible forms (Hockey and Bosman 1986). This exploitation also increases diversity, in line with the predictions of the Intermediate Disturbance Hypothesis (Connell 1978). However, from the standpoint of conservation ethics, changes in the functioning of ecosystems and decreases in the availability of preferred food items are in no way offset by the putative benefits of increased diversity (Siegfried et al 1985; Hockey and Bosman 1986; Hockey et al 1988).

FIGURE 19.1 Geographic patterns of diversity in isopods (from Kensley 1978), bivalves (from Day 1974) and patellid limpets (from Day 1974; Robson 1986; pers obs), and dispersion of patellid limpet species in southern Africa.
Codes for limpet species are: 1 *Patella compressa*, 2 *P granatina*, 3 *P argenvillei*, 4 *Helcion dunkeri*, 5 *H pectunculus*, 6 *H pruinosus*, 7 *P barbara*, 8 *P granularis*, 9 *P miniata*, 10 *P cochlear*, 11 *P oculus*, 12 *P longicosta*, 13 *P tabularis*, 14 *P concolor*, 15 *Cellana capensis*, 16 *P obtecta*, 17 *P aphanes*. *P safiana* and *P pica* occur in the subregion to the north of South Africa on the west and east coasts, respectively.

In Natal, exploitation of marine invertebrates is controlled by licence, and approximately 462 tonnes are removed annually under licence. The issuing of licences is managed by the Natal Parks Board on the advice of the Oceanographic Research Institute at Durban. In 1985, 33 654 bait licences were issued, nearly twice the number issued in 1974. Brown mussels *Perna perna* are among the most popular species for collection, and the number of mussel licences issued increased nearly elevenfold between 1974 and 1985, although the total number of mussels landed only trebled. Concerns about potential overexploitation along this coast have been expressed (de Freitas and Martin 1988).

Although the west coast is probably the most productive of southern Africa's intertidal regions, subsistence exploitation here is minimal. However, there is extensive commercial exploitation of rock lobsters *Jasus lalandii* and abalone *Haliotis midae*. The relatively low intensity of shellfish collection reflects the distribution of subsistence communities. Historically, however, indigenous people on the west coast relied heavily on the shellfish resource (Buchanan et al 1984). An application for a commercial licence for harvesting wild stocks of the limpets *Patella argenvillei* and *P granatina* on the Namaqualand coast is pending, awaiting research findings.

LEGISLATION

Control of the seashore is embodied in diverse and complex legislation, some of it dating back more than 50 years. What follows is a brief summary of the existing legislation.

South Africa

The Seashore Act (No 21 of 1935) afforded ownership of the coastal zone below the high-water mark to the State President, subject to traditional common law rights of the public. This Act controls aspects of coastal development and, in 1987, its implementation was transferred from the Department of Public Works and Land Affairs to the Department of Environment Affairs.

The Environmental Conservation Act (No 100 of 1982) proclaims a 'limited area' extending 1 000 m inland from the high-water mark along the entire South African coast. This Act is designed to control development, and certain activities are prohibited within this 'limited area', except under permit. Under Section (12)(2)(C) the Minister is empowered to exempt areas from the provisions of the bill: in the Cape Province, all coastal areas under the control of local authorities have been exempted (R2587 of Government Gazette 10546, 12 December 1986).

Exploitation of coastal living resources in the Cape Province is subject to legislative control under the Sea Fisheries Act (No 58 of 1973) and the accompanying Sea Fisheries Regulations, which were amended no fewer than 62 times between September 1974 and October 1987. In Natal, the Sea Fisheries Act is substituted by the Natal Parks, Game and Fish Preservation Ordinance (No 35 of 1947), and the Natal Fisheries Licensing Board Regulations (under Ordinance 15 of 1974). Both sets of legislation govern aspects of exploitation such as bag limits, sizes and seasons, but they differ in that Natal requires the issue of an *ad hominem* licence for which a fee is payable. Further protection to coastal biota is provided at national level by the Sea Birds and Seals Protection Act (No 46 of 1973), which protects both the animals themselves and the majority of their breeding sites (particularly offshore islands) from interference. In addition, the public may be excluded from the shore under the conditions of certain other Acts, including the Defence Act (1957) and the Precious Stones Act (1964).

Nature reserves are proclaimed at national level, provincial level or by local authorities, but state ownership of the sea and shore precludes nature reserves from extending below the high-water mark unless the power to allow this is delegated by the Minister to the relevant province. As a consequence, the jurisdiction of all nature reserves does not extend below the high-water mark.

Provision is made in the National Parks Act (No 57 of 1976) for declaring reserves in the sea or on the seashore, as defined in the Seashore Act, and protection of the shore can be afforded under the former Act, although exploitation can be permitted at the discretion of the administering authority. At present the Tsitsikamma Coastal National Park is the only marine reserve proclaimed under the National Parks Act where Sea Fisheries Regulations have officially been withdrawn. The Lake Areas Development Act (No 39 of 1975) provides for the establishment of Lake Areas under the control of the National Parks Board, and under the auspices of the Department of Environment Affairs. Within Lake Areas, the National Parks Board is empowered to modify the controls on exploitation specified by the Sea Fisheries Regulations subject to written notice in the Government Gazette. In November 1987, the state-owned land within the Wilderness National Lake Area was proclaimed a national park. This includes the intertidal region, but the Sea Fisheries Regulations have not yet been withdrawn.

Marine reserves are proclaimed at ministerial level, in both the Cape Province and Natal, under section 13 of the Sea Fisheries Act. Normally, no exploitation of marine resources (sometimes with the exception of fish) is permitted within these reserves, but the Minister can provide for exceptions. In Natal, for example, collection of specified intertidal invertebrates for food is permitted, under licence, within certain sections of marine reserves. Authority to issue such licences has been transferred from ministerial to provincial level.

Although much legislation exists governing the use of the coastal zone, not all of it is directed at conservation. The proposed Marine Affairs Bill (Government Gazette 9996, November 1985) would have repealed and superceded several Acts and ordinances, including the Sea Fisheries Act, Sea Birds and Seals Protection Act, Territorial Waters Act (No 87 of 1963), and the Fishing Industry Development Act (No 56 of 1978). Although this draft bill aimed to consolidate the existing diverse legislation, several recommendations made by commissions of inquiry (eg the Alant Commission) were not incorporated (Grindley et al 1986). It appears, however, that the Marine Affairs Bill has been abandoned. A new Sea Fisheries Act (No 12 of 1988) has been passed, but not promulgated. Under section (56)(1) provision is made for it to come into operation on a date (as yet unknown) to be determined by the State President. This Act includes some significant changes from the existing Act. Section 2 states that the Minister can determine policy in terms of optimizing the utilization of South African marine living resources. The Act also introduces a Quota Board for the allocation of quotas, as recommended by the Diemont Commission; the total allowable catch, however, remains at the discretion of the Minister.

Transkei and Ciskei

When Transkei and Ciskei became independent, they both took over the South African legislation *en bloc*. Since then they have repealed some Acts, amended others, and added their own new Acts.

Transkei retains both the South African Seashore Act and the Sea Fisheries Act and Regulations as amended by the time of independence in October 1976. The Transkeian Nature Conservation Act does not include the seashore, but nature reserves in Transkei are the only sections of coast where effective policing and implementation of the Sea Fisheries Regulations occurs. Elsewhere the level of exploitation far exceeds legal limits (Hockey et al 1988). New legislation is being formulated.

The Ciskeian Nature Conservation Act (Act 10 of 1987) effectively repealed all extant legislation governing the seashore, including the Seashore, Sea Fisheries, Sea Birds and Seals Protection and National Parks Acts. As such it is the most compact and comprehensive legislation governing any section of the southern African coast. In addition to vesting ownership of the sea and seashore in the President, and prescribing legal limits to exploitation, the Act establishes a

TABLE 19.1   Coastal conservation areas in southern Africa, their administering authorities, status and shoreline lengths. Private nature reserves, forest reserves and military areas are excluded.

	Conservation area	Administering authority[1]	Conservation status[2]	Shore length (km)	
				Rock	Sand
	West-coast province				
1	McDougall's Bay Rock Lobster Sanctuary	3	2	0,00	2,25
2	Rocher Pan Nature Reserve	3	1	0,00	2,75
3	St Helena Bay Rock Lobster Sanctuary[3]	3	2	9,00	11,25
4	Columbine Nature Reserve	5,3	3	8,75	1,25
5	Saldanha Bay Rock Lobster Sanctuary	3	2	13,50	14,50
6	Oude Post Nature Reserve[4,5]	1	1	4,50	3,00
7	Sixteen Mile Beach[4,5]	1	1	0,00	26,50
8	Table Bay Rock Lobster Sanctuary[3]	3	2	34,00	23,25
9	Cape of Good Hope Nature Reserve	5,3	3	13,25	0,00
10	Cape of Good Hope Marine Reserve[5]	5,3	1[6]	10,00	5,25
11	Miller's Point Marine Reserve[3]	3	1	2,75	0,50
12	Glencairn Marine Reserve[3]	3	1	2,50	0,50
13	Kalk Bay Marine Reserve[3]	3	1	1,00	0,00
14	Muizenberg Marine Reserve[3]	3	1	1,75	0,00
15	Wolfgat Nature Reserve	5,3	3	2,00	1,33
16	Rooi-Els Nature Reserve	5,3	3	0,33	0,00
17	H F Verwoerd Coastal Reserve	3	1	4,50	2,00
18	Kleinmond Coastal Nature Reserve	5,3	3	4,75	0,25
	South-coast province				
19	De Hoop Marine Reserve[5]	3	1	32,25	13,50
20	Wilderness National Park[5]	1	3	1,00	24,25
21	Goukamma Marine Reserve[5]	3	1	2,00	8,50
22	Knysna National Lake Area	1	3	10,25	5,75
23	Robbeberg Marine Reserve[5]	3	2	8,00	1,00
24	Keurboomsrivier Nature Reserve	3	3	0,00	2,50
25	Tsitsikamma Forest National Park[7]	1	3	6,00	2,00
26	Tsitsikamma Coastal National Park[5]	1	1	61,00	4,75
27	Cape St Francis Nature Reserve	3	3	2,25	0,00
28	Sylvic Nature Reserve	5,3	3	0,50	3,50
29	Sardinia Bay Marine Reserve[5]	5,3	1	6,50	5,75
30	Cape Recife Nature Reserve	5,3	3	2,00	0,75
31	Joan Muirhead Nature Reserve	5,3	3	1,00	1,75
32	Great Fish River Mouth Bird Sanctuary	5,3	3	0,00	0,50
33	Gxulu Marine Reserve[3]	3	4	8,25	5,75
34	Gonubie Marine Reserve[3]	3	4	8,50	2,50
35	Nyara Marine Reserve[3]	3	4	17,25	8,50
36	Dwesa Nature Reserve	4	3[8]	14,00	9,50
37	Hluleka Nature Reserve	4	3	4,00	1,00
38	Silaka Nature Reserve	4	3	1,75	0,50
	East-coast province				
39	Trafalgar Marine Reserve	2	1	2,25	2,50
40	Mlalazi Nature Reserve	2	2	0,00	19,00
41	St Lucia Park	2	2	0,25	8,25
42	St Lucia Marine Reserve 1 (south)	2	2	9,25	14,75
43	St Lucia Marine Reserve 2	2	1	6,40	16,80
44	St Lucia Marine Reserve 1 (north)[5]	2	2	7,50	23,10
45	Maputaland Marine Reserve 1 (south)	2	2	6,20	35,70
46	Maputaland Marine Reserve 2	2	1	1,30	9,70
47	Maputaland Marine Reserve 1 (north)[5]	2	2	4,40	13,10

Notes to Table 19.1

1   Administering authority: 1 = National Parks Board; 2 = Natal Parks Board; 3 = Cape Chief Directorate of Nature and Environmental Conservation; 4 = Transkei Department of Agriculture (Division of Nature Conservation); 5 = local authority

2   Conservation status: 1 = intertidal biota fully protected, angling allowed in some instances; 2 = modified Sea Fisheries Regulations (Cape, Transkei), limited collecting licences issued (Natal), 3 = Sea Fisheries Regulations (Cape, Transkei), collecting licences issued (Natal); 4 = controls on angling only

3   Marine reserves/rock lobster sanctuaries which lack an officially designated name

4   Areas fall within the West Coast (formerly Langebaan) National Park. Although the Sea Fisheries Regulations have not been officially withdrawn, the National Parks Board enforces full protection of the intertidal region

5   Marine reserves in which part or all of the area above high-water mark is included within a conservation area

6   Rock lobsters may be caught within a limited area of the reserve

7   This small section of the Tsitsikamma Forest National Park was proclaimed in January 1988 by the annexation of the De Vasselot Nature Reserve. The adjacent coastline is not a marine reserve

8   Some sections of the reserve are closed to all exploitation

---

'Coastal Conservation Area' which extends one kilometre inland from the coast. Any development, including that by any state department, within this strip must be sanctioned by the Ciskeian Department of Agriculture, Forestry and Rural Development.

   The key to successful management is simplicity although governmental systems tend to be complex (Heydorn 1983). In terms of simplifying the legislative framework for coastal conservation, Ciskei has progressed further than either South Africa or Transkei.

TABLE 19.2   An analysis of the conservation status of the coastline in the three marine biogeographical provinces, and calculation of the potential shore length conserved if all conservation areas on the coast included the region below high-water mark. Conservation status codes as in Table 19.1.

	West coast Rock[2]	Sand	South coast Rock	Sand	East coast Rock	Sand
Total shore length (km)[1]	339	461	519	681	136	464
Shore length with status 1	36,0	40,6	101,8	32,5	10,0	29,0
% of shore with status 1	10,6	8,8	19,6	4,8	7,4	6,3
Shore length with status 2 – 4	85,6	54,2	4,8	69,8	27,6	113,9
% of shore with status 2 – 4	25,3	11,8	16,3	10,3	20,3	24,6
Potential % of shoreline protected	35,9	20,6	35,9	15,1	27,7	30,9

1   Derived from Hockey et al 1983
2   Relative proportions of sand and rock in the Cape Province, Transkei and Ciskei were derived from 1:50 000 topographical maps of the South African Department of Surveys and Mapping. Details for Natal reserves were supplied by the Natal Parks Board, and were based on analysis of aerial photographs and on ground surveys.

WHAT IS CONSERVED?

As discussed above, legislation exists to control development and exploitation along the entire southern African coast, but total protection of intertidal biota is provided only in two national parks and in certain marine reserves (Table 19.1). Excluding offshore islands, where the intertidal zone is effectively conserved under the provisions of the Sea Birds and Seals Protection Act, total protection (excluding fish in some instances) is afforded to a relatively small proportion of the coastline. The least protection is provided to sandy beaches of the south coast (4,8%) and rocky shores of the east coast (4,6%), and the most to the rocky shores of the south coast (19,6%), the size of the latter figure being due largely to the existence of the Tsitsikamma Coastal National Park. It should be noted, however, that the high-energy shoreline of the Tsitsikamma Coastal National Park is atypical of most of the south coast. If all other coastal conservation areas (excluding private nature reserves, forest reserves and military areas) were extended to include the intertidal zone, at least 9%, and up to 35,1%, of each major habitat type (rock/sand) within each biogeographical province would be protected (Table 19.2). Only the east-coast rocky habitat would have less than 15% of its length conserved.

ARE CURRENT CONSERVATION MEASURES ADEQUATE?

Two extensive sections of coast are currently inadequately conserved. Between the Great Kei River and Richards Bay, a heavily exploited stretch of coast, there is only one small marine reserve (Figure 19.2): and in the marine reserves around East London, only angling is controlled. On the west coast there are no coastal conservation areas north of Rocher Pan (Figure 19.2). Proclamation of the proposed national park on the Namaqualand coast between the Groen and Spoeg rivers (32,8 km of rocky shore, 21 km of sandy shore) would help to redress this imbalance.

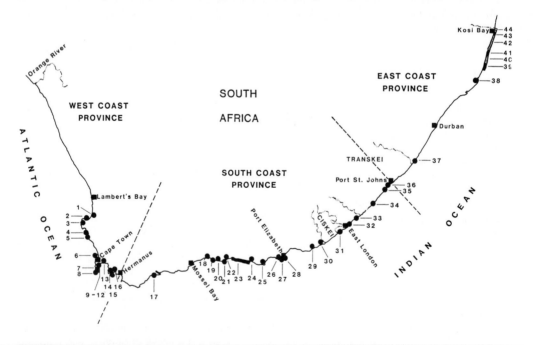

FIGURE 19.2 Distribution of coastal conservation areas between the Orange River and Kosi Bay. Reserve names, sizes and conservation status are detailed in Table 19.1

Resources are always defined in terms of a given economic system, and conservation should be a means of adjusting short-term uses and needs so as not to jeopardize longer-term values and future options (Ray 1984). Conservation objectives must both maintain essential ecological processes and ensure sustainable utilization of species and ecosystems. However, conservation issues cannot be expected to enjoy a high rate of concern among people whose basic economic and nutritional needs are not being met (Johannes and Hatcher 1986).

The need for a network of reserves, or refuges, along the coast to balance the processes of immigration and local extinction has a conceptual basis in island biogeography theory. Unfortunately, however, island biogeography does not answer the critical questions of reserve size and spacing. These decisions rest to a large extent on the dispersive capabilities of the biota, which vary dramatically among intertidal organisms. For example, limpets and mussels are broadcast spawners with external fertilization and a pelagic veliger stage, ostensibly promoting long-distance dispersal. Clinid fish, by contrast, are viviparous and have limited potential for dispersal over long distances. It has been suggested that viviparity in these species is a reproductive specialization to limit dispersal away from local rock pools (Bennett 1987).

Many problems attend the study of marine invertebrates with a planktonic larval dispersive phase. For example, some patellid limpets that reportedly have planktonic larvae (Grahame and Branch 1985) may, on the basis of macro- and mesospatial distributions, disperse over considerably shorter distances than predicted (P A R Hockey unpublished data). Holt and Talbot (1978) advocate that management decisions should include a safety factor to allow for the fact that knowledge is limited and institutions are imperfect. This is equivalent to the 'adaptive management' strategy outlined by Heydorn (1983).

Potentially, the conservation status of the southern African coast is good, with the exceptions outlined above. However, the realization of this potential is hampered by unwieldy legislation in that the administration of conservation generally is decentralized, but the coastal zone is legislated at national level. This leads to an anachronistic situation in which nature reserves on the coast do not extend seaward of the high-water mark. Ciskei has consolidated conservation legislation, but accounts for only a small fraction of the southern African coastline. If the authority to allow protection of the seashore within nature reserves in South Africa were transferred to provincial level, adequate protection would probably be achieved along much of the coast. Such protection should ideally extend seawards in the manner of a general marine reserve.

The evidence of rapid degradation of intertidal resources in parts of Natal and particularly in Transkei is cause for immediate concern. Shellfish are an important source of nutrition for coastal Transkeians (Hockey et al 1988) who view the seashore as a commonage to which they have traditional rights; they believe these rights transcend and render ineffective any legislation designed to safeguard the flora and fauna. The Transkeian coast is of particular interest in that much of it falls within a biogeographical transition zone. Research is urgently required into developing a strategic conservation plan to ensure not only the viability of Transkei's coastal ecosystems, but also the continued provision of food to the local people. This clearly will involve considering more than just the intertidal zone; a 'Critical Habitat Evaluation' of the type suggested by Ray (1984) may provide the most effective means of tackling the problem.

PRIORITIES FOR EFFECTIVE COASTAL CONSERVATION

The priorities for the conservation of southern Africa's seashore may be divided along two axes, administrative and research.

Administrative priorities
— Deregulating legislation in the Seashore Act to allow for provincial designation of marine reserves, in line with the regionalization policy, and consolidating existing diverse legislation;
— Proclaiming a national park on the Namaqualand coast which would include a marine reserve;
— Establishing marine reserves in Transkei and southern Natal pending development of a strategic coastal conservation plan;

— Increasing control over the use of off-road vehicles on beaches and coastal dunes; and

— Ensuring adequate advertisement and policing of marine reserves.

Research priorities

— Establishing guidelines, in a South African context, for the sizes and spacing of a 'backbone' of marine reserves to ensure conservation and judicious utilization of coastal resources;

— Establishing dispersal patterns and immigration rates of exploited organisms with a view to optimizing the effectiveness of a reserve network;

— Investigating alternative avenues, specifically mariculture, for meeting the nutritional needs of coastal people in subsistence economies;

— Monitoring the dispersion and ecological impacts of alien species on southern African shores, particularly the alien mussel *Mytilus galloprovincialis* ;

— Investigating the influence of geology on intertidal community structure, particularly in areas where coastal geology is complex (eg Transkei), with a view to optimizing site selection for reserves;

— Long-term monitoring of intertidal communities in selected undisturbed sites (eg national parks) to detect natural cyclic phenomena, their limits, and changes which elsewhere may be attributed erroneously to the effects of exploitation or disturbance. Such monitoring exercises should be designed in such a way that they can be carried out by resident on-site personnel;

— Assessing the ecological impact of recreational activities such as the use of off-road vehicles, bait collection and shore angling on coastal biota, including mobile elements such as birds and mammals which have not been discussed elsewhere in this review;

— Investigating the impact of marine diamond mining on the biota of the Namaqualand coast; and

— Monitoring the levels and spatial distribution of toxins resulting from industrial or other pollution, which accumulate in the tissues of (particularly) filter-feeding invertebrates, and investigating the effects of these toxins on growth, reproduction and survivorship of the affected species.

Research programmes to address some of the above points are already under way, but they require active co-ordination on a national level to ensure that results are comparable and robust, and are replicated over a geographic range wide enough to be relevant to the objectives of a national coastal conservation strategy and the establishment of a viable reserve network.

ACKNOWLEDGEMENTS

I am grateful to Renate Blaich for assistance with data compilation and for preparing the figures. Funding was provided by the University of Cape Town and the South African National Committee for Oceanographic Research. Charles Griffiths and George Branch provided helpful comments on an earlier draft.

REFERENCES

BECKLEY L E (1985). The fish community of East Cape tidal pools and an assessment of the nursery function of this habitat. *South African Journal of Zoology* **20**, 21 – 27.

BENNETT B A (1987). The rock-pool fish community of Koppie Alleen and an assessment of the importance of Cape rock pools as nurseries for juvenile fish. *South African Journal of Zoology* **22**, 25 – 32.

BENNETT B A and GRIFFITHS C L (1984). Factors affecting the distribution, abundance and diversity of rock-pool fishes on the Cape Peninsula, South Africa. *South African Journal of Zoology* **19**, 97 – 104.

BOLTON J J (1987). Seaweed biogeography in Southern Africa. *Veld and Flora* **73**, 147 – 148.

BOSMAN A L and HOCKEY P A R (1988a). The influence of primary production rate on the population dynamics of *Patella granularis*, an intertidal limpet. *PSZNI: Marine Ecology* **9**, 181 – 198.

BOSMAN A L and HOCKEY P A R (1988b). Life-history patterns of populations of the limpet *Patella granularis:* the dominant roles of food supply and mortality rate. *Oecologia* **75**, 412 – 419.

BOSMAN A L, HOCKEY P A R and SIEGFRIED W R (1987). The influence of coastal upwelling on the functional structure of rocky intertidal communities. *Oecologia* **72**, 226 – 232.

BRANCH G M (1985). Competition: its role in ecology and evolution in intertidal communities. In *Species and Speciation* (ed Vrba E S) Transvaal Museum, Pretoria. pp 97 – 104.

BRANCH G M and GRIFFITHS C L (1988). The Benguela ecosystem. Part V. The coastal zone. *Oceanography and Marine Biology Annual Review* **26**, 396 – 486.

BROWN A C and JARMAN N (1978). Coastal marine habitats. In *Biogeography and ecology of southern Africa* (ed Werger M J A) Junk, The Hague. pp 1239 – 1277.

BUCHANAN W F, PARKINGTON J E, ROBEY T S and VOGEL J C (1984). Shellfish, subsistence and settlement: some western Cape Holocene observations. In Frontiers: southern African archaeology today. (eds Hall M, Avery G, Avery D M, Wilson M L and Humphreys A J B) *British Archaeology Records International Series* **207**, 121 – 130.

BUXTON C D, SMALE M J, WALLACE J H and COCKROFT V G (1984). Inshore small-mesh trawling survey of the Cape south coast. Part 4. Contributions to the biology of some Teleostei and Chondrichthyes. *South African Journal of Zoology* **19**, 180 – 188.

CHRISTENSEN M S and WINTERBOTTOM R (1981). A correction factor for, and its application to, visual censuses of littoral fish. *South African Journal of Zoology* **16**, 73 – 79.

CLARKE J R (1988). Aspects of the biology of the musselcracker, *Sparodon durbanensis* and the bronze bream, *Pachymetopon grande*, (Pisces: Sparidae), with notes on the Eastern Cape recreational rock-angling and spear fisheries. MSc thesis, Rhodes University, Grahamstown.

CONNELL J H (1978). Diversity in tropical rain forests and coral reefs. *Science* **199**, 1302 – 1310.

DAY J H (1967). *A monograph of the Polychaeta of southern Africa.* 2 vols. British Museum (Natural History), London.

DAY J H (1974). *A guide to marine life on South African shores.* Balkema, Cape Town.

DE FREITAS A J and MARTIN L K P (1988). Long-term data on exploitation of the mussel *Perna perna* in Natal. In Proceedings of the conference on long-term data series relating to southern Africa's renewable natural resources. (eds Macdonald I A W and Crawford R J M) *South African National Scientific Programmes Report* **157**, 123 – 125.

DENNY M W (1987). Life in the maelstrom: the biomechanics of wave-swept rocky shores. *Trends in Ecology and Evolution* **2**, 61 – 66.

GIBSON R N (1982). Recent studies on the biology of intertidal fishes. *Oceanography and Marine Biology Annual Review* **20**, 363 – 414.

GRAHAME J and BRANCH G M (1985). Reproductive patterns of marine invertebrates. *Oceanography and Marine Biology Annual Review* **23**, 373 – 398.

GRIFFITHS C L (1974). The gammaridean and caprellid Amphipoda of southern Africa. PhD thesis, University of Cape Town, Cape Town.

GRINDLEY J R, GLAZEWSKI J I and RABIE M A (1986). The control of South Africa's living marine resources. In *The law of the sea.* (eds Bennett T W, Dean W H B, Hutchinson D B, Leeman I and van Zyl Smit D) Juta, Cape Town. pp 133 – 159.

HEYDORN A E F (1983). Management of sandy coastlines — report on review and workshop. In *Sandy beaches as ecosystems.* (eds McLachlan A and Erasmus T) Junk, The Hague. pp 703 – 708.

HOCKEY P A R and BOSMAN A L (1986). Man as an intertidal predator in Transkei: disturbance, community convergence and management of a natural food supply. *Oikos* **46**, 3 – 14.

HOCKEY P A R, BOSMAN A L and SIEGFRIED W R (1988). Patterns and correlates of shellfish exploitation by coastal people in Transkei: an enigma of protein production. *Journal of Applied Ecology* **25**, 353 – 364.

HOCKEY P A R, SIEGFRIED W R, CROWE A A and COOPER J (1983). Ecological structure and energy requirements of the sandy beach avifauna of southern Africa. In *Sandy beaches as ecosystems.* (eds McLachlan A and Erasmus T) Junk, The Hague. pp 507 – 521.

HOLT S J and TALBOT L M (1978). New principles for the conservation of wild living resources. *Wildlife Monographs* **59**, 3(2), 1 – 33.

JACKSON L F and LIPSCHITZ S (1984). *Coastal sensitivity atlas of southern Africa.* Government Printer, Pretoria.

JOHANNES R E and HATCHER B G (1986). Shallow tropical marine environments. In *Conservation biology: the science of scarcity and diversity.* (ed Soule M E) Sinauer, Sunderland, Mass. pp 371 – 382.

KENSLEY B (1978). *Guide to the marine isopods of southern Africa.* Trustees of the South African Museum, Cape Town.

LASIAK T A (1981). Nursery grounds of juvenile teleosts: evidence from the surf zone at King's beach, Algoa Bay. *South African Journal of Science* **77**, 388 – 390.

MCLACHLAN A, WOOLDRIDGE T and DYE A H (1981). The ecology of sandy beaches in southern Africa. *South African Journal of Zoology* **16**, 219 – 231.

MCQUAID C D and BRANCH G M (1984). The influence of sea temperature, substratum and wave exposure on rocky intertidal communities: an analysis of faunal and floral biomass. *Marine Ecology Progress Series* **19**, 145 – 151.

MCQUAID C D and BRANCH G M (1985). Trophic structure of rocky intertidal communities: response to wave action and implication for energy flow. *Marine Ecology Progress Series* **22**, 153 – 161.

MENGE B A (1976). Organization of the New England rocky intertidal community, role of predation, competition and environmental heterogeneity. *Ecological Monographs* **46**, 355 – 393.

RAY G C (1984). Conservation of marine habitats and their biota. In *Conservation of threatened habitats.* (ed Hall A V) *South African National Scientific Programmes Report* **10**, 109 – 134.

ROBSON G (1986). A new species of South African limpet *Patella aphanes*, (Mollusca: Gastropoda: Patellidae), with a discussion of *P. obtecta* Krauss, 1848. *Durban Museum Novitates* **13**, 305 – 320.

SIEGFRIED W R, HOCKEY P A R and CROWE A A (1985). Exploitation and conservation of brown mussel stocks by coastal people of Transkei. *Environmental Conservation* **12**, 303 – 307.

SMALE M J and BUXTON C D (1989) The subtidal gulley fish community of the eastern Cape and the role of this habitat as a nursery area. *South African Journal of Zoology* **24**, 58 – 67.

STEPHENSON T A and STEPHENSON A (1972). *Life between tidemarks on rocky shores.* Freeman, San Francisco, Ca.

CHAPTER 20

The conservation status of the pelagic ecosystems of southern Africa

P A Shelton

INTRODUCTION

Prior to the intensification of exploitation of the marine resources off southern Africa after World War II there was little information on the conservation status of the pelagic ecosystem off this coast. Much of the subsequent information collected is a direct consequence of exploitation, either through information on harvests, or through research related to management of fisheries. It follows that most often management has related more closely to a policy of maximizing economic yield than to an adherence to principles of conservation biology.

This policy has led to overexploitation, and in some cases severe population declines and range contractions, stopping short of actual species extinction, particularly when short time horizons have been dominant in decision-making. In recent years there has been an increasing tendency for decision-makers to take a longer term view in the management of fish populations, and for fishermen to move away from the 'if I don't catch it someone else will' attitude which prevailed in the 1950s and 1960s. This change is partly a consequence of the declaration of the 200 nautical-mile exclusive economic zone off South Africa in 1977. But experience of social and ecological implications of overexploitation and species depletion, and difficulty encountered in rebuilding depleted populations, has also influenced the attitude of fishermen and resource managers. The interdependence of populations and the need to maintain diversity, both in the ecosystem and in catches, are now given much more serious consideration than before, and the time horizon for evaluating management decisions is lengthening from one or two years to decades.

In this chapter the conservation status of some important species in the pelagic ecosystem off southern Africa are reviewed. Consideration is given to managing for biological diversity, given the effect of largely unpredictable physical forcing and complex interactions between system variables.

BOUNDARIES AND PHYSICAL FORCING

The pelagic habitat off southern Africa is influenced by several water masses of tropical, subtropical, south Atlantic and Antarctic origins (Shannon 1985), making the region one of the most oceanographically heterogeneous in the world. However, for simplicity the region between the Angolan border on the west coast and the Mozambique border on the east coast can be classified into four major water 'masses', distinguishable on the basis of surface temperature — Benguela Current, Agulhas Current, Angolan Current and oceanic water (Figure 20.1).

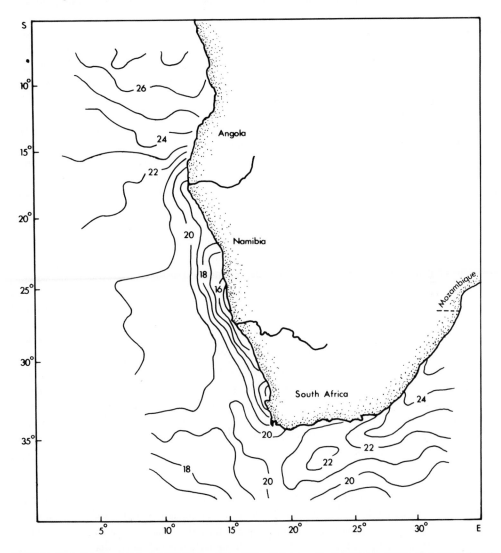

FIGURE 20.1 Summer sea surface temperature (°C) off the coast of southern Africa. The warm Angolan Current and Agulhas Current waters to the north and south enclose cooler Benguela Current water originating from upwelling on the west coast. Warmer water offshore corresponds to 'oceanic water'. Modified from Parrish et al (1983).

The area dominated by the Benguela Current, commonly termed the Benguela Ecosystem (Shannon 1985), varies, but it generally extends longshore from the vicinity of Cape Agulhas (southernmost tip) to about 16°S (just north of the Angolan border). The offshore boundary of the Benguela Ecosystem is not clearly defined and varies seasonally (Shannon 1985), but a useful definition is that it includes the area of decreased temperature and elevated production arising from coastal and shelf-edge upwelling. Offshore of the Benguela Ecosystem warm subtropical water associated with the South Atlantic Gyre, commonly referred to as 'oceanic water', occurs.

On the east coast of southern Africa, the southward flowing warm Agulhas Current of tropical origin is a dominant influence. It mixes with water from the Benguela Current on the Agulhas Bank (the shallow water region off the southern tip of Africa) and extends seasonally some distance up the west coast. The area off northern Namibia is seasonally influenced by the southward movement of warm tropical Angolan Current water.

Each of the four water masses constituting the pelagic habitat off southern Africa have endemic species (eg some species of copepods (De Decker 1984)), whereas other species, especially some of the larger, more mobile forms, may migrate between water masses (eg many of the cetacean species (Ross and Best 1989), and some of the predatory fish species such as snoek *Thyrsites atun* (Crawford et al 1987)). High species diversity is characteristic of the entire area as a result of the presence of several water masses, but is often highest in association with the mixing areas between water masses (eg a high number of species in the ichthyoplankton recorded in the mixing area between Agulhas Current and Benguela Current waters, Shelton 1986). A local decrease in species diversity may therefore reflect changes in the distribution of water masses, and not extinction.

The pelagic habitat can be divided vertically into epipelagic, mesopelagic and bathypelagic zones, together covering the whole water column from the sea surface to just above the bottom. Many pelagic species, while found predominantly in one or other depth zone, may migrate vertically in search of food, to spawn, to escape predators, or to avail themselves of the vertical shear in current velocities in order to facilitate longshore or onshore-offshore movements.

Given this heterogeneity, a strong argument could be made to treat the pelagic habitat as several ecosystems and to review their conservation status separately. However, the variable influence of the different water masses in the region, the occurrence of mixing, the movement of some of the organisms between water masses, and the paucity of detailed distributional data largely precludes such a treatment. Political boundaries could also be used as system demarcators, but again these are not adhered to, either by the water masses or their biota. For simplicity therefore the sea up to 200 nautical miles off the coast of southern Africa has been treated as a single pelagic 'ecosystem' for the purpose of reviewing the conservation status of the biota.

The pelagic ecosystem off southern Africa is characterized by variable physical forcing at time scales from hours to decades. Although winds are perennially favourable along the west coast for the upwelling of cold, nutrient-rich water, which stimulates plankton production and drives the food chain, seasonal changes in the position of the South-east Atlantic high-pressure anticyclone influences the intensity of upwelling. Higher frequency variability in wind stress, generated by the passage of easterly moving cyclones south of the continent, imparts a three to six day pulsing in upwelling strength (Nelson and Hutchings 1983).

Seasonal changes in atmospheric forcing control the penetration of warm Angolan Current and Agulhas Current waters into the region, but this in turn is amplified or suppressed by interannual variations in larger-scale climatic processes. For example there have been several periods of abnormally warm sea temperatures since the 1980s, termed Benguela Niños (Shannon et al 1986). Analysis of the data indicates a possible, but weak, 10-year cycle in ocean warming (Taunton-Clark and Shannon 1988) similar to, and in phase with, the 10 to 12-year oscillation in southern South African rainfall described in Vines (1980) and Tyson (1986), see Shelton et al (1985). This may be evidence that both responses are associated with the same large-scale changes in the Southern Oscillation and Walker Circulation. Superimposed on these processes is a much longer-term warming trend in sea temperature off the west coast of southern Africa, with the post-World War II era being typically 0,8°C warmer than earlier periods (Shannon and Taunton-Clark 1988).

In combination, the three to six day pulsing, the seasonal cycle, interannual variability, and the long-term warming trend contribute to form complicated, and largely unpredictable, environmental forcing. Organisms that persist have evolved a variety of 'bet hedging' life-history traits to cope with the variability. In some instances these traits form a biological filter, allowing population size to be more constant than the environment (Shelton 1987). However, prolonged unfavourable conditions must eventually lead to a decrease in numbers. The rapidity of a subsequent recovery will depend on the potential for the population to grow, determined by age to maturity, fecundity and survival of the premature stages. Some species such as anchovy *Engraulis capensis* and pilchard *Sardinops ocellatus*, which have a low tolerance to more than a few years of adverse conditions, but which have the ability for rapid numerical increase given favourable conditions, may be adapted to a 'boom-and-bust' existence rather than to constancy. Dramatic population declines in such species would therefore not have implications for conservation, unless they were prolonged. Indices of historic change in pelagic fish population size (such as guano yield from pelagic fish-eating seabirds on near-shore islands (Crawford and Shelton 1981; Figure 20.2) and fish-scale deposits in the bottom sediments off Namibia (Shackleton 1987; Figure 20.3)) do indeed suggest that these kinds of fluctuations were the norm for anchovy and pilchard populations even prior to exploitation. When species declines do occur in an ecosystem characterized by variable environmental forcing, it is important to determine whether the cause is natural, or whether it is man-induced and therefore potentially controllable.

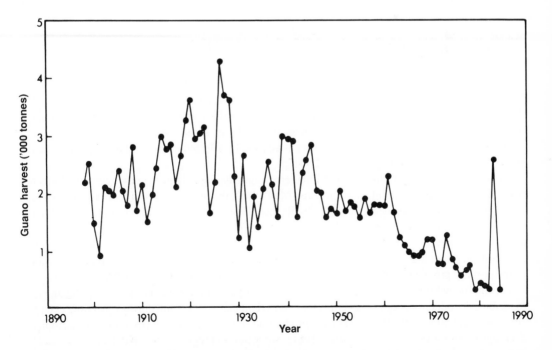

FIGURE 20.2   Time series of seabird guano deposited on islands off South Africa for the period 1897 to 1984

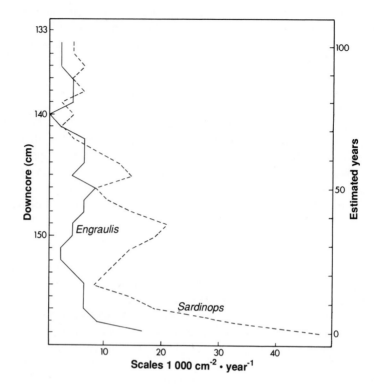

FIGURE 20.3 Time series of anchovy and pilchard scales deposited in bottom sediments off Namibia for the approximate period 1865 to 1960 (from Shackleton 1987)

SPECIES RICHNESS AND FOOD-CHAIN STRUCTURE

The biotic components of the southern African pelagic ecosystem are diverse, ranging from micro-organisms such as bacteria and protozoa to whales and man, and the interactions between them are complex. Between the primary producers and the top carnivores there is a minimum of three and a maximum of six steps in the food chain (Figure 20.4), even without taking into account juveniles feeding at lower trophic levels than the adults in some species, or cannibalism. By comparison, 62 terrestrial and aquatic foodwebs analysed by Briand (1983) and Briand and Cohen (1984) had a mean chain length of only 2,9 steps.

Although food chains in three-dimensional environments (eg pelagic) are thought to be longer than those in two-dimensional environments (May 1986), the high productivity of pelagic fish in upwelling ecosystems has been attributed by Ryther (1969) to the potential for just one step (and therefore little energy loss) between the high levels of primary production and harvestable fish (eg anchovy). However, in the pelagic ecosystem off southern Africa anchovy feed selectively on zooplankton rather than on phytoplankton (James 1987), and zooplankton are inefficient grazers of the primary production in newly upwelled water (Borchers and Hutchings 1986).

This inefficient transfer is most likely to be a consequence of the lagged response of consumers to increased production, both seasonally (Jones and Henderson 1987) and during upwelling events (Borchers and Hutchings 1986). Primary production not consumed directly enters a microbial decomposer pathway (Newell and Turley 1987). Some nutrient regeneration within this 'microbial loop' may contribute to the maintenance of primary production, and bacteria and protozoa may

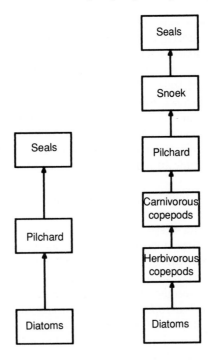

FIGURE 20.4 Diagram showing a short food chain leading to pelagic fish and a longer food chain leading to seals

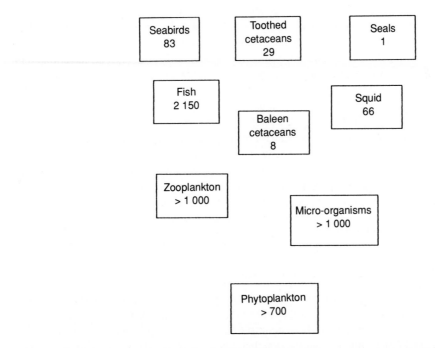

FIGURE 20.5 Some of the more important species groups in the pelagic ecosystem off southern Africa showing species richness, arranged roughly according to trophic level. The species richness data for squid is from J Augustyn, and the plankton is from L Hutchings (pers comm).

contribute directly to zooplankton production, although there will be considerable loss through respiration. Although the energy transfer from primary production to zooplankton is inefficient, there is nevertheless sufficient energy transfer to allow several steps, and therefore a diverse assemblage of consumers, in the food chain leading to top predators.

The species richness of some of the more important groups of organisms in the pelagic ecosystem, arranged roughly according to trophic level, are given in Figure 20.5. The 2 150 marine fish species described by Smith and Heemstra (1986) represent 15% of the total number of marine fish species in the world, and include representatives of 83% of the world's marine fish families. About 13% of the species described are endemic to the region. However, some of the fish species have strong inshore or benthic affinities and cannot therefore be considered as part of the pelagic ecosystem.

Of the 83 seabird species found off southern Africa, only seven are endemic and 14 breed locally (Cooper and Berruti this volume). A total of 37 cetacean species occur in the region, a relatively high number compared with 40 species for the entire North Atlantic, or 41 species for the whole North Pacific (Ross and Best 1989). Many are migratory, for example right whales *Eubalaena glacialis* visit the southern African coast each spring to bear and raise their young and possibly mate, while humpback whales *Megaptera novaeangliae* use the coastline as a migratory corridor to and from their tropical breeding areas (Best and Ross 1989). Only one species, the Heaviside's dolphin *Cephalorhynchus heavisidii*, is endemic to southern Africa waters.

Food chains in three-dimensional environments are considered to be generally 'thinner' (fewer species at the same trophic level) than those in two-dimensional environments (May 1986), and therefore potentially more sensitive to species depletions or extinctions. Although this generalization does not appear to hold for all trophic levels in the pelagic ecosystem, there do appear to be only a few dominant species at the level of zooplankton consumers (eg pilchard, anchovy, horse mackerel *Trachurus trachurus capensis*, mackerel *Scomber japonicus*, round herring *Etrumeus whiteheadi*, lantern fish *Lampanyctodes hectoris* and light fish *Maurolicus mulleri*), and the depletion or removal of one or more of these could be expected to have profound effects on ecosystem functioning. Crawford and Shelton (1981) and Shelton et al (1984a) have linked the deteriorating conservation status of the jackass penguin *Spheniscus demersus* to pilchard collapses off Namibia and South Africa. However, where alternative prey have been locally available, predators have displayed considerable opportunism, often switching to whichever prey is abundant (Crawford et al 1985, 1988). In contrast, switching occurs rather more slowly in commercial fisheries where there is considerable inertia with regard to fishing systems and markets.

The large number of species, especially lower down in the food chain, and the complicated linkages that exist between them, constitute a complex ecosystem. The generalization that complex ecosystems are more stable than simple ones, and therefore potentially less sensitive to the activities of man, does not appear to have a strong theoretical or empirical basis (May 1981, 1986). In fact, theoretical studies have suggested that under certain conditions more complex ecosystems may even be less stable (May 1973).

EXPLOITATION AND INTERFERENCE BY MAN

The observation that some populations in the pelagic ecosystem have been exploited for many years without decreasing to extinction implies the existence of some form of density dependence in rates governing population size. If there is simple linear density dependence, then equilibrium (or sustainable) yield (the amount of production that can be harvested at constant current biomass) is a quadratic function of biomass, with maximum sustainable yield (MSY) occurring at half the unexploited biomass. Using the biomass at MSY as a reference point, any fish population can be assigned to one of six states, depending on the degree of exploitation (Figure 20.6): (1) unexploited; (2) underexploited; (3) fully exploited; (4) overexploited; (5) collapsed; and (6) collapsed but recovering. This provides a useful classification of the conservation status of some of the resources in the pelagic ecosystem off southern Africa (Table 20.1).

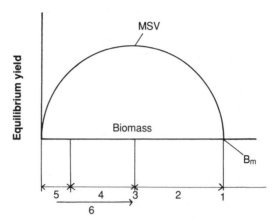

FIGURE 20.6 Diagram illustrating six possible stages in the exploitation of a fish population: (1) unexploited; (2) underexploited; (3) fully exploited; (4) overexploited; (5) collapsed; and (6) collapsed but recovering. B_m is the maximum (unexploited) biomass.

It is significant that of the 23 resources currently being exploited (listed in Table 20.1), only two can be classified as underexploited, whereas many are either overexploited or collapsed. Most resources listed in Table 20.1 gave peak yields in the 1960s and 1970s, and only Namibian horse mackerel, squid and anchovy yields have peaked in the 1980s.

Of the collapsed resources, none can yet be classified as recovered although there is evidence that some of the overexploited hake resources are recovering to the fully exploited state. Although many individual populations have decreased through overfishing, the combined yield of the major commercially exploited species has remained fairly constant since reaching a peak in 1968 (Crawford et al 1988; Figure 20.7).

Generally, data on recreational marine linefish cannot be subject to the same kind of classification as commercial species because of the absence of appropriate stock assessments. Of the 205 species of linefish for which recreational anglers report statistics, 90% of the total catch by number or by weight is made up of fewer than 10 species (Van der Elst 1989). There is evidence to show that several of these favoured angling species, especially sparids, have been depleted through overfishing, and in some instances have been replaced, often by less desirable species (Van der Elst 1989).

Exploitation of marine resources off southern Africa has not been restricted to fish alone. Products from marine mammals, seabirds and turtles have also been highly valued. Harvesting of the Cape fur seal *Arctocephalus pusillus pusillus* commenced in the seventeenth century. By the twentieth century seals were no longer to be found at 23 localities at which they had previously bred or had regularly occurred, and the overall population size had been severely reduced (Shaughnessy and Butterworth 1981; Shaughnessy 1984). Controls on harvesting were introduced in 1893 and since then the population has been recovering (Butterworth et al 1987), allowing an increase in the pup harvest over this century (Figure 20.8). The cessation of the international market for seal pelts in 1983 through conservation action has further facilitated the recovery of the population.

TABLE 20.1   Status of some important pelagic resources off southern Africa (Newman 1977; De Villiers 1985, Crawford et al 1987; Anonymous 1987; J Augustyn pers comm, Sea Fisheries Research Institute, unpublished data). Start refers to the earliest available catch statistics for the resource. Overexploited+ indicates recovering. Divisions are those used by the International Commission for the Southeast Atlantic Fisheries. TAC refers to total allowable catch.

Species	Locality	Start	Max catch	Year Max	1985 Catch	Status	Management
Hake							
Merluccius polli	Mainly Angola	1973	60 185	1978	3	Unknown	None
M capensis & M paradoxus	Northern Namibia(Div. 1.3 & 1.4)	1964	606 000	1972	212 154	Overexploited	TAC, mesh size
M capensis & M paradoxus	Southern Namibia (Div 1.5)	1964	290 000	1973	201 481	Overexploited	TAC, mesh size
M capensis & M paradoxus	South Africa west coast	1955	244 000	1972	71 958	Overexploited+	TAC, limited entry, mesh size
M capensis & M paradoxus	South Africa east coast	1964	101 000	1974	56 223	Fully exploited	TAC, limited entry, mesh size
Horse mackerel							
Trachurus trecae (Cunene)	Angola and northern Namibia	1956	380 150	1978	56 766	Collapsed	TAC, mesh size
Trachurus trachurus (Cape)	Namibia	1966	659 994	1982	428 140	Overexploited	TAC, mesh size
Trachurus trachurus (Cape)	South Africa west coast	1950	118 100	1954	1 689	Collapsed	TAC, limited entry, mesh size
Trachurus trachurus (Cape)	South Africa south coast	1968	93 000	1977	25 588	Overexploited	TAC, limited entry, mesh size
Chub mackerel							
Scomber japonicus	Namibia	1972	199 088	1978	29 797	Overexploited	Limited entry, TAC
Scomber japonicus	South Africa	1954	128 200	1967	2 047	Collapsed	Limited entry, TAC
Snoek							
Thyrsites atun	Southern Africa	1960	81 688	1978	39 806	Fully exploited	Season, TAC
Yellowtail							
Seriola dorsalis	South Africa	1968	1 265	1975/6	414	Fully exploited	Limited entry
Tuna							
Thunnus etc.	Southern Africa	1960	43 200	1971	37 294	Unknown	None
Squid							
Loligo etc.	Southern Africa	1973	11 866	1985	11 866	Fully exploited	Limited entry off SA
Pilchard							
Sardinella spp	Angola	1956	240 600	1978	211 431	Fully exploited	None?
Sardinops ocellatus	Namibia	1950	1 400 100	1968	56 800	Collapsed	TAC, limited entry, season
Sardinops ocellatus	South Africa	1950	410 200	1962	29 600	Collapsed	TAC, limited entry, season
Anchovy							
Engraulis capensis	Namibia	1964	380 400	1987	51 866	Fully exploited	TAC, limited entry, season
Engraulis capensis	South Africa	1963	596 015	1987	277 012	Fully exploited	TAC, limited entry, season
Round herring							
Etrumeus whiteheadi	South Africa	1958	67 000	1978	37 700	Underexploited	Limited entry, season
Lanternfish							
Lampanyctodes hectoris	South Africa	1968	42 400	1973	32 200	Underexploited	Limited entry, season

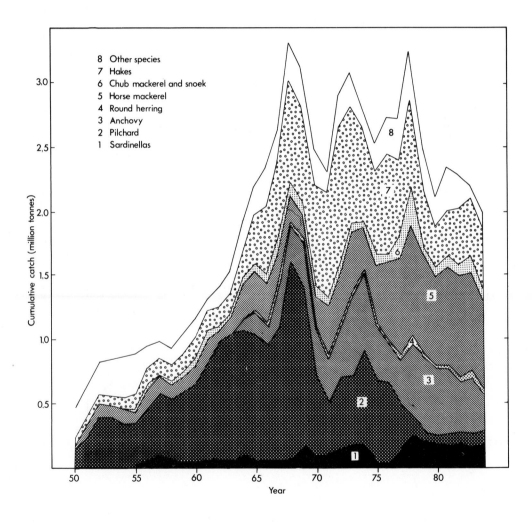

FIGURE 20.7 The combined yield of some important commercially exploited species off southern Africa showing the fairly constant yield since the peak in 1968 (from Crawford et al 1988)

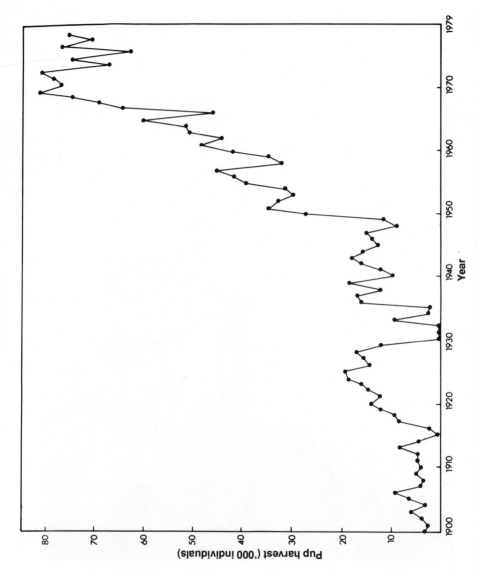

FIGURE 20.8 Annual harvest of Cape fur seal bulls and immatures from 1900-1979 (from Shaughnessy and Butterworth 1981). The increase in the pup harvest is taken to be an index of the recovery in the seal population during the twentieth century.

Exploitation of whales off southern Africa had commenced by 1770 but intensified in the 1900s with the advent of modern whaling techniques and the extension of activities to the Antarctic. This led to the sequential over-harvesting or collapse of one population after the other before whaling ceased off South Africa in 1975 (Best and Ross 1989; Table 20.2). Species such as the Minke whale *Balaenoptera acutorostrata* and Bryde's whale *Balaenoptera edeni* escaped comparatively unscathed, and there is evidence that the right whale *Eubalaena glacialis* and the humpback whale *Megaptera novaeanglia* are now beginning to recover as a result of a long period of protection (Best and Ross 1989).

TABLE 20.2   Status of southern hemisphere stocks of large whales (from Best and Ross 1989)

Species	Date intensive exploitation began	Initial population size (exploitable)	Date legally protected	Population size at protection	Degree of depletion at protection
Right	±1770	no data	1935	no data	>0,90
Humpback	1910	100 000	1963	3 000	0,97
Blue	1915	180 000	1967	5 000	0,97
Pygmy blue	1960	10 000	1967	5 000	0,50
Fin	1915	400 000	1976	84 000	0,79
Sei	1960	64 000	1979	11 000	0,83
Sperm (males)	1946	257 000	1982	128 000	0,52
(mature females)	1946	330 000	1982	259 000	0,22
Minke	1971	no data	1986	287 000	?

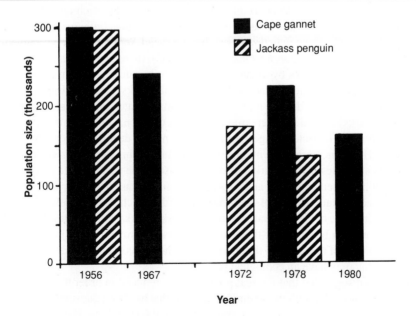

FIGURE 20.9 Changes in the population size of Cape gannets *Morus capensis* and jackass penguins breeding on islands off southern Africa between 1956 and 1980. The data is from Shelton et al (1984a) and Crawford et al (1983).

Disturbance and exploitation of jackass penguins (for meat and eggs) had led to their extinction at three colonies by the beginning of the twentieth century (Shelton et al 1984a). The decline from an estimated 300 000 adult birds in 1956 to about 130 000 in 1978/79 (Figure 20.9) can be attributed to guano scraping, the collapse of the pilchard population and competition by seals for space (Shelton et al 1984a). The population size of the Cape gannet has also decreased from about 150 000 breeding pairs in 1956, to about 80 000 in 1980 (Figure 20.9), again attributed largely to the collapse of the pilchard (Crawford et al 1983). The Cape cormorant population has fluctuated without a clear trend (Cooper et al 1982) and the bank cormorant population has increased (Cooper 1981, see also Cooper and Berruti this volume).

Two species of sea turtle, the loggerhead *Caretta caretta* and the leatherback *Dermochelys coriacea*, nest on the Natal coast of South Africa, and the green turtle *Chelonia mydas* is a non-breeding resident (Hughes 1989). Turtles off South Africa appeared to be becoming scarce through overexploitation by the early 1900s, and in 1916 the killing of turtles was prohibited in Natal, although subsistence exploitation continued and aggravated the decrease (Hughes 1989). The leatherback female population was critically low by 1963 and only five females nested in Natal in 1966, but serious conservation measures adopted since the 1960s have led to an improvement and the populations at studied and protected nesting grounds are stable or improving (Hughes 1989).

Man's impact on marine ecosystems has not been confined to direct commercial exploitation of marine animal populations. Before the 1960s it was the policy of the then Guano Islands Division of the Department of Industries to destroy kelp gull *Larus dominicanus vetula* eggs and chicks and to poison or shoot adult kelp gulls in order to increase the yield of jackass penguin eggs (Crawford et al 1982).

Gill nets have been employed off Natal's bathing beaches since 1952 to reduce the incidence of shark attacks, and in data covering the period 1966 to 1975 there is evidence of a decrease in catch per net (Van der Elst 1979). Concurrent with the decrease in the catch rate of sharks large enough to be captured by the gill nets (mainly blackfin *Carcharhinus limbatus*, dusky *Carcharhinus obscurus*, bronze *Carcharhinus brachyrus*, hammerhead *Sphyrna zygaena*, ragged-tooth *Carcharias taurus* and Zambezi *Carcharhinus leucas* adults), there has been an increase in the catch rate of juvenile dusky sharks by anglers. The increase in catch rate of juveniles may be a consequence of decreased predation by larger sharks on small sharks (preferred food item). It may be implicated in aggravating the decline in the sport fisherman's catch rates of teleosts (food item of juvenile dusky sharks) caused by overexploitation by recreational anglers off Natal (Van der Elst 1979). However, the expected longer-term consequence of decreased mortality on young dusky sharks would be an increase in the number of older sharks. This would be reflected by either an increase in the catch rates by the shark nets or a decrease in the survival rate of young dusky sharks as a result of increased predation, and consequently an increase in the catch rates of teleosts caused by decreased predation. Part of the decrease in the catch rates of teleosts may be a consequence of the increase in the catch rates of juvenile dusky sharks under finite effort. Although no strong conclusions can be drawn at this stage, the data do provide some justification for a quantitative examination of the effect of the shark nets on the status of shark and teleost biotas off Natal. Safe swimming may have an ecological price tag.

It has recently been suggested that the seal population be reduced in order to increase the fish yield (Anonymous 1986), although the scientific basis for this has been questioned (Butterworth et al 1988). A disturbance programme was introduced on Seal Island, False Bay, in early 1987 to reduce the breeding success of seals, and culling has been considered as a method for limiting population size. The increasing seal population is also seen as a threat to seabird populations, already stressed by overfished food resources, through competition for breeding space (Crawford et al 1989). A programme of disturbance was carried out at Mercury Island off Namibia in 1987 to limit the competition by seals for jackass penguin, bank cormorant *Phalacrocorax neglectus* and gannet breeding areas.

Decreases in fish population size cannot always be attributed primarily to overexploitation (Shelton 1981). In the case of anchovy, pilchard, horse mackerel, chub mackerel *Scomber japonicus* and hake *Merluccius* spp the increased harvests have apparently followed the entry of one or more strong year-classes into the fishery (Crawford et al 1988), a consequence of favourable environmental conditions for reproduction and/or early-stage survival. These strong year-classes often sustain a build-up in fishing effort and large catches result for a few years. When the fishery again returns to being dependent on average-year class strength, fishing effort is not decreased or redirected to other species quickly enough, and the effort overshoot causes the population to become severely depleted. When demand for a sought-after species exceeds supply and the product price is high, any further slight increase in population size is immediately countered by increased harvests (provided the fishery can remain viable by catching large amounts of less valuable species between these increases), which prevents or delays recovery — the so-called 'predator pit'. The South African pilchard fishery may be a good example of these events.

In terms of conservation status, an insidious consequence of harvesting is the effect increased mortality has on eroding the age structure of a population, and consequently on the ability of a population to withstand or recover from adverse environmental conditions. Increased mortality reduces the numbers surviving to older age and may consequently reduce the number of age classes in the mature population. Because adverse environmental anomalies that persist for longer than the reproductive life span of a species have severe effects on population size, there is an increased probability of population decline. Also, there are often age-dependent spatial patterns of abundance, for example in pelagic fish (Crawford 1980), so that range contractions often follow overexploitation (eg South African and Namibian pilchard populations). This results in a reduction in the diversity of environments being encountered by the population, again increasing the probability of decline. Fecundity is normally a function of weight so that younger fish, which weigh less than older fish, may produce fewer eggs or fewer batches of eggs, or even smaller eggs with smaller yolk reserves. Together these factors reduce the ability of a population to withstand variability in the environment or to recover from a decline. To some extent these changes may be compensated for by density-dependent increases in the growth rate, reduced age at first maturity, or increased fecundity at age (eg Shelton and Armstrong 1983).

MANAGING FOR BIOLOGICAL DIVERSITY

Commercial utilization of pelagic marine populations off southern Africa has in many instances had a detrimental effect on their conservation status, although no species extinctions have been recorded. The reason why those using the resource do not themselves conserve the resource has been clearly illustrated by Hardin (1968) as the 'tragedy of the commons'. The removal of one more unit of fish has a direct benefit to the person removing it, whereas the loss through the change in conservation status is spread over all the users.

In the absence of control, more and more people will enter the fishery, providing the catch per unit effort is high enough for the revenue generated to exceed the cost of fishing. However, as the biomass declines under fishing pressure, the catch per unit effort also declines until the revenue generated from using the resource balances the cost, at which point new entrants will not be attracted because fishing is no longer profitable. This 'bionomic equilibrium' occurs at a low biomass, corresponding to overexploitation, for any open-access fishery with a sufficiently low cost:revenue ratio (Clark 1985). A lag effect in the response of effort to profitability may cause effort to overshoot the equilibrium in a rapidly expanding fishery, with detrimental effects to both the resource and the economy of the fishery. State assistance in developing high-cost fisheries essentially lowers the cost:revenue ratio, and may be instrumental in precipitating overexploitation, and consequently reduced conservation status. In the short term this generates food and jobs, but the longer-term consequences are likely to be adverse.

In contrast to open-access, the apportioning of rights to resource use, which can be bought and sold at market value, prevents the competitive scramble between resource users and could be expected to reduce the incentive for overexploitation. This is indeed so, provided the discount rate (related to the real or non-inflationary rate of return that investors expect to earn on their investments) is low relative to the growth rate of the population being exploited. If not, it may make better economic sense to collapse the resource and invest the profit at the current interest rate (Clark 1985). The implication is that the longer-lived species with the lowest population growth rates will be the first to decline (whales, seals etc), as has occurred in the pelagic ecosystem off southern Africa and elsewhere, leaving the shorter-lived and perhaps less valuable species (eg gobies *Suff logobius bibarbatus* off Namibia; Crawford et al 1985). Also of relevance is that a deteriorating economy may adversely affect the discount rate and lead to worsening attitudes towards conserving resources. The development of economic infrastructure, such as large corporations with fishing companies as minor subsidiaries, may also have a detrimental effect on resource conservation.

Conservation of populations and the maintenance of ecological processes may not, therefore, be compatible with prevailing economic practice in exploiting renewable marine resources. Some form of management towards a broader objective may be necessary, assuming that population depletions and the resultant altered ecological processes are detrimental to the long-term good of mankind.

Management of exploited marine pelagic resources off southern Africa has traditionally been based on the analysis of catch data within the context of single-species models. This has three major shortcomings: (i) the influence of man and his related economic and social structures are relegated to an outside forcing function, normally harvest, rather than treated as dynamic state variables which are updated within the model based on rates that are influenced by other variables; (ii) self-defeating attempts are made simultaneously to obtain the maximum benefit to society from several valued variables of the system or attributes based on these variables, while ignoring the trade-offs determined by the interactions between variables; and (iii) the status of less valued variables is allowed to deteriorate while the focus is on attempts to maximize the benefit derived from more valued variables.

To some extent, it may be argued, these broader issues are taken into account implicitly in the debates by people who influence the final decision and who modify the output of the single-species models, based on their perception of how the variables in the system interact. Immediately, however, questions must be raised about time horizons and value judgements. It is preferable that a structured approach be developed which deals explicitly with the individual attributes (Shelton et al 1984b, Stewart and Brent 1988; Stewart in press; Stewart et al in prep). While the need to base management on an analysis of the multiple variables which comprise the system is now recognized, there are seldom sufficient data of adequate quality to parameterize more than the simplest single-species models; problems of structural specification and parameterization generally prevent complex models from performing better than simple ones (eg Shelton 1988 and references therein).

Although the modelling and management of multispecies systems remains a major unresolved problem in fisheries (see for example May et al 1979), several small, decision-driven speculative models (inadequate data to determine structure or parameters properly) of the interaction between variables, and between variables and controls, can be used to bring a variety of insights to bear on decision problems, such as those involving the culling of seals to increase fish yield (Starfield et al 1988). These should be built and used where possible to isolate critical data that need to be collected in order to distinguish between alternative models, or to reduce surprises when, inevitably, decisions are made. The relationships between the catch and the economic and social variables of the system are potentially more observable than those between populations of animals, and may be more easily incorporated into models to improve management.

In conclusion, the conservation of the biotic integrity of the pelagic ecosystem has to do with the management of a complex system of interacting physical, biotic, social, economic and political variables. In the past, inadequately controlled exploitation of renewable resources has had profound implications, not only for the exploited populations but also for dependent biotic, economic and social variables in the system. The understanding of how these variables interact is still minimal. It is imperative that this understanding be improved, despite the absence of a clear paradigm for analysing complex systems. It must be taken into account in management decisions in an objective manner if the biological diversity of the ecosystem is to be preserved and if future options for the utilization by man are not to be foreclosed.

REFERENCES

ANONYMOUS (1986). *Report of the commission of Inquiry into the Allocation of Quotas for the Exploitation of Living Marine Resources.* South African Government Report RP 91/1986. 107 pp.

ANONYMOUS (1987). International Commission for the Southeast Atlantic Fisheries. *Statistical Bulletin* **15**, 375 pp.

BEST P B and ROSS G J B (1989). Whales and whaling. In *Oceans of Life off Southern Africa.* (eds Payne A I L and Crawford R J M) Vlaeberg, Cape Town. pp 315 – 338.

BORCHERS P and HUTCHINGS L (1986). Starvation tolerance, development time and egg production of *Calanoides carinatus* in the southern Benguela Current. *Journal of Plankton Research* **8**, 855 – 874.

BRIAND F (1983). Environmental control of food web structure. *Ecology* **64**, 253 – 263.

BRIAND F and COHEN J E (1984). Community food webs have scale-invariant structure. *Nature* **307**, 264 – 267.

BUTTERWORTH D S, DAVID J M, MCQUAID L and XULU S S (1987). Modeling the population dynamics of the South African Fur Seal *Arctocephalus pusillus pusillus.* In *Status, biology, and ecology of fur seals; Proceedings of an international symposium and workshop, Cambridge, England, 23-27 April 1984.* (eds Croxall J P and Gentry R L) *National Oceanic and Atmospheric Association Technical Report. National Marine Fisheries Service* **51**, 141 – 164.

BUTTERWORTH D S, DUFFY D C, BEST P B and BERGH M O (1988). On the scientific basis for reducing the South African seal population. *South African Journal of Science* **84**, 179 – 188.

CLARK C W (1985). *Bioeconomic modelling and fisheries management.* Wiley, New York, NY.

COOPER J (1981). Biology of the bank cormorant 1. Distribution, population size, movement and conservation. *Ostrich* **52**, 208 – 215.

COOPER J, BROOKE R K, SHELTON P A and CRAWFORD R J M (1982). Distribution, population size and conservation of the Cape cormorant *Phalacrocorax capensis. Fisheries Bulletin of South Africa* **16**, 121 – 143.

CRAWFORD R J M (1980). Seasonal patterns in South Africa's Western Cape purse-seine fishery. *Journal of Fish Biology* **16**, 649 – 664.

CRAWFORD R J M, COOPER J and SHELTON P A (1982). Distribution, population size, breeding and conservation of the kelp gull in southern Africa. *Ostrich* **53**, 164 – 177.

CRAWFORD R J M, CRUIKSHANK R A, SHELTON P A and KRUGER I (1985). Partitioning of a goby resource amongst four avian predators and evidence for altered trophic flow in the pelagic community of an intense, perennial upwelling system. *South African Journal of Marine Science* **3**, 215 – 228.

CRAWFORD R J M, DAVID J H M, WILLIAMS A J and DYER R M (1989). Competition for space: recolonising seals displace endangered, endemic seabirds off Namibia. *Biological Conservation,* **59** – 72.

CRAWFORD R J M, SHANNON L V and POLLOCK D E (1987). The Benguela ecosystem. Part IV. The major fish and invertebrate resources. *Oceanography and Marine Biology Annual Review* **25**, 353 – 505.

CRAWFORD R J M, SHANNON L V and SHELTON P A (1988). Characteristics and management of the Benguela as a large marine ecosystem. In *Proceedings of the American Association for the Advancement of Science Conference on Biomass and Geography of Large Marine Ecosystems, Chicago, 1987.* (eds Sherman K and Alexander L M) 171 – 221.

CRAWFORD R J M and SHELTON P A (1981).Population trends for some southern African seabirds related to fish availability. In *Proceedings of the Symposium on Birds of the Sea and Shore, 1979.* (ed Cooper J) Cape Town, African Seabird Group. pp 15 – 41.

CRAWFORD R J M, SHELTON P A, COOPER J and BROOKE R K (1983). Distribution, population size and conservation of the Cape gannet. *Morus capensis, South African Journal of Marine Science* **1**, 153 – 174.

DE DECKER A (1984). Near surface copepod distribution in the southwestern Indian and southeastern Atlantic Ocean. *Annals of the South African Museum* **93**, 303 – 370.

DE VILLIERS G (1985). Living resources of the Benguela current region. In *Simposio Internacional Sobres Las Areas de Afloramiento Mas importantes del Oeste Africano (Cabo Blanco Y Benguel)* **2**, (eds Bas C, Margalef R and Rubies P) 1005 – 1039.

HARDIN G (1968). The tragedy of the commons. *Science* **162**, 1243 – 1248.

HUGHES G R (1989). Turtles. In *Oceans of Life off Southern Africa*. (eds Payne A I L and Crawford R J M) Vlaeberg, Cape Town. pp 230 – 243.

JAMES A G (1987). Feeding ecology, diet and field-based studies on feeding selectivity of the Cape anchovy *Engraulis capensis* Gilchrist. In Benguela and Comparable Ecosystems. (eds Payne A I L, Gulland J A and Brink K H) *South African Journal of Marine Science* **5**, 673 – 692.

JONES R and HENDERSON E W (1987). The dynamics of energy transfer in marine food chains. In Benguela and Comparable Ecosystems. (eds Payne A I L, Gulland J A and Brink K H) *South African Journal of Marine Science* **5**, 447 – 465.

MAY R M (1973). *Stability and complexity in model ecosystems*. Princeton University Press, Princeton, NJ.

MAY R M (1981). Patterns in multi-species communities. In *Theoretical Ecology: Principles and Applications*. (ed May R M) Blackwell, Oxford.

MAY R M (1986). The search for pattern in the balance of nature: advances and retreats. *Ecology* **67**, 1115 – 1126.

MAY R M, BEDDINGTON J R, CLARKE C W, HOLT S J and LAWS R M (1979). Management of multispecies fisheries. *Science* **205**, 267 – 177.

NELSON G and HUTCHINGS L (1983). The Benguela Current. *Progress in Oceanography* **12**, 333 – 356.

NEWELL R C and TURLEY C M (1987). Carbon and nitrogen flow through pelagic microheterotrophic communities. In Benguela and Comparable Ecosystems. (eds Payne A I L, Gulland J A and Brink K H). *South African Journal of Marine Science* **5**, 717 – 734.

NEWMAN G G (1977). The living marine resources of the Southeast Atlantic. *Food and Agricultural Organization, Fisheries Technical Paper* **178**, 59 pp.

PARRISH R H, BAKUN A, HUSBY D M and NELSON Ç S (1983). Comparative climatology of selected environmental processes in relation to eastern boundary current pelagic fish reproduction. In *Proceedings of the Expert Consultation to Examine Changes in Abundance and Species Composition of Neritic Fish Resources, San José, Costa Rica, April 1983*. (eds Sharp G D and Csirke J) *FAO Fisheries Report* **291**, 731 – 777.

ROSS G J B and BEST P B (1989). Smaller whales and dolphins. In *Oceans of Life off Southern Africa*. (eds Payne A I L and Crawford R J M) Vlaeberg, Cape Town. pp 303 – 314.

RYTHER J H (1969). Photosynthesis and fish production in the sea. *Science* **171**, 72 – 76.

SHACKLETON L Y (1987). A comparative study of fossil fish scales from three upwelling regions. In Benguela and Comparable Ecosystems. (eds Payne A I L, Gulland J A and Brink K H) *South African Journal of Marine Science* **5**, 79 – 84.

SHANNON L V (1985). The Benguela Ecosystem Part 1. Evolution of the Benguela, physical features and processes. *Oceanograph and Marine Biology Annual Review* **23**, 105 – 182.

SHANNON L V, BOYD A J, BRUNDRIT G B and TAUNTON-CLARK J (1986). On the existence of an El Niño-type phenomenon in the Benguela System. *Journal of Marine Research* **44**, 495 – 520.

SHANNON L V and TAUNTON-CLARK J (1988). Interannual and decadal changes in the sea surface temperature and relative wind stress in the south-east Atlantic this century. In Long-term data series relating to South Africa's renewable natural resources. (eds Macdonald I A W and Crawford R J M) *South African National Scientific Programmes Report* **157**, CSIR, Pretoria. 49 – 51.

SHAUGHNESSY P D (1984). Historical population levels of seals and seabirds on islands off southern Africa, with special reference to Seal Island, False Bay. *Investigational Report of the Sea Fisheries Institute of South Africa* **127**, 61 pp.

SHAUGHNESSY P D and BUTTERWORTH D S (1981). Historical trends in the population size of the Cape fur seal (*Arctocephalus pusillus*). In *The Worldwide Furbearer Conference Proceedings*. Donnelly, Falls Church, Virginia. (eds Chapman J A, Pursley D). 1305 – 1327.

SHELTON P A (1981). Research on pelagic fish in the southern Benguela region. *Transactions of the Royal Society of South Africa* **44**, 365 – 371.

SHELTON P A (1986). Fish spawning strategies in the variable southern Benguela Current region. PhD thesis. University of Cape Town, Cape Town, 327 pp.

SHELTON P A (1987). Life-history traits displayed by neritic fish in the Benguela Current ecosystem. In Benguela and Comparable Ecosystems. (eds Payne A I L, Gulland J A and Brink K H) *South African Journal of Marine Science* **5**, 235 – 242.

SHELTON P A (1988). Models for the management of renewable marine resources embedded in complex marine ecosystems. In *Ecodynamics. Contributions to Theoretical Ecology*. Research Reports in Physics. (eds Soeder C-J, Drepper F R and Wolff W) Springer, Heidelberg. pp 235 – 242.

SHELTON P A and ARMSTRONG M J (1983). Variations in parent stock and recruitment of pilchard and anchovy populations in the southern Benguela system. In *Proceedings of the Expert Consultation to Examine Changes in Abundance and Species Composition of Neritic Fish Resources, San José, Costa Rica, April 1983*. (eds Sharp G D and Csirke J) *FAO Fisheries Report* **291**, 1113 – 1132.

SHELTON P A, BOYD A J and ARMSTRONG M J (1985). The influence of large-scale environmental processes on neritic fish populations in the Benguela Current system. *California Cooperative Fisheries Investigations Reports* **26**, 72 – 92.

SHELTON P A, CRAWFORD R J M, COOPER J and BROOKE R K (1984a). Distribution, population size and conservation of the jackass penguin *Spheniscus demersus*. *South African Journal of Marine Science* **2**, 217 – 257.

SHELTON P A, DE VILLIERS G and CRAWFORD R J M (1984b). Southern African pelagic fish resources — towards a structured management approach. *Papers of the International Conference on Operations Research in Resources and Requirements in Southern Africa, Pretoria 2–5 April 1984*.

SMITH M M and HEEMSTRA P C (1986). *Smith's Sea Fishes*. Macmillan, South Africa. 1047 pp.

STARFIELD A M, SHELTON P A, FIELD J G, CRAWFORD R J M and ARMSTRONG M J (1988). A note on a modelling schema for renewable resource problems. *South African Journal of marine Science* **7**, 299 – 303.

STEWART T J (1988). Experience with prototype multicriteria decision support systems for pelagic fish quota determination. *Naval Research Logistics* **35**, 719 – 734.

STEWART T J and BRENT M (1988). Decision support system for pelagic fish management policy generation. In *Operational Research '87*. (ed Rand G K) Elsevier, BV (North-Holland). pp 119 – 129.

STEWARD T J, SHELTON P A and CRAWFORD R J M (in prep). Application of multicriteria decision support systems for pelagic fish quota determination.

TAUNTON-CLARK J and SHANNON L V (1988). Annual and interannual variability in the South-east Atlantic during the 20th century. *South African Journal of Marine Science* **6**, 97 – 106.

TYSON P D (1986). *Climatic Change and Variability in Southern Africa*. Oxford University Press, Cape Town. 220 pp.

VINES R G (1980). Analysis of South African rainfall. *South African Journal of Science* **76**, 404 – 409.

VAN DER ELST R P (1979). A proliferation of small sharks in the shore-based Natal sport fishery. *Environmental biology of fishes* **4**, 349 – 362.

VAN DER ELST R P (1989). Marine recreational angling in South Africa. In *Oceans of Life off Southern Africa*. (eds Payne A I L and Crawford R J M) Vlaeberg, Cape Town. pp 164 – 176.

PART 6

Policies to protect biotic diversity

CHAPTER 21

Outlines of a national environmental policy for South Africa

P Roelf Botha, B J Huntley

INTRODUCTION

South Africa has developed a very strong and internationally respected tradition in terms of the establishment and scientific management of protected areas such as national parks, wilderness areas and nature reserves. A consequence of this success is the rather limited perception of environmental conservation developed in the minds of most South Africans. Environmental conservation is often seen as synonymous with preserving spectacular landscapes and populations of large mammals in rigorously policed nature reserves. This concept is remote from the current dimensions of environmental needs and aspirations of both rural and urban communities, and a new and wider perspective of the real problems facing the South African environment is required.

The establishment of the Council for the Environment in 1982 brought with it the opportunity to take a fresh look at environmental issues in South Africa. The council set about formulating a series of policy documents on topics such as environmental education, coastal-zone management, protected areas, noise control, solid-waste management, open-space planning, and integrated environmental management. In addition, the council initiated the preparation of an overall 'National Environmental Policy and Strategy' which would provide the conceptual basis for and integrate the different facets of environmental matters in South Africa (Council for the Environment 1988). Biotic diversity is one such facet that must be seen within the context of the multitude of priorities and issues of the day. The council views the maintenance of biotic diversity as a prerequisite to improving the quality of life for all South Africans.

This paper provides a brief outline of the key goals and principles elucidated in the policy.

BACKGROUND

Attitudes to nature conservation and protected areas entered an era of change in the 1960s and 1970s, with classic texts such as Rachel Carson's 'Silent Spring' and Paul Ehrlich's 'The Population Bomb' introducing a new dimension to our view of the world. The International Union for the Conservation of Nature and Natural Resources (IUCN), traditionally concerned with the management of protected areas and the survival of threatened plants and animals, broadened the scope of its concern to include the whole environment on a world scale. In 1980, with the World Wildlife Fund and the United Nations Environment Programme, IUCN published the World Conservation Strategy (WCS) which seeks to provide a practical outline to nations throughout the world on problems and principles relating to the sustained use of natural resources (IUCN 1980). The WCS set off a chain reaction throughout the environmental movement, resulting in a series of national, regional, and local conservation strategies of varying cover and quality. From the outset, the WCS addressed a far wider spectrum of landuse options and issues than merely the protection of nature. Yet many still criticize it for reflecting too restricted a view. It is therefore pertinent to state that the WCS and National Conservation Strategies that have emerged subsequently view the protection of nature as but a single component of environmental conservation, which is dependent as much on social, economic and industrial factors as it is on wild living resources.

In South Africa, the first attempt to develop a National Environmental Conservation Policy and Strategy was that of the Wildlife Society of Southern Africa which published its document in 1981, the first non-governmental organization in the world to do so (Wildlife Society of Southern Africa 1981). The document set broad environmental goals and the principles and policies needed to attain them. It also provided outlines of the approaches and actions needed to realize specific policy objectives. During the eight years since the publication of the Wildlife Society and the WCS documents, enormous advances have been witnessed in conservation thinking and action in South Africa. Notable developments have included the promulgation of the Environment Conservation Act of 1982 (whereby the Council for the Environment was established), the Reports of the Planning Committee of the President's Council, the preparation of White Papers or Guide Plans on Forestry Resources, Agricultural Policy, Water Resources, Industrial Development and Energy Policy, the Agricultural Resources Act, and the National Grazing Strategy. These developments are without parallel in South Africa's conservation history and their impact on environmental resources may still be timely enough to reverse many of the negative trends that have developed during the last half century.

A need exists, however, for an updated and comprehensive statement of environmental needs and the guiding principles for their attainment in South Africa. Such a report should recognize but be independent of political, geographic or organizational constraints, it should be concise without being superficial, and it should be presented in a format suitable for communication to a very wide audience. The policy and strategy document of the Council for the Environment attempts to meet these objectives. It draws heavily from the WCS, national conservation strategies already produced by a number of countries, white papers submitted to the South African Parliament, and the Report on Nature Conservation by the President's Council. At all phases of the preparation of the document, comment from the full membership of the Council for the Environment, and from a wide spectrum of the South African population, has been sought to ensure balance in cover and opinion.

GOALS

The national policy addresses the problem of arresting and reversing current resource and environmental deterioration while at the same time promoting approaches to attaining a better quality of life for all South Africans. The policy identifies the following broad goals:

— the preservation of genetic diversity;
— ensuring the sustainable utilization of species and ecosystems;
— ensuring that non-renewable resources are not depleted before transition to the use of other materials has been achieved; and
— the maintenance and enhancement of cultural, spiritual and other qualities which enrich South African life.

The achievement of these goals is dependent largely on a recognition of the interdependence of conservation and development, activities that have traditionally been regarded as in conflict, rather than as mutually beneficial components of integrated resource management. The policy also recognizes the growing interdependence of international economic and political systems as a major influence on environmental issues. Demography, migration, agriculture, communication, energy, industry, technology, finance and socio-political evolution are key determinants in the creation and solution of environmental problems.

PRINCIPLES AND OBSTACLES

Seven fundamental principles and strategic approaches form the basis of the proposed policy:

— the equitable distribution of and access to resources;
— the integration of planning for conservation and development;
— retaining options for future use;
— focusing on causes as well as symptoms of environmental deterioration;
— the development of environmental knowledge and a predictive capacity for future application;
— the development of procedures for integrated environmental management; and
— the education of and active participation by the public in environmental matters.

The achievement of the policy's goals is dependent on the above principles being widely adopted by decision-makers and the population at large in South Africa. They are relevant to all environmental components and should serve as the unifying concepts on which the policy is implemented.

The document recognizes that effective environmental management requires both popular and political will to bring it about. Popular response to the policy and strategy will be dependent on the perceived equitability of access to resources and on reducing conditions of poverty and hunger. Political implementation will require both social and economic stability. The attainment of environmental goals is therefore closely related to the emergence of commonly accepted new political and social dispensations in South Africa.

Human actions have complex motivations and are directed towards many and varied goals. The environmental policy will therefore operate most effectively in a context in which individuals and groups have the opportunity to express their needs and aspirations, and in which open negotiation concerning the use of resources is facilitated.

THE ROAD AHEAD

In terms of a new Bill on Environmental Conservation to be submitted to Parliament, the Council for the Environment will advise the Minister of Environment Affairs on the formulation of an environment policy for South Africa. The approach to a policy and strategy outlined here will therefore be a useful point of departure for the council to execute its responsibilities under the new Act. For this reason it is important that the council's draft document be disseminated to as large an audience as possible to obtain comment from a wide range of opinion.

The development of an environmental policy is seen as an ongoing process. The philosophy and broad strategy must, after review, lead to the formulation of detailed operational guidelines which will provide decision-makers with the mechanisms to implement policy. The multidisciplinary nature of nearly all environmental issues implies that their solution can be found only through interaction between technical specialists and the end user — very often the individual taxpayer. Thus the working groups charged with the responsibility of drafting operational guidelines must include an appropriate representation of interest groups.

Once accepted for use, the guidelines should furthermore be subjected to review at least every five years to accommodate changing conditions and new circumstances. In this sense no policy document can ever be final or complete, and the Council for the Environment would like to have its policy and strategy seen as a contribution to the dynamic process of environmental conservation and management in South Africa.

REFERENCES

CARSON R (1962). *Silent Spring.* Houghton Mifflin, Boston, Mass.
COUNCIL FOR THE ENVIRONMENT (1989). An approach to a National Environmental Policy and Strategy for South Africa. Pretoria. 76 pp.
EHRLICH P R (1968). *The Population Bomb.* Ballentine, New York, NY.
IUCN (1980). *World Conservation Strategy: Living Resource conservation for Sustainable Development.* IUCN-UNEP-WWF, Gland, Switzerland. 44 pp.
WILDLIFE SOCIETY OF SOUTHERN AFRICA (1981). A strategy for survival. *African Wildlife* **35**,(5) 3 – 44.

CHAPTER 22

The role of large corporations and the private sector

G L Smuts, J C A Hobbs

INTRODUCTION

Threats to the diversity of life forms on earth are essentially due to loss of habitats, changes in habitats, and occasionally direct overexploitation of a species by man. Virtually all recent plant and animal extinctions have been caused through the indirect actions of humankind (Myers 1984).

The need to conserve biotic diversity (intraspecific genetic diversity, species richness or the diversity of interactions and ecological processes) has been recognized in many local and international programmes geared not only to protect wild living resources and their habitats but equally to implement reforms that influence man's impact on the earth's resources generally, for example the World Conservation Strategy (IUCN 1980), the Global 2000 Report (US Council on Environmental Quality 1980) and the Brundtland Report (WCED 1987). What is significant about these initiatives is that threats to our environment, and particularly the world's climate and everything it supports, are recognized to be inextricably linked to man's actions whether these be of an economic, social, political or ecological nature.

Large corporations have been at the forefront in altering our environment by both improvement and destruction. As the human population has increased and material standards and expectations have risen, demands have been unleashed for the services and products provided by such organizations, and thus the scale and complexity of their operations and impacts have increased.

Of particular concern to us in this paper is the implication that corporate organizations have been instrumental in reducing biotic diversity through the pursuit of short-term economic goals. It has been suggested that the (corporate) bulldozer has been a far more effective agent of extinction than the (oft-cited) pioneer's gun (US Council on Environmental Quality 1980).

However, the concept of biotic diversity should not be foreign to the corporate business community (Clarke 1983). Genetic diversity, for example, has its parallel in the business environment. An organization's chances of withstanding competition are served by its ability to diversify and adapt. In a similar way genetic diversity maximizes the likelihood of a species surviving in the face of environmental change. Ecological (or economic) diversity, on the other hand, enhances the resilience and sustained productivity of a total ecosystem (or economy).

Besides this, the prospects of an enterprise's survival are enhanced by planning for long-term goals, which is precisely what the maintenance of biotic diversity is all about. It is an investment to ensure the survival of those living resources that drive vital life-support systems on which human survival and development depend (IUCN 1980), while at the same time protecting those wild species which are of value as current or future suppliers of food, fibre, energy, medicines, industrial chemicals and the like (Prescott–Allen and Prescott–Allen 1982).

As stated in the World Conservation Strategy, the protection of biotic diversity is both a matter of insurance and investment (IUCN 1980). In this paper we hope to introduce some of the means whereby South Africa's large corporations and even smaller companies can develop and implement environmental management policies, an integral part of which must be the protection of biotic diversity. We shall also be touching on how socio-political and socio-economic issues impact on natural resources, and how these impacts can be influenced by a corporate organization's social responsibility programme and its policies towards its employees, shareholders, customers and the public at large. In short, we shall be sketching a blueprint for organizations to consider on the road to protecting biotic diversity and ensuring sustainable economic development and human progress.

With the apparent imminence of revised environmental legislation in South Africa, many organizations will be faced with the need to develop this ability in a short time. For some responsible organizations this will not alter much; for others it will be seen as an unwelcome revolution in the way they conduct their activities. For most, we hope, it will be seen as a challenge.

FROM RESOURCE DEPLETION TO ENVIRONMENTAL MANAGEMENT

We turn first to a contemporary paradigmatic shift that has been identified by O'Riordan and Turner (1983). Until recently few corporate organizations exhibited any sense of basic ecological understanding of, or responsibility for, the environments and resources they have impacted. Initially the impressive economic growth and material benefits that western affluent societies have received as a consequence has overshadowed their failure to calculate the social and environmental costs. 'Pernicious damage to environmental quality subsidised low resource prices' (O'Riordan and Turner 1983). Over the past three decades society has begun to realize the real costs of this growth, and this has frequently engendered an uncompromising adversarial relationship between industrial, environmental and community interests.

O'Riordan and Turner (1983) have developed a classification of ideologies that constitute the spectrum of the far from unitary concept of modern 'environmentalism'. They place 'ecocentrists' and 'technocentrists' at the two extremes of the environmentalist continuum. The radical pessimistic ecocentrists believe in a morality of limits to growth and in the intrinsic importance of nature for the sanity of man. Endangered species, in their view, have an inherent right to remain unmolested by our activities. At the other extreme, the optimistic technocentrist has faith in growth and the ability of man and his technological and scientific ability to get us out of any (self-imposed) difficulties.

We suggest that southern African corporate society has been dominated by the latter technocentrist ideology but that there is a discernible and needed shift away from this pole towards the ideology of environmental management, a median position on the continuum.

While believing in the value of economic growth, the 'environmental manager' qualifies this by recognizing the need for development to be based on thorough project appraisal and development control. He recognizes the need for decision-making to be balanced so that environmental factors are incorporated alongside technical and financial considerations, and are accorded equal weight. He believes that 'so long as man designs his activities with due care, man and nature can live in productive harmony' (O'Riordan and Turner 1983), a philosophy clearly reflected in the World Conservation Strategy (IUCN 1980) and the Brundtland Commission's report (WCED 1987).

It is this ideology that is taking root in enlightened southern African business, promoted in no small measure by public pressure and the creation of internal structures which now combine ecological and environmental expertise with the usual specialist inputs from, for example, engineers, architects and soil scientists. In essence a more holistic learning process has been initiated and the ecocentrist and technocentrist are losing ground to the 'environmental manager'.

The tools by which the 'environmental manager' initially gained credibility were the various procedures and methods of environmental impact appraisal (EIA). EIA has become more centred (yet more streamlined) in the concept of 'adaptive environmental assessment and management' (Holling 1978), or what is being re-labelled in South Africa as 'integrated environmental management' (Council for the Environment 1988). The prospects for protecting biotic diversity will be greatly improved by this change of 'conceptual spectacles'.

REASONS FOR ENVIRONMENTAL MANAGEMENT

Any corporate organization's principal operating ethos is to develop a resource(s), to provide a service, and to earn a suitable return for its shareholders — it is not to protect the environment. It would, however, be a foolhardy (and short-lived) organization if it simply provided a service without considering how that service was provided (Spalthoff 1987).

Some organizations have carried on regardless, safe in the belief that government agencies lack the finance and technical ability to enforce environmental protection legislation. They characteristically intimidate authorities with threats of unemployment and economic decline if attempts are made to enforce controls or refuse permits.

Profit-orientated entrepreneurial corporations have been reluctant to become involved in day-to-day environmental management for fear of the perceived implications of increased product or service costs to their customers and loss of competitive advantages. This puts them in a Catch-22 situation because it is those very same customers who are encouraging policy-makers not to allow the old-guard technocentrist organizations to operate with impunity.

Today, more and more people are appreciating their dependence on the environment and the reasons why it needs to be managed wisely. In the following sections we list some of the more salient reasons for applying sound environmental management.

Ethical

The great achievements of science and technology have not only been unable to meet all the requirements of the human species, but many of those that have been met have been gained at a terrible price — an unprecedented destruction of the natural environment based upon materialist philosophies which deny the spiritual dimension to Mother Earth and look upon it merely as a material substance to be manipulated at will (Singh 1987). Since ethical values are in reality those that should be conducive to the welfare of all humans, it must be recognized that every person, business or government needs to plan and manage their behaviour in an environmentally responsible way.

Although several ethical bases for environmental conservation exist, all recognize that individual desires must be regulated and limited, and that expediency or self-interest are not necessarily appropriate bases for judging human actions (Fuggle 1983). What are required are utilitarian ethical values that attempt to ensure wise stewardship of the earth's resources so that these will produce the greatest benefit for the greatest number of people for the greatest period of time.

Aesthetic

Natural environments are valued for material and spiritual qualities. Beautiful landscapes, plants and animals are assets which make South Africa a pleasant place in which to live, as well as a sought-after destination for tourists. The aesthetic qualities of a country, from a clean city or factory to a natural wonder such as a mountain or desert wilderness, help to maintain and engender cultural, spiritual and other qualities which make a proud and strong community. Any development must consider the aesthetic qualities of natural resources and how these can best be protected so that the proposed development can be blended into the environment without blighting it.

Scientific and educational

The pursuit of wisdom and knowledge (philosophy) by observation, experimentation and critical assessment (science) has taught man to utilize his environment and improve his standard of living. The natural environment is a valuable laboratory, an outdoor classroom and a source of research material. Learning about plants and animals (biology) and how they interact with their total environment (ecology) has enabled man to develop the agricultural, livestock, forestry, fishing and wildlife industries of the world. Wild species of plants and animals form the basis of these industries.

Virtually all branches of science have benefited from living resource conservation. Some modern medicines, foodstuffs, renewable sources of energy and many industrial and household chemicals come from plants and animals (Prescott-Allen and Prescott-Allen 1982).

Apart from satisfying man's desire for knowledge, science and education have enabled us to appreciate the extent of our dependence on the earth's living resources and the need to work towards sustainable resource use.

Social, political and legal

The slow but purposeful growth in environmental literacy during the past decade has pushed the weight of public opinion towards a desire for explicit, sustainable resource utilization. Public concern has followed close on the heels of demonstrations by environmental scientists of how habitats are being destroyed, various plant and animal species lost, and life-sustaining processes damaged. This concern cannot be ignored. It is, after all, the public who ultimately bears the cost of maintaining environmental quality or alternatively the loss of jobs and services, so it is they who legitimately must assist in making choices regarding environmental quality objectives (US Chamber of Commerce 1987).

Environmental protection is not a passing fashion; it is gaining ground as a major public issue. Politicians are listening, and no longer fear the risk of committing political suicide when raising and supporting environmental concerns. Inevitably, this will be (and has been) followed by enactment of protective legislation and regulations. Corporate organizations will have to face up to environmental issues and take a leading role in resolving the most persistent. By keeping their own houses in order, such organizations are precluding the need for authorities to impose draconian and sometimes ill-conceived legislation.

Economic

In the business environment, economic considerations are a fact of survival. Reference has been made earlier to the economic advantages of incorporating environmental management into business practice. We now explore some direct and indirect advantages in greater detail.

A corporation's image will be enhanced if it demonstrates a willingness to compromise on environmental concerns. Conflicts are time-consuming and may result in legal fees, lost production, delayed or cancelled projects, and a damaged corporate image. A stable, co-operative future is in the best interests of any organization.

If a responsible environmental track record is demonstrated, the chances of conflicts and opposition to future proposals is reduced. Potential antagonists will trust the organization, even on contentious issues, to act with responsibility. The benefit of a credible reputation in the environmental arena has been recognized by several corporate organizations who not only meet statutory requirements but often exceed them. They have also developed a company-wide awareness of their own high internal standards.

Too often environmental protection has been a reactive response to unexpected consequences. Corporate business behavior must shift from simply reacting to environmental problems *post facto* to playing an active part in anticipating and therefore reducing and solving them.

In short, environmental management makes good business sense, as long as it is pre-emptive whenever possible. Planned investment at the start of the production process to avoid or at least minimize impacts is far preferable to forced investment to remedy damage at the end, or retrofitting environmental controls once a problem has arisen. Similarly, by considering the potential environmental consequences of a range of options at the earliest design and planning stages, an organization can improve prospects for identifying and managing potential conflicts, thus ensuring a cheaper and more effective decision-making process.

Polluting industries are generally inefficient and the generation of waste is often an indication of non-productive utilization of resources. Operating costs may be reduced and profits increased when organizations address waste-marketing and resource-recovery technologies.

Environmental management does not hinder or burden development, it enhances the prospects for sustainable development. Maintaining the productivity of a natural resource base is of real importance to corporations for business reasons (eg forestry, fisheries and agriculture), since they depend on these resources for survival. Through environmental management the probability that natural resources of potential commercial value will continue to be available is increased.

Corporate responsibility

Today no industry ... can afford to put environmental concerns at the bottom of the Corporate agenda ... (we) ... are all "environmentalists" ... after all ... we all drink the same water and breathe the same air ... long after (the) oil and gas wells run dry the results of (our) handiwork will be evident on the land, in the waters and in the air. We will pass on to future generations the legacy of our care ... or neglect. Faced with a choice between environmental disregard or environmental concern, there is really no choice (Hair 1986).

The need is 'to factor environmental protection into corporate planning and to elevate it to a primary corporate objective. For example in the energy sector this means taking the commitment to clean atmosphere just as seriously as the obligation to supply power at acceptable cost' (Spalthoff 1987).

It is a primary management obligation to rank environmental protection alongside traditional corporate objectives such as profitability, quality and competitiveness (Spalthoff 1987). It is not an easy exercise to convince management of this need because they are inclined to consider environmental management as divorced from the productive process. Nor is it easy to demonstrate that environmental management is beneficial when it has saved an organization money it did not need to spend, for example, through avoiding environmental clean-ups or costly legal battles.

We now turn to the means whereby an organization can incorporate environmental management into its operations, after a brief consideration of what constitutes effective environmental management.

EFFECTIVE ENVIRONMENTAL MANAGEMENT

Environmental protection is too often naively interpreted as being concerned solely with the biological environment, particularly so in southern Africa where impressive large mammals and national parks are for many people 'the environment'.

As a consequence some organizations appear to think environmental conservation is sufficiently served by taking out corporate membership of non-government wildlife groups or sponsoring and being associated with wildlife-related themes on the pages of wildlife publications. These may be valid contributions to the coffers of non-government organizations (NGOs) but they may also run the risk of being interpreted as facades of 'conscience money', marketing strategies or public relations exercises which, when scrutinized, demonstrate a lack of understanding of the real problems facing the future of our planet. The danger is that such contributions may be totally divorced from the management of the environmental consequences of the day-to-day operations of the organization concerned.

These strategies are unfortunately not pursued only by some members of the business community; in many instances, conservation groups employ professional sales people who attempt to convince the business community to support their cause, a cause which admittedly is usually much easier to sell than one of sometimes obscure, yet more essential, environmental need.

These problems apply not only to the terrestrial environment but equally to the marine environment where the whale and other mammals are recognized as important symbols but where 'the survival of whales is less tied, in the long term, to direct exploitation than to understanding how anthropogenic stresses alter oceanic and coastal ecological processes' (Ray 1986).

It is imperative that corporate organizations invest in environmental management (with adequate financial, material and human resources) as an integral part of their daily activities of modifying and using land, water and atmospheric and biological resources, and disposing of wastes. Ideally, a significant percentage of an organization's turnover should be earmarked for environmental management. This does not mean that species-orientated projects must not be supported but rather that environmental investments deserve higher priority and motivation than public relations benefits or tax incentives. We certainly do not wish to denigrate the philanthropy of corporate organizations; we simply caution them against the dangers of regarding this kind of support as an adequate interpretation of what constitutes effective environmental management.

IMPLEMENTATION OF ENVIRONMENTAL MANAGEMENT

Foremost in meeting the objectives of effective environmental management for any corporate or private-sector organization is the establishment of a formal, concise and unambiguous public statement on environmental policy and strategy (BP 1985; Eskom 1988). This will illustrate the commitment of the highest echelons within the organization and enhance the prospects for across-the-board acceptance and compliance. We submit the following an an example of a statement of environmental intent:

This company recognizes that sustained economic growth is possible only in a managed and protected environment, and we commit ourselves to encouraging a greater awareness of the environment and the impacts the company has on it in the interests of sustainable human progress and human survival.

Without a commitment by top management, the tendency to interpret environmental protection as a burden and hindrance to the organization's business activities will remain.

Accountability and motivation

An accountability matrix should be a part of a policy or directive. This identifies senior managers who will be responsible for implementation of the policy and who will be obliged to enforce the objectives throughout the organization. They will also have to report back on the effectiveness of actions taken. Ideally, a board member should shoulder the ultimate accountability for environmental performance, thereby facilitating communication among major decision-makers. However, in the final analysis, the chief executive should accept overall responsibility for the environmental performance of the organization.

Traditional management tasks of informing, motivating and supervising personnel should become important means of protecting the environment (Spalthoff 1987). In order for personnel to be motivated to serve the environmental policy objectives, it is fundamental that open two-way communication channels exist. For employees to make the right decisions in their areas of responsibility, they must have access to appropriate information such as the priorities for environmental protection. On the other hand, senior management must be open to recommendations from operational personnel concerning opportunities and constraints with regard to implementation.

Structure and role of an environmental management division

Environmental management is a multi-disciplinary activity, one that can be effectively performed only by different specialists working together. Co-ordination, however, should be controlled by an environmental management division (EMD) established in an independent, service role in the organization. The EMD should comprise persons of sufficiently broad-based training and integrative abilities who can draw on the expertise of natural and social scientists but, most importantly, who can interact regularly with the engineers and architects planning a project. In many large corporations the establishment of an EMD will require only minor restructuring, with the size and composition of the unit being determined by the organization's specific requirements. In small companies environmental needs could be met by employing outside consultants whenever necessary. Here, however, the need still exists for a particular employee to manage the environmental portfolio. If this does not happen, information will tend to flow in one direction, there will be insufficient interactions between the parties concerned, and the process of learning, which precedes changes in interpretation of environmental needs, will not take place effectively.

The role of an EMD is multi-faceted. It has to advise, guide, train, track and audit environmental performance and compliance while anticipating trends in environmental issues that may affect the organization. It will assist management to achieve sufficiently high standards of environmental awareness and protection. It is important, however, that the responsibility for carrying out impact appraisal and impact control measures rest with 'line' or operational personnel. It should not usually be the task of the EMD to budget for and directly solve the environmental impacts of the organization's 'line' managers.

An EMD will act as a data bank on environmental information, and establish a network of contacts with environmental experts. It will continually liaise with government, NGOs, consultants, professional institutes and other interested parties. It is essential for the organization to understand the environment whose standards, externally imposed or internally formulated, it aims to safeguard. The EMD, through its understanding of the complexities of relevant biophysical and socio-economic environments, will facilitate the rational assessment of the potential effects the organization may have on these complex environments. Thus it will advise on steps to be taken to prevent, minimise or correct harmful impacts, and to maximise positive impacts.

It follows that the EMD will also be equipped with experts in the procedures, methods and techniques of project appraisal, impact control, monitoring and auditing. Given that environmental assessment is in its embryonic stages in South Africa (Hobbs 1985), the division will be engaged in devising and improving (particularly the predictive capabilities of) impact assessment methods and techniques, and recommending and updating internal standards, procedures, objectives and guidelines for all areas of the organization's activities likely to have significant environmental impacts.

A further function of an EMD will be to train all relevant employees in the skills of environmental management. This will ensure that standards of environmental literacy are raised and that staff develop the necessary skills to discharge their functions effectively. Contractors working under the control of an organization must also be advised of relevant standards to ensure their compliance. This is best achieved by incorporating environmental requirements into contracts and establishing compliance guidelines.

We agree with the statement made by Tomlinson (1987): 'Corporate commitment to sound environmental management is increasingly common among larger companies and it is hoped that their example will act as a model to medium and smaller companies.'

Environmental auditing and contingency planning

In order to check that measures designed to minimize adverse effects are effective, and that corrections can be made as early as possible to problem areas, the functions of monitoring and environmental auditing are necessary. Any organization prefers an audit to be an internal self-assessment rather than an external agency review and, understandably, the private sector has taken the lead in this new initiative, probably motivated by fear of potential liabilities (Bleiweiss 1987). The US giant General Motors, for example, has been conducting internal environmental audits since 1972.

In America, auditing has now been adopted as Environmental Protection Agency policy (United States of America Federal Register 9 July 1985). It actively encourages the use of environmental auditing by regulated organizations to help achieve and maintain compliance with environmental laws and regulations, as well as to help identify and correct unregulated environmental hazards.

There are several different kinds of auditing. They all, however, represent forms of 'structured feedback mechanisms' (Tomlinson 1987) designed to check that mitigatory recommendations are working as they are supposed to. If this is not the case, problems are identified and recommendations are made to the manager responsible. Audits are also a valid way of checking the accuracy of predictions made at the appraisal stage, thereby improving this skill for future project assessments. This variation is known as 'post development audits' (Bisset 1986).

Audits have been demonstrated to be effective tools in sensitizing personnel to environmental issues. They have also helped to make upper management aware of environmental impacts associated with the organization's activities (Bleiweiss 1987).

Finally, a further need exists for contingency planning to ensure fast and effective responses to emergency situations. For example, an oil or chemical company will need a clearly defined action plan to deal with spillage risks. As an example, the Sullom Voe oil terminal in the Shetland Islands has more than US $ 20 million worth of spillage control equipment, as well as the availability of a dedicated pollution-control vessel (BP 1986). The recent experiences of Union Carbide at Bhopal, in India, where 2 500 people were killed by a release of MIC gas (a phosgene derivative), and its demise are an example of the consequences of not having such a contingency plan (Weir 1987).

CORPORATE ENVIRONMENTAL RESPONSIBILITY IN PRACTICE

In addition to the appointment and training of appropriate staff and the implementation of policies and procedures within a private company or corporation, meaningful environmental protection also includes specific landuse practices and the implementation of enlightened socio-political and socio-economic strategies. It is to these considerations that we now turn.

Landuse considerations

Before discussing acceptable landuse practices, it is necessary to adopt a sound form of land tenure. This can be achieved by practising the concept of land stewardship (Fisher 1983) 'in which we may discern the notions of making full use of the land and taking all its fruits without degrading it, of active possession without denying possession to others, of nurturing a proper respect for this fragile and limited resource so that those who come after us will benefit equally.'

Organizations that own and/or utilize land for any purpose can promote sustainable landuse and conservation of biotic diversity by actions which include first, the conservation and management of appropriate pieces of land together with their indigenous plants and animals. The status of such land can be improved by having it proclaimed a private nature reserve, a wetland of international importance (IUCN 1984a), or by having it registered under the South African Natural Heritage Programme. Land-owners should also co-operate with their neighbours and conservation authorities, particularly with respect to the formation of conservancies (Markham 1986) or other ventures geared to facilitate the management of groups of farms.

Secondly, land-owners should develop landuse practices, especially in agriculture and forestry, that allow for the economic and sustainable use of land as well as provide living space for indigenous plants and animals within the man-managed system. This requires setting aside some marginal land, corners of fields and wetlands for wildlife, rather than putting everything to the plough, to plantations or to livestock production.

Thirdly, organizations should avoid the temptation to buy up farms or tracts of undeveloped land purely for purposes of property speculation, unless the means and the determination exist to manage the land in an environmentally acceptable way. Too often 'spec' property is left to become the local rubbish dump, to be burnt unseasonally, overgrazed, invaded with alien plants, or abused by squatters or poachers.

Fourthly, wherever land is disturbed, developers must ensure safe and productive re-instatement — whether the land in question was used for waste disposal, mining or any other purpose.

Fifthly, organizations should support meaningful landuse and related research programmes, and enter into joint sponsorship of such programmes with the State or other, private, organizations.

Finally, it would be in the best interests of all if a special effort were made to encourage better land stewardship and the development of a land ethic by 'examining each question in terms of what is ethically and aesthetically right, as well as what is economically expedient' (Leopold 1966).

Socio-political and socio-economic issues

It is well established today that economic, social, political and ecological systems are interlinked (IUCN 1980; Barbier 1987; Myers 1987; WCED 1987). We are firmly committed to the belief that economic development will increase the prospects for better resource management. Poverty and underdevelopment force people to disregard sound resource management practices. Sustainable economic development cannot occur unless the strategies being formulated are ecologically sustainable in the long term. They also need to be consistent with social values and institutions and encourage 'grassroots' participation in the development process (Barbier 1987). Moreover, it has been established that 'economic instability, political turmoil or outright conflict is aggravated when it is accompanied by rapid population growth — especially when its impact is amplified through environmental impoverishment' (Myers 1987). This situation will never change. Nonetheless, it is imperative that the policies and practices of all institutions concerned with these matters change. Large corporations and the private sector can influence the impact people and their demands have on biotic diversity by implementing actions such as the following in their social-responsibility programmes and in their interactions with their staff, shareholders, customers and the public at large:

— Help to improve educational opportunities (Collings 1982) for disadvantaged people by providing study grants, bursaries, additional appropriate schools, transport to schools, feeding schemes or any other incentives to encourage school attendance.
— Support or initiate environmental education programmes and training courses which emphasise the importance of an integrated approach to conservation, development and fertility reduction for all levels of society. Not only will the degree of environmental literacy of future generations determine the quality of future environments, but if the money is not spent now, the costs to the economy of correcting environmental problems in the future may be beyond our financial resources (Hurry 1984).
— Assist employees to improve their physical environment at home (and at work) by encouraging and assisting with land- and home-ownership schemes. It is now generally recognized that the physical environment in which an individual is reared has a profound effect on his or her growth and intellectual development (Hurry 1984), and as Royston (1985) says: 'A good environment and a motivated people lead to a physically and mentally healthy community.'
— To break the cycle of poverty and associated high birth and death rates, improved economic, social and environmental conditions are needed. These could be achieved by supporting sustainable rural and urban development programmes (Collings 1982; IUCN 1984b), by improving the position of women in society (more and equal job and education opportunities will reduce birth rates), and by working for administrative practices that advance normal family life and socio-economic equality (WCED 1987).
— Develop a corporate health policy and programme which includes access to safe and effective methods of family planning and contraception (IUCN 1984b).
— Work towards a political dispensation and an economic system in which all are able to participate fairly, and put this into practice in the organization through genuine equal opportunities. In this respect, salary structures should also be market-related and based on demands of the free-enterprise system. This will ensure that more, rather than fewer, people are drawn into the economic system and that salaries are related to an individual's productivity and job grading.
— Encourage the decentralization of funds, political power and personnel to local authorities which are best placed to appreciate and manage local needs (WCED 1987).
— Participate in initiatives by private organizations and governments to discuss and rationalize the use of natural resources. An example of such an initiative is the so-called Corporate Conservation Council established between the National Wildlife Federation and sixteen senior-level American corporate executives (The Environmental Forum 1983). The idea behind this council is to foster a closer working relationship between industrial leaders and the Wildlife Federation, America's largest private conservation organization (Hair 1986).
— Private-sector organizations and companies should attempt to remain up to date with local and international developments in environmental matters. This will help them make the right decisions with regard to support for or investment in major projects which have environmental

impacts. The World Bank, International Monetary Fund and regional development banks warrant special attention because of their influence on development throughout the world (WCED 1987). The World Bank in particular has been severely criticized for supporting environmentally destructive projects in various part of the world (IUCN 1987).

— Private companies or shareholders involved in farming ventures should encourage farmers to adopt practices that are ecologically sustainable and should resist the temptation to overprotect farmers through subsidies, price controls or tax relief (WCED 1987). Interventions such as these not only encourage the degradation of the agricultural resource base and the overuse of harmful chemicals, but produce surpluses and their associated financial burdens for poorer producers. Whole communities eventually suffer as is currently the case in the Karoo and the western Transvaal of South Africa.

— Pursue profit-seeking objectives within a framework of long-term sustainable development and avoid the asymmetry in bargaining power often prevalent between large corporations and small or poor countries without equivalent access to relevant information (WCED 1987).

— Of particular relevance to multi-nationals is that they should share managerial and technological skills with host country nationals (WCED 1987).

— The same standards of environmental and health protection must be applied, irrespective of the areas where development takes place (WCED 1987). These enterprises usually have considerable technical skills and economic backing, and they should not only adopt high safety and health standards but also give trans-boundary pollution issues adequate attention (WCED 1987).

Whereas one does expect private corporations, and governments for that matter, to promote the development of an environmental ethic, the role of the individual should not be forgotten. People, after all, consume resources, generate waste, and create demands for material goods. It would be fair to say that no community ethic should ever be used as a convenient screen behind which individuals can plunder the earth's resources for selfish gain. Groups and individuals, therefore, have a joint responsibility to ensure that the environmental ethic becomes enshrined in the traditions, folk tales and customs of all people, thereby ensuring its sustainability (Royston 1985).

DISCUSSION AND CONCLUSION

In a developing world in which ecosystems are being radically altered to provide man with resources, it is not possible to protect all forms of biotic diversity. Thousands of species have already been lost and this process is continuing. One can at least attempt to retard the process by protecting life forms in specially declared areas. It is better still to ensure that land-owners, developers and industrialists give due consideration to the way resources are used and take the necessary precautions when planning and operating processes that have impacts on local and more distant ecosystems.

The reduction of biotic diversity is an unwitting, secondary consequence and not an intended result of the activities of a corporate organization. There is little to be gained in wagging a finger at 'culprits'. Their impact on biotic diversity is a 'spillover effect' or 'externality' resulting from activities directed towards other goals (Myers 1979).

As Myers points out, an organization that misuses a 'common resource' is not necessarily blind to humankind's needs. It is pursuing its immediate self-interests within 'the rules of the game laid down by society's inadequate institutions'. Society needs to devise measures to ensure that incentives for looking after the immediate needs of an organization are matched by incentives to conserve what the Brundtland Commission (WCED 1987) labelled 'our common future'. We feel this will be realized only if large corporations establish permanent, multi-disciplinary teams with a strong biological/environmental staff base.

When producing a document geared to influence the policy of large corporations and the private sector on general conservation, and specifically on environmental management, one should not lose sight of the fact that positive action depends ultimately on the attitude and behaviour of individual people — everyone needs to be environmentally literate. If people do not understand environmental problems, they are unlikely to support the need for management or, at best, the final task will not be done very well. Clearly, then, environmental awareness and education are vital if progress is to be made in the critical years that lie ahead for southern Africa.

The earth's living resources are an incomparably rich natural and global bank account. If they are managed wisely we can draw on them in perpetuity without depleting it. Protecting biotic diversity is the highest form of thrift (USCEQ 1980). To this end the peoples of the world must co-operate more freely at all levels of society. If this is not achieved, environmental resources will continue to be squandered, species extinction rates will escalate, and life-support systems will be irreparably damaged. In such a situation it would be possible to ensure neither sustainable human progress nor, in the final analysis, human survival.

ACKNOWLEDGEMENTS

The authors wish to thank Brian Clark (University of Aberdeen, Scotland), Lynn Pirrozzoli and Dale Manty (Environmental Protection Agency USA) for providing valuable information. The paper also benefited from the comments and support provided by colleagues at Anglo America Corporation (Messrs G W H Relly, A W Lea, G R Pardoe, Dr Z J de Beer) and Eskom (Messrs O F Graupner, G Isaacs, D McLeod, D J Strydom, H van Tonder, G A Visser, D Willemse). Rosemary Galbraith (AAC) kindly undertook all the typing, and Arend Hoogervorst commented on the final draft.

REFERENCES

BARBIER E B (1987). The concept of sustainable economic development. *Environmental conservation* **14**, 101 – 110.

BISSET R (1986). Role of monitoring and auditing in EIA. Unpublished paper presented at the international seminar on environmental impact assessment, University of Aberdeen, Scotland, July 1986. 18 pp.

BLEIWEISS S J (1987). Legal considerations in environmental audit decisions. *Chemical Engineering Progress,* January 1987, 15 – 19.

BP (1985). Petroleum development (North-West Europe). Health, safety and environmental policy statement, September 1985. 8 pp.

BP (1986). BP and the environment. Environmental services branch, BP Petroleum Development Ltd, Aberdeen. Publicity document.

CLARKE H G (1983).Business ecology — a new science. *South African Journal of Business Management* **14**, 66 – 74.

COLLINGS J (1982). Social responsibility in South Africa. The work of The Anglo American and De Beers Chairman's Fund. *Supplement to Optima* **31**, 20 pp.

COUNCIL FOR THE ENVIRONMENT (1988). *Integrated environmental management in South Africa.* Department of Environment Affairs, Pretoria. 48 pp.

ESKOM (1988). *Corporate directive on environmental impact management EV1011.* Eskom, January 1988. 4 pp.

FISHER R C (1983). Land and land tenure. In *Environmental concerns in South Africa.* (eds Fuggle R F and Rabie M A) Juta, Cape Town, 435 – 444.

FUGGLE R F (1983). Nature and ethics of environmental concerns. In *Environmental concerns in South Africa.* (eds Fuggle R F and Rabie M A) Juta, Cape Town. 8 pp.

HAIR J D (1986). Better living through chemistry. Unpublished paper presented at Monsanto Worldwide Environmental Conference. National Wildlife Federation. 8 pp.

HOBBS J C A (1985). EIA a must for South Africa — from philosophical platitudes to potent planning? *Journal of the Institute of Landscape Architects of Southern Africa,* 16 – 23.

HOLLING C S (1978). *Adaptive environmental assessment and management. International series on applied systems analysis.* Wiley and Sons, Chichester. 377 pp.

HURRY L (1984). Saving spaceship earth. *Leadership SA* **3**, 131 – 141

IUCN (1980). The world conservation strategy. International Union for the Conservation of Nature and Natural Resources, Gland, Switzerland.

IUCN (1984a). Gronigen 1984 — a bigger splash or a drop in the ocean. *IUCN bulletin* **15**, 42 – 45.

IUCN (1984b). *Population and natural resources. A supplement to the world conservation strategy.* Revised edition, August 1984. Commission on Ecology (IUCN) and International Planned Parenthood Federation. 57 pp.

IUCN (1987). The greening of the World Bank. *IUCN Bulletin* **18**, 7 – 8.

LEOPOLD A (1966). *A sand county almanac*. Oxford University Press, new York, NY. 269 pp.

MARKHAM R (1986). *Establishing a wildlife conservancy. Wildlife Management. Technical guides for farmers.* Natal parks Board, Pietermaritzburg. 4 pp.

MYERS N (1979). *The sinking ark: a new look at the problem of disappearing species.* Pergamon Press, Oxford. 307 pp.

MYERS N (1984). The mega-extinction of animals and plants. In *Ecology 2000. The changing face of Earth.* (ed Hillary E) Michael Joseph, London, pp 82 – 107.

MYERS N (1987). Population environment and conflict. *Environmental Conservation* **14**, 15 – 22.

O'RIORDAN T and TURNER R K (1983). An annotated reader in environmental planning and management. *Urban and regional planning series* **30**. Pergamon Press, Oxford. 460 pp.

PRESCOTT-ALLAN R and PRESCOTT-ALLEN C (1982). *What's wildlife worth? Economic contributions of wild plants and animals to developing countries.* Earthscan/PA data. IIED, London. 92 pp.

RAY G C (1986). Conservation concepts for the seas and coasts. *Environmental Conservation* **13**, 95 – 96.

ROYSTON M G (1985). Local and multi-national corporations: reappraising environmental management. *Environment* **27**, 13 – 43.

SINGH K (1987). The ethics of conservation. *Environmental Conservation* **14**, 1 – 3.

SPALTHOFF F J (1987). Managing environmental protection. *Siemens Review* **54**, 4 – 6.

THE ENVIRONMENTAL FORUM (1983). Wildlife's "corporate detente". February 1983, 41 – 43.

TOMLINSON P (1987). Environmental monitoring and assessment. Special Issue "Audits in Environmental Assessment" 8, 183 – 261.

US CHAMBER OF COMMERCE (1987). Policy statement on environmental quality. USCC, Washington DC. 5 pp.

US COUNCIL ON ENVIRONMENTAL QUALITY AND US DEPARTMENT OF STATE. *The Global 2000 report to the President: Entering the twenty-first century*, 1 (summary report). US Government Printing Office, Washington DC, 1980.

WCED (1987). *Our common future. World Commission on Environment and Development.* Oxford University Press, Oxford. 383 pp.

WEIR D (1987). *The Bhopal syndrome — pesticides, environment and health.* Earthscan, London. 208 pp.

INDEX

QH
77
.A356
B56
1989

QH
77
.A356
B56

1989

35.00